Patrick Moore's Yearbook of Astronomy 2016

Patrick Moore's Yearbook of Astronomy 2016

EDITED BY

John Mason

MACMILLAN

First published 2015 by Macmillan
an imprint of Pan Macmillan
20 New Wharf Road, London N1 9RR
Associated companies throughout the world
www.panmacmillan.com

ISBN 978-1-4472-8708-7

3 5 7 9 8 6 4 2

A CIP catalogue record for this book is available from the British Library.

Typeset by Ellipsis Digital Limited
Printed and bound by CPI Group (UK) Ltd, Croydon, CR0 4YY

Contents

Part Two

Article Section

Part Three

Miscellaneous

Editor's Foreword

We seem to be living in a golden age for astronomy, astrophysics and space exploration. Within the past year, we have seen the Hubble Space Telescope celebrate its twenty-fifth birthday – in April 2015. Hubble has undergone a series of essential repairs and upgrades, but it is still beaming back cutting-edge science and breathtaking images of the universe from Earth orbit. Scientists expect it to continue functioning through to at least 2018, when its successor, the James Webb Space Telescope, is scheduled for launch.

Meanwhile, the Kepler spacecraft's K2 observing campaign continues as the number of confirmed exoplanets (planets it has discovered orbiting other stars) exceeds 1,000 and the number of planetary candidates (those awaiting confirmation) approaches 4,700. Many of these confirmed exoplanets are less than twice Earth-size, with a handful found to be orbiting within the so-called habitable zone of their parent star, i.e., the region around a star within which planetary-mass objects with sufficient atmospheric pressure can support liquid water at their surfaces; liquid water is considered essential for many forms of life.

Elsewhere NASA's Dawn spacecraft – which orbited the asteroid Vesta from July 2011 to September 2012 – started circling the dwarf planet Ceres, the largest object in the asteroid belt between Mars and Jupiter, in March 2015. Following its remarkable success in dropping a small probe, Philae, onto the surface of Comet 67P/Churyumov-Gerasimenko in November 2014, the European Space Agency's Rosetta spacecraft has accompanied the comet during its closest approach to the Sun in August 2015, monitoring how the comet has changed over time and beaming back groundbreaking science as it goes. The Rosetta mission marked the first time a spacecraft had been placed in orbit around a comet, and Philae was the first probe to soft-land on a comet. Also in August 2015, NASA's Curiosity Rover celebrated its third anniversary on Mars; it continues its exploration of the base of Mt Sharp in Gale Crater.

The undoubted highlight of the past year was the long-awaited fly-by of the dwarf planet Pluto by NASA's New Horizons spacecraft in

July 2015, after a nine-and-a-half year journey to reach its target. Pluto turned out to be an even more remarkable world than had been expected. I can only imagine how excited my former co-Editor of this Yearbook, the late Sir Patrick Moore, would have been by the images sent back by New Horizons! Patrick had a lifelong interest in Pluto and was a good friend of Clyde Tombaugh, Pluto's discoverer. Tombaugh died in 1997 and a small sample of his ashes were on board the spacecraft as it hurtled past the dwarf planet.

We have tried to capture some of this excitement within the pages of the 2016 *Yearbook of Astronomy* while continuing to give readers a wide choice, both of subject and of technical level. We have the usual mix of contributions both for those who want to know exactly what is going on in the night sky, month-by-month, and for those who enjoy reading the longer articles.

Clearly there have been many exciting and significant advances in astronomy and space exploration in recent times. Natalie Starkey recounts the amazing comet-chasing exploits of the Rosetta mission, including the landing on the comet by Philae, and the many important scientific discoveries that are being made. Stephen Webb looks at an unusual experiment on-board the International Space Station, the Alpha Magnetic Spectrometer, which is detecting energetic cosmic-ray electrons and positrons from Earth orbit and is providing a deeper understanding of the nature of high-energy cosmic rays and of the existence of mysterious dark matter. The Square Kilometre Array (SKA) is an international project to build the world's largest radio telescope, with, eventually, over a square kilometre (one million square metres) of collecting area. As Lisa Harvey-Smith explains, designing and constructing such a unique instrument represents a huge leap forward in both engineering and research and development and, as one of the largest scientific endeavours in history, the SKA project is bringing together a wealth of the world's finest scientists, engineers and policy-makers to make it a reality. David Harland examines one of the Hubble Space Telescope's greatest achievements: helping astronomers to refine the value of the Hubble Constant, the unit of measurement which describes the rate of expansion of the Universe. David describes the enormous efforts made to measure the Hubble Constant over the years and reviews its cosmological significance. As mentioned earlier, NASA's New Horizons spacecraft flew past Pluto and its moons in mid-July 2015 and the first scientific results were only just appearing as this Yearbook went to press, but we have done our best to present 'the story so far' in this edition, with a more detailed summary to follow next year.

For the amateur observer, Martin Mobberley describes how one goes about choosing an astrograph, the name usually given to telescope systems designed for imaging the night sky. Martin looks at the wide range of such equipment currently on offer and provides a wealth of helpful hints and tips, based on his own extensive experience, not only on what to buy but also on how to get the best out of your astrograph once you have obtained it.

For those with an interest in the history of astronomy, we have Richard Baum's fascinating story of the bright object seen quite near the Sun on 21 December 1882 from Broughty Ferry, near Dundee in Scotland. Was it the brilliant planet Venus, which had been in transit across the face of the Sun earlier that month, or perhaps a sungrazing comet, or even a supernova? The arguments have continued for over a century. Finally, another of our regular contributors, Allan Chapman, provides an enthralling biography of the life and work of Sir John Frederick William Herschel, the eminent astronomer, cosmologist and 'natural philosopher' of the Victorian Age, a man who in every way embodied the 'Grand Amateur' tradition of self-funded scientific research.

I am, as always, most grateful to all of the regular contributors for their support in completing this edition. Martin Mobberley has supplied the notes on eclipses, comets and minor planets, and Nick James has produced the data for this year's transit of Mercury, the phases of the Moon, longitudes of the Sun, Moon and planets, and details of lunar occultations. As usual, John Isles and Bob Argyle have provided the information on variable stars and double stars, respectively. Wil Tirion, who produced our stars maps for the Northern and Southern Hemispheres, has again drawn all of the line diagrams showing the positions and movements of the planets to accompany the Monthly Notes. These provide a detailed guide to what's happening in the night sky throughout the year, on a month-by-month basis.

I very much hope that you will enjoy reading the range of articles that have been prepared for you this year.

John Mason
Barnham, August 2015

Preface

New readers will find that all the information in this Yearbook is given in diagrammatic or descriptive form; the positions of the planets may easily be found from the specially designed star charts, while the Monthly Notes describe the movements of the planets and give details of other astronomical phenomena visible in both in the Northern and Southern Hemispheres. Two sets of star charts are provided. The **Northern Star Charts** (pp.7 to 31) are designed for use at latitude 52°N, but may be used without alteration throughout the British Isles, and (except in the case of eclipses and occultations) in other countries of similar northerly latitude. The **Southern Star Charts** (pp.33 to 57) are drawn for latitude 35°S, and are suitable for use in South Africa, Australia and New Zealand, and other locations in approximately the same southerly latitude. The reader who needs more detailed information will find *Norton's Star Atlas* an invaluable guide, while more precise positions of the planets and their satellites, together with predictions of occultations, meteor showers and periodic comets, may be found in the *Handbook of the British Astronomical Association*. Readers will also find details of forthcoming events given in the American monthly magazine *Sky & Telescope* and the British periodicals *The Sky at Night* and *Astronomy Now*.

Important note

The times given on the star charts and in the Monthly Notes are generally given as local times, using the twenty-four-hour clock, the day beginning at midnight. All the dates, and the times of a few events (e.g. eclipses) are given in Greenwich Mean Time (GMT), which is related to local time by the formula:

Local Mean Time = GMT – west longitude

In practice, small differences in longitude are ignored, and the observer will use local clock time, which will be the appropriate Standard (or

Zone) Time. As the formula indicates, places in west longitude will have a Standard Time slow on GMT, while places in east longitude will have a Standard Time fast on GMT. As examples we have:

Standard Time in	
New Zealand	GMT + 12 hours
Victoria, NSW	GMT + 10 hours
Western Australia	GMT + 8 hours
South Africa	GMT + 2 hours
British Isles	GMT
Eastern ST	GMT – 5 hours
Central ST	GMT – 6 hours, etc.

If Summer Time is in use, the clocks will have been advanced by one hour, and this hour must be subtracted from the clock time to give Standard Time.

Part One

Monthly Charts and Astronomical Phenomena

Notes on the St[illegible]

[illegible]

Notes on the Star Charts

The stars, together with the Sun, Moon and planets, seem to be set on the surface of the celestial sphere, which appears to rotate about the Earth from east to west. Since it is impossible to represent a curved surface accurately on a plane, any kind of star map is bound to contain some form of distortion.

Most of the monthly star charts which appear in the various journals and some national newspapers are drawn in circular form. This is perfectly accurate, but it can make the charts awkward to use. For the star charts in this volume, we have preferred to give two hemispherical maps for each month of the year, one showing the northern aspect of the sky and the other showing the southern aspect. Two sets of monthly charts are provided, one for observers in the Northern Hemisphere and one for those in the Southern Hemisphere.

Unfortunately, the constellations near the overhead point (the zenith) on these hemispherical charts can be rather distorted. This would be a serious drawback for precision charts, but what we have done is to give maps which are best suited to star recognition. We have also refrained from putting in too many stars, so that the main patterns stand out clearly. To help observers with any distortions near the zenith, and the lack of overlap between the charts of each pair, we have also included two circular maps, one showing all the constellations in the northern half of the sky, and one showing those in the southern half. Incidentally, there is a curious illusion that stars at an altitude of 60° or more are actually overhead, and beginners may often feel that they are leaning over backwards in trying to see them.

The charts show all stars down to the fourth magnitude, together with a number of fainter stars which are necessary to define the shapes of constellations. There is no standard system for representing the outlines of the constellations, and triangles and other simple figures have been used to give outlines which are easy to trace with the naked eye. The names of the constellations are given, together with the proper names of the brighter stars. The apparent magnitudes of the stars are indicated roughly by using different sizes of dot, the larger dots representing the brighter stars.

The two sets of star charts – one each for Northern and Southern Hemisphere observers – are similar in design. At each opening there is a single circular chart which shows all the constellations in that hemisphere of the sky. (These two charts are centred on the North and South Celestial Poles, respectively.) Then there are twelve double-page spreads, showing the northern and southern aspects for each month of the year for observers in that hemisphere. In the **Northern Star Charts** (drawn for latitude 52°N) the left-hand chart of each spread shows the northern half of the sky (lettered 1N, 2N, 3N . . . 12N), and the corresponding right-hand chart shows the southern half of the sky (lettered 1S, 2S, 3S . . . 12S). The arrangement and lettering of the charts is exactly the same for the **Southern Star Charts** (drawn for latitude 35°S).

Because the sidereal day is almost four minutes shorter than the solar day, the stars appear to rise and set about four minutes earlier each day, and this amounts to two hours in a month. Hence the twelve pairs of charts in each set are sufficient to give the appearance of the sky throughout the day at intervals of two hours, or at the same time of night at monthly intervals throughout the year. For example, charts 1N and 1S here are drawn for 23 hours on 6 January. The view will also be the same on 6 October at 05 hours; 6 November at 03 hours; 6 December at 01 hours and 6 February at 21 hours. The actual range of dates and times when the stars on the charts are visible is indicated on each page. Each pair of charts is numbered in bold type, and the number to be used for any given month and time may be found from the following table:

Local Time	**18h**	**20h**	**22h**	**0h**	**2h**	**4h**	**6h**
January	11	12	1	2	3	4	5
February	12	1	2	3	4	5	6
March	1	2	3	4	5	6	7
April	2	3	4	5	6	7	8
May	3	4	5	6	7	8	9
June	4	5	6	7	8	9	10
July	5	6	7	8	9	10	11
August	6	7	8	9	10	11	12

September	7	8	9	10	11	12	1
October	8	9	10	11	12	1	2
November	9	10	11	12	1	2	3
December	10	11	12	1	2	3	4

On these charts, the ecliptic (the plane of the Earth's orbit around the Sun) is drawn as a broken line on which longitude is marked every 10°. The positions of the planets are then easily found by reference to the table on page 64. It will be noticed that on the **Southern Star Charts** the ecliptic may reach an altitude in excess of 62.5° on the star charts showing the northern aspect (5N to 9N). The continuations of the broken line will be found on the corresponding charts for the southern aspect (5S, 6S, 8S and 9S).

Northern Star Charts

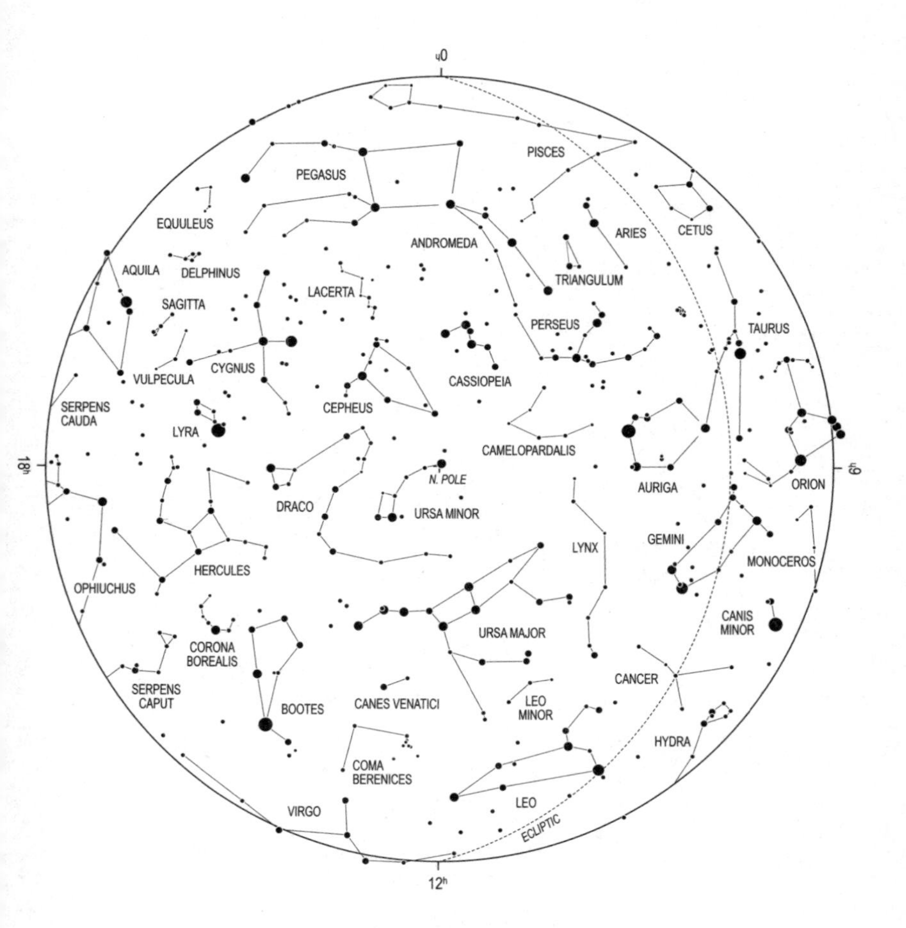

Northern Hemisphere

Note that the markers at 0^h, 6^h, 12^h and 18^h indicate hours of Right Ascension.

1N

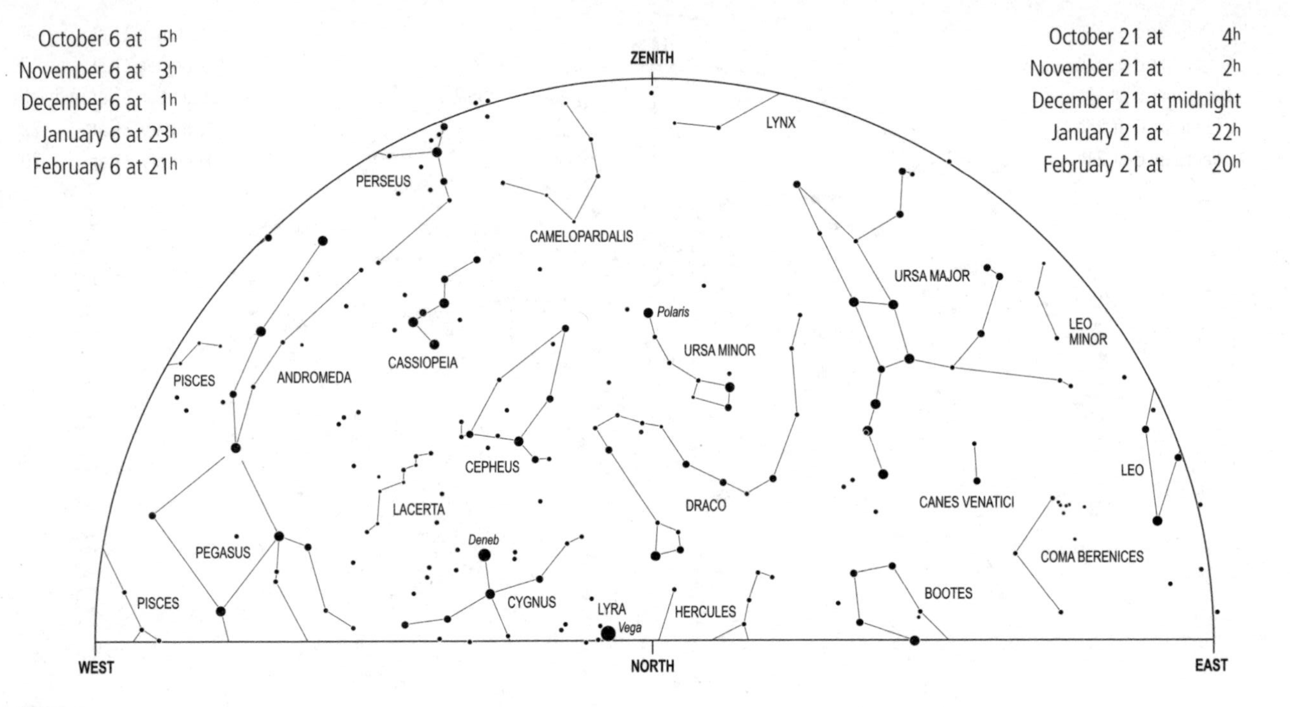

October 6 at 5h
November 6 at 3h
December 6 at 1h
January 6 at 23h
February 6 at 21h
October 21 at 4h
November 21 at 2h
December 21 at midnight
January 21 at 22h
February 21 at 20h
ZENITH
LYNX
PERSEUS
CAMELOPARDALIS
URSA MAJOR
LEO
MINOR
Polaris
URSA MINOR
CASSIOPEIA
PISCES
ANDROMEDA
CEPHEUS
LEO
DRACO
LACERTA
CANES VENATICI
Deneb
PEGASUS
COMA BERENICES
BOOTES
CYGNUS
PISCES
LYRA
HERCULES
Vega
WEST
NORTH
EAST

1S

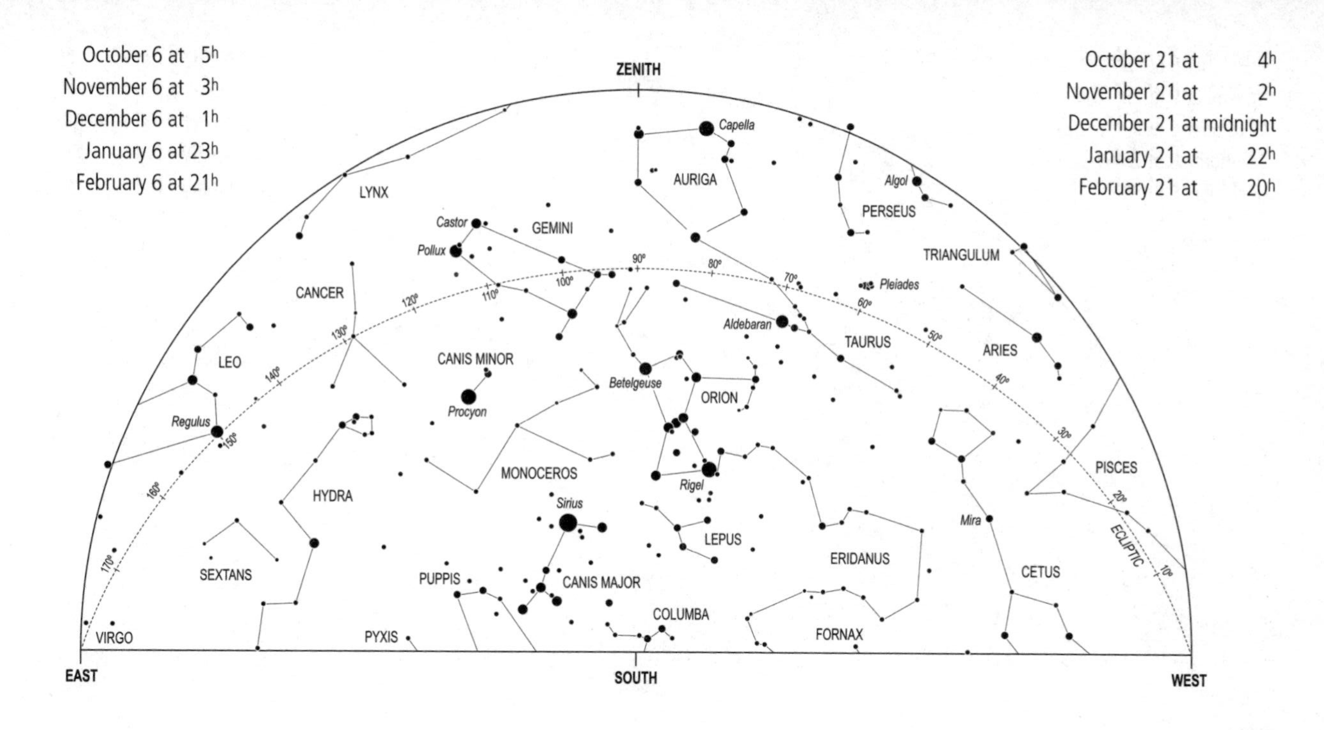
October 6 at 5h
November 6 at 3h
December 6 at 1h
January 6 at 23h
February 6 at 21h
October 21 at 4h
November 21 at 2h
December 21 at midnight
January 21 at 22h
February 21 at 20h
ZENITH
EAST
SOUTH
WEST
LYNX
AURIGA
Capella
PERSEUS
Algol
TRIANGULUM
Pleiades
GEMINI
Castor
Pollux
CANCER
LEO
Regulus
CANIS MINOR
Procyon
Aldebaran
TAURUS
ARIES
Betelgeuse
ORION
MONOCEROS
Rigel
Sirius
HYDRA
LEPUS
ERIDANUS
Mira
CETUS
PISCES
ECLIPTIC
SEXTANS
PUPPIS
CANIS MAJOR
COLUMBA
FORNAX
PYXIS
VIRGO

2N

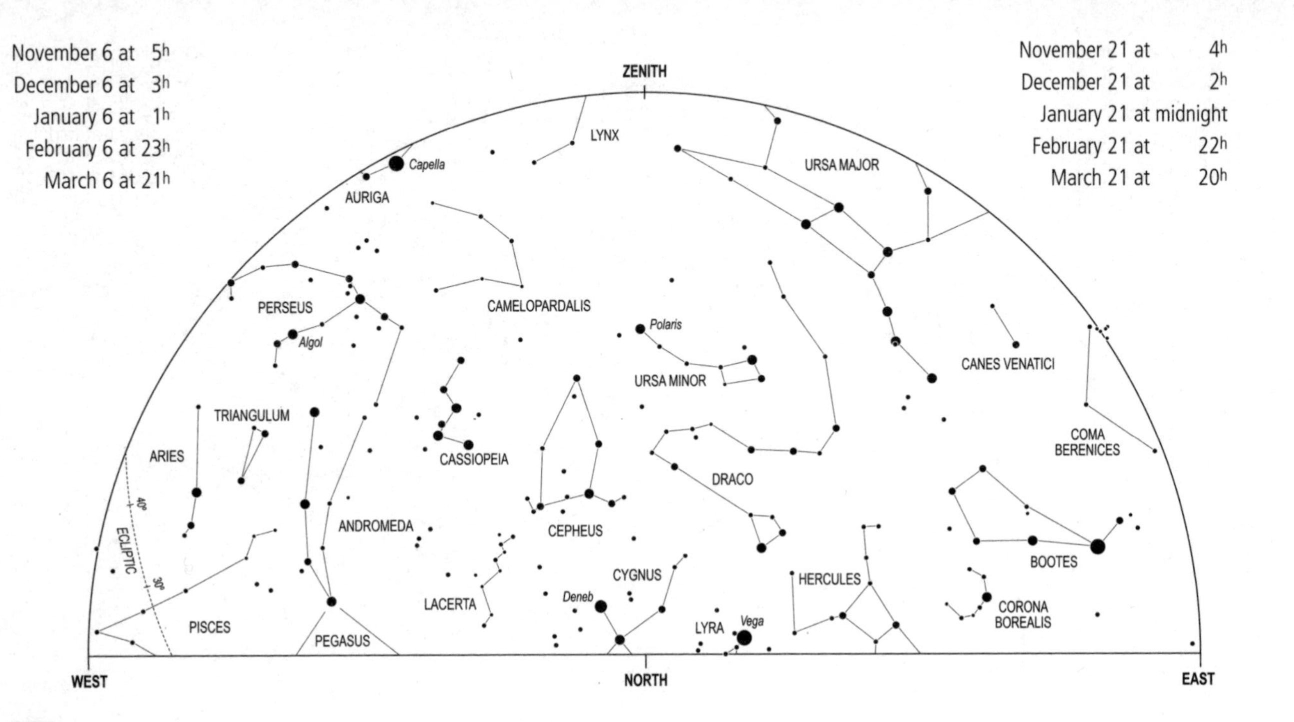
November 6 at 5h
December 6 at 3h
January 6 at 1h
February 6 at 23h
March 6 at 21h
November 21 at 4h
December 21 at 2h
January 21 at midnight
February 21 at 22h
March 21 at 20h
ZENITH
WEST
NORTH
EAST
ECLIPTIC
40°
30°
LYNX
Capella
AURIGA
URSA MAJOR
PERSEUS
Algol
CAMELOPARDALIS
Polaris
URSA MINOR
CANES VENATICI
TRIANGULUM
ARIES
CASSIOPEIA
DRACO
COMA
BERENICES
ANDROMEDA
CEPHEUS
BOOTES
CYGNUS
HERCULES
Deneb
LACERTA
CORONA
BOREALIS
PISCES
PEGASUS
LYRA
Vega

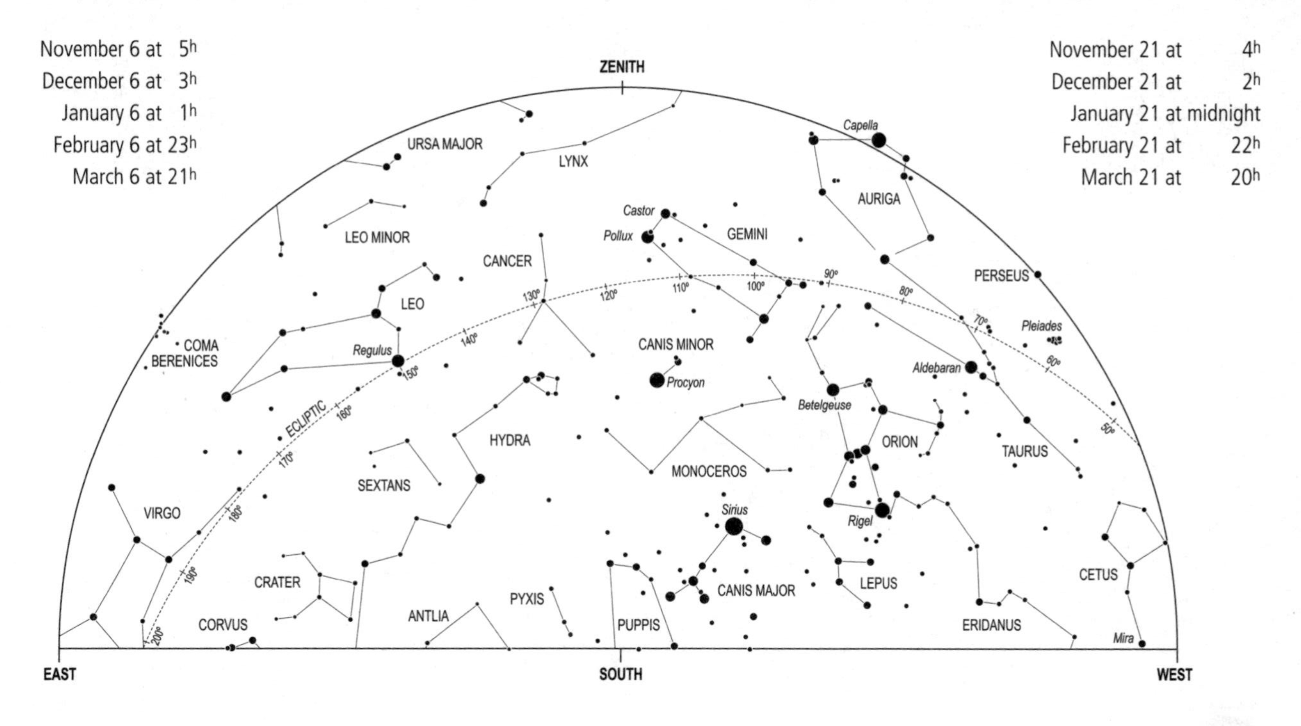
November 6 at 5h
December 6 at 3h
January 6 at 1h
February 6 at 23h
March 6 at 21h
November 21 at 4h
December 21 at 2h
January 21 at midnight
February 21 at 22h
March 21 at 20h
ZENITH
EAST
SOUTH
WEST
URSA MAJOR
LYNX
Capella
AURIGA
LEO MINOR
CANCER
Castor
Pollux
GEMINI
PERSEUS
LEO
COMA
BERENICES
Regulus
CANIS MINOR
Procyon
Pleiades
Aldebaran
Betelgeuse
ECLIPTIC
HYDRA
ORION
TAURUS
MONOCEROS
SEXTANS
VIRGO
Sirius
Rigel
CETUS
CRATER
CANIS MAJOR
LEPUS
PYXIS
ANTLIA
CORVUS
PUPPIS
ERIDANUS
Mira
50°
60°
70°
80°
90°
100°
110°
120°
130°
140°
150°
160°
170°
180°
190°
200°

2S

3N

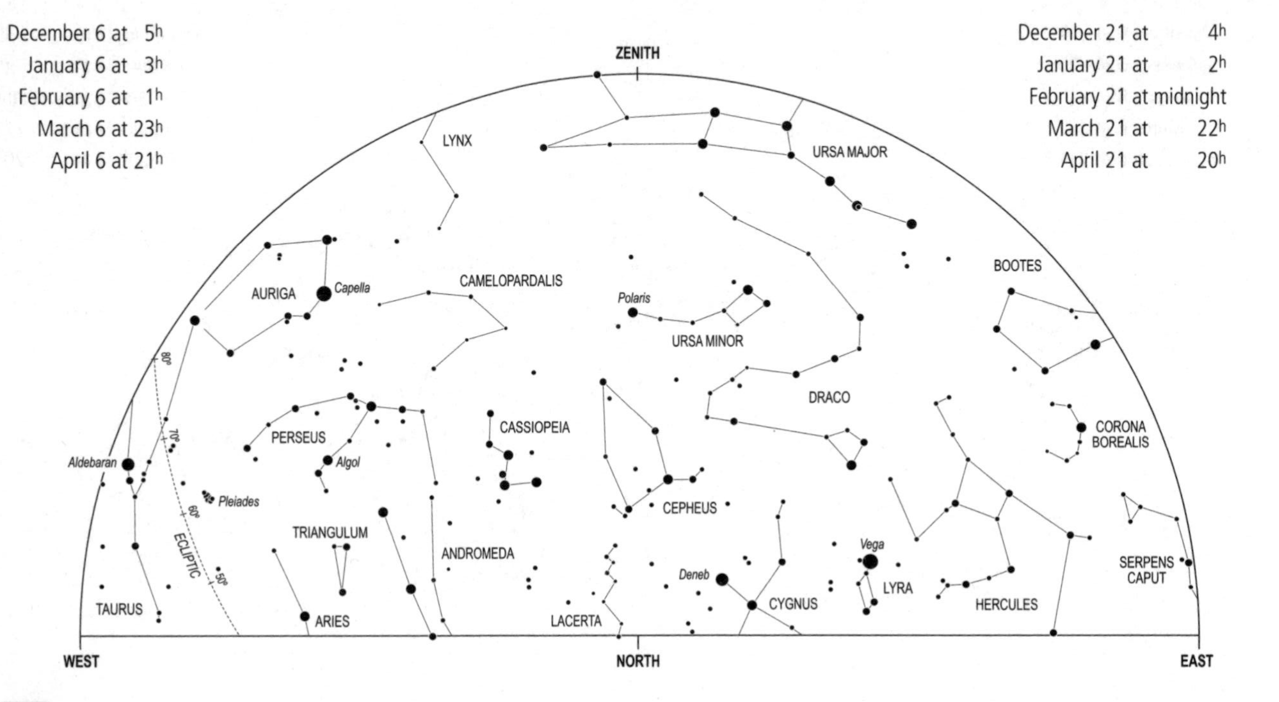
December 6 at 5h
January 6 at 3h
February 6 at 1h
March 6 at 23h
April 6 at 21h
December 21 at 4h
January 21 at 2h
February 21 at midnight
March 21 at 22h
April 21 at 20h
ZENITH
WEST
NORTH
EAST
LYNX
URSA MAJOR
AURIGA
Capella
CAMELOPARDALIS
Polaris
URSA MINOR
BOOTES
80°
70°
60°
50°
ECLIPTIC
PERSEUS
Algol
Aldebaran
Pleiades
CASSIOPEIA
DRACO
CORONA BOREALIS
TRIANGULUM
ANDROMEDA
CEPHEUS
Vega
Deneb
LYRA
SERPENS CAPUT
TAURUS
ARIES
LACERTA
CYGNUS
HERCULES

December 6 at 5^h
January 6 at 3^h
February 6 at 1^h
March 6 at 23^h
April 6 at 21^h

December 21 at 4^h
January 21 at 2^h
February 21 at midnight
March 21 at 22^h
April 21 at 20^h

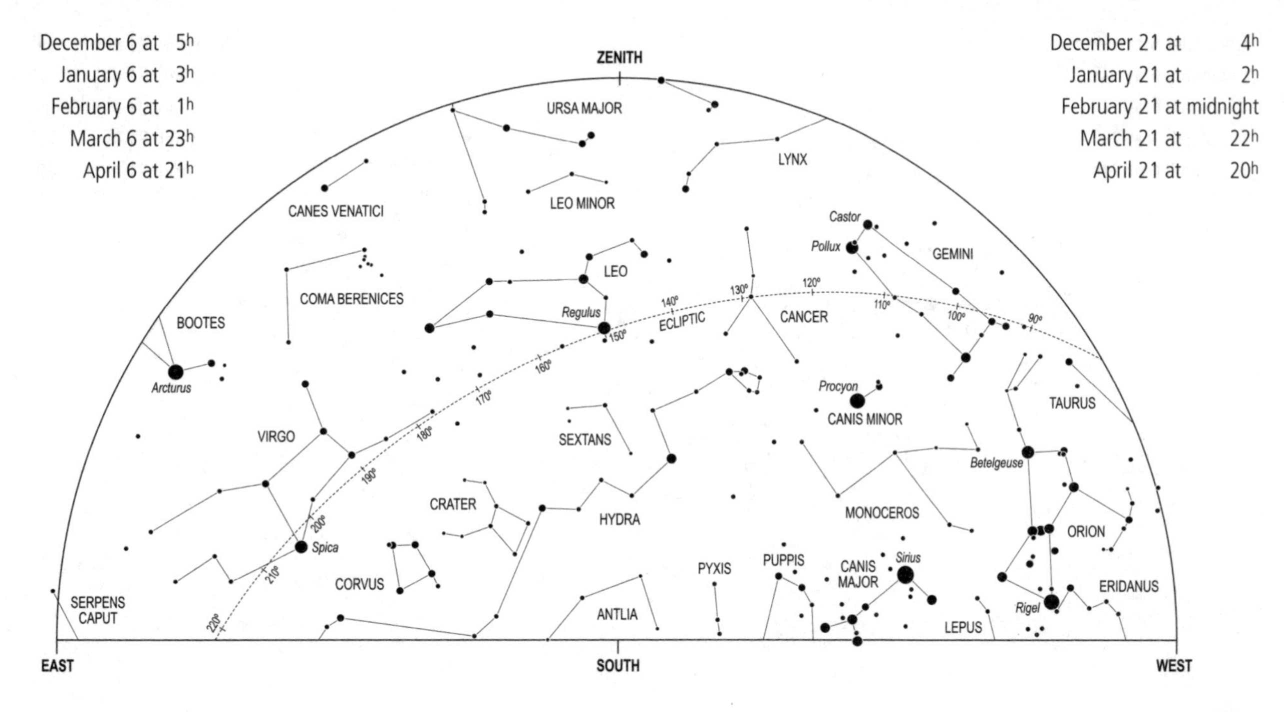

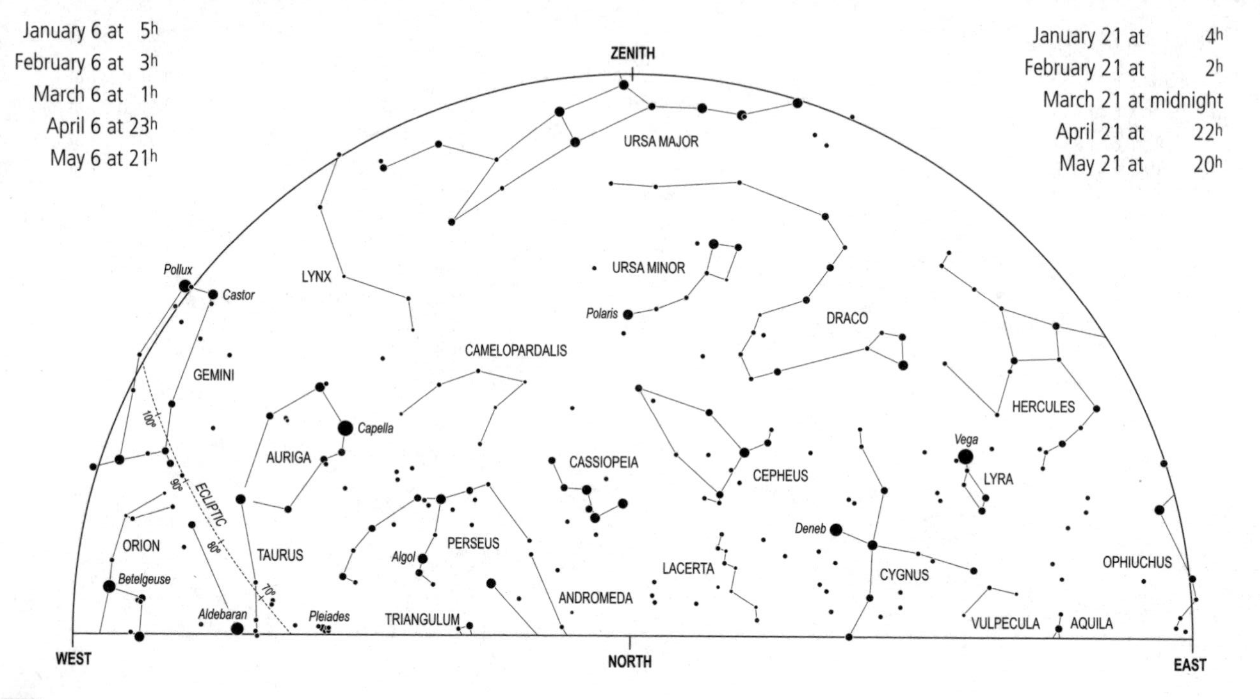
4N
January 6 at 5h
February 6 at 3h
March 6 at 1h
April 6 at 23h
May 6 at 21h
January 21 at 4h
February 21 at 2h
March 21 at midnight
April 21 at 22h
May 21 at 20h
ZENITH
WEST
NORTH
EAST
URSA MAJOR
URSA MINOR
Polaris
LYNX
Pollux
Castor
GEMINI
CAMELOPARDALIS
DRACO
HERCULES
Capella
AURIGA
CASSIOPEIA
CEPHEUS
Vega
LYRA
100°
90°
80°
70°
ECLIPTIC
ORION
Betelgeuse
TAURUS
Aldebaran
Pleiades
Algol
PERSEUS
TRIANGULUM
ANDROMEDA
LACERTA
Deneb
CYGNUS
OPHIUCHUS
VULPECULA
AQUILA

4S

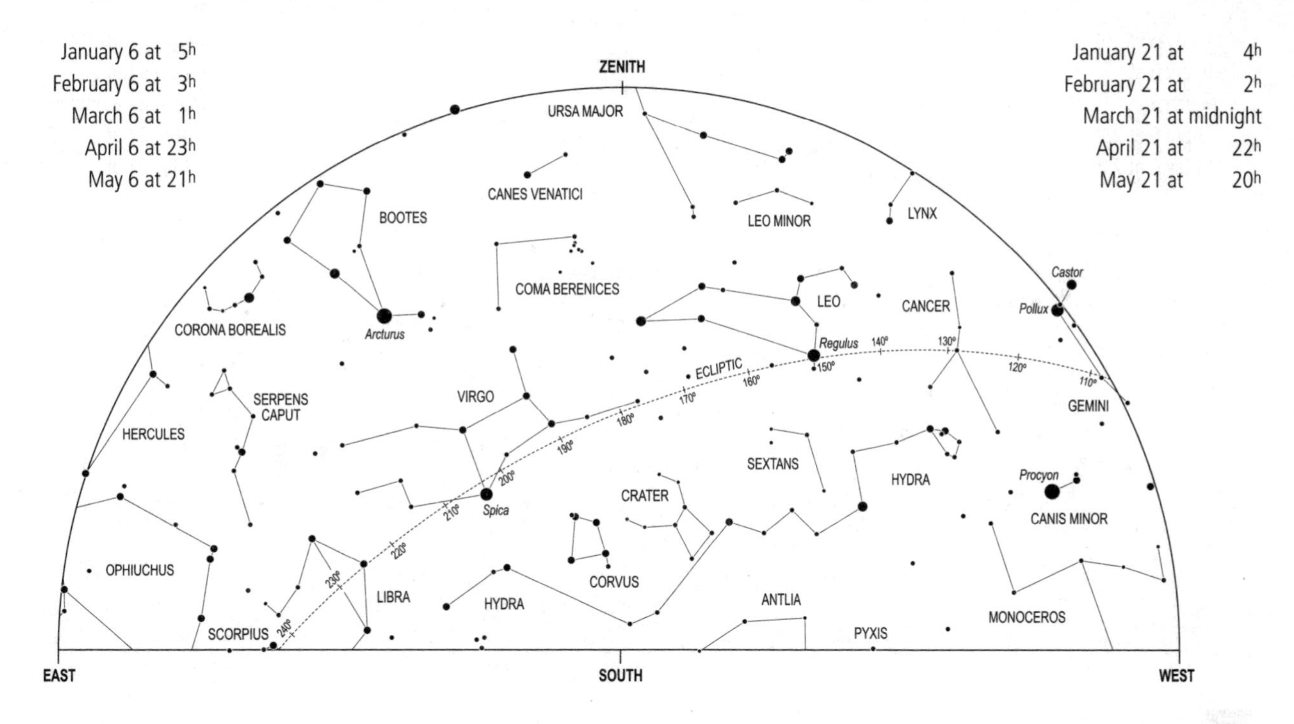

January 6 at 5h
February 6 at 3h
March 6 at 1h
April 6 at 23h
May 6 at 21h
January 21 at 4h
February 21 at 2h
March 21 at midnight
April 21 at 22h
May 21 at 20h
ZENITH
EAST
SOUTH
WEST
URSA MAJOR
CANES VENATICI
BOOTES
LEO MINOR
LYNX
COMA BERENICES
LEO
CANCER
Castor
Pollux
CORONA BOREALIS
Arcturus
Regulus
ECLIPTIC
110°
120°
130°
140°
150°
160°
170°
180°
190°
200°
210°
220°
230°
240°
GEMINI
SERPENS CAPUT
VIRGO
HERCULES
SEXTANS
HYDRA
Procyon
CRATER
Spica
CANIS MINOR
OPHIUCHUS
CORVUS
LIBRA
HYDRA
ANTLIA
MONOCEROS
SCORPIUS
PYXIS

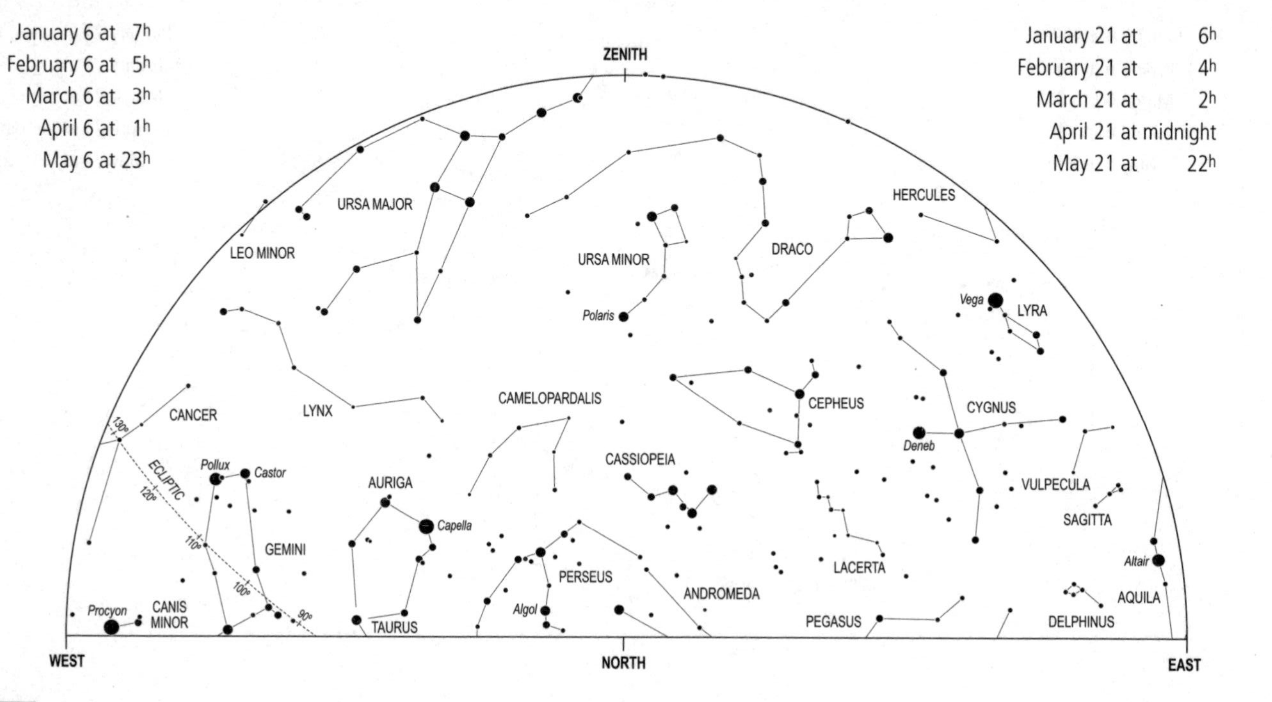
5N
January 6 at 7h
February 6 at 5h
March 6 at 3h
April 6 at 1h
May 6 at 23h
January 21 at 6h
February 21 at 4h
March 21 at 2h
April 21 at midnight
May 21 at 22h
ZENITH
WEST
NORTH
EAST
ECLIPTIC
130°
120°
110°
100°
90°
URSA MAJOR
LEO MINOR
URSA MINOR
DRACO
HERCULES
Polaris
Vega
LYRA
CANCER
LYNX
CAMELOPARDALIS
CEPHEUS
CYGNUS
Deneb
Pollux
Castor
AURIGA
CASSIOPEIA
VULPECULA
Capella
SAGITTA
GEMINI
PERSEUS
LACERTA
Altair
ANDROMEDA
AQUILA
Procyon
CANIS MINOR
Algol
PEGASUS
DELPHINUS
TAURUS

5S

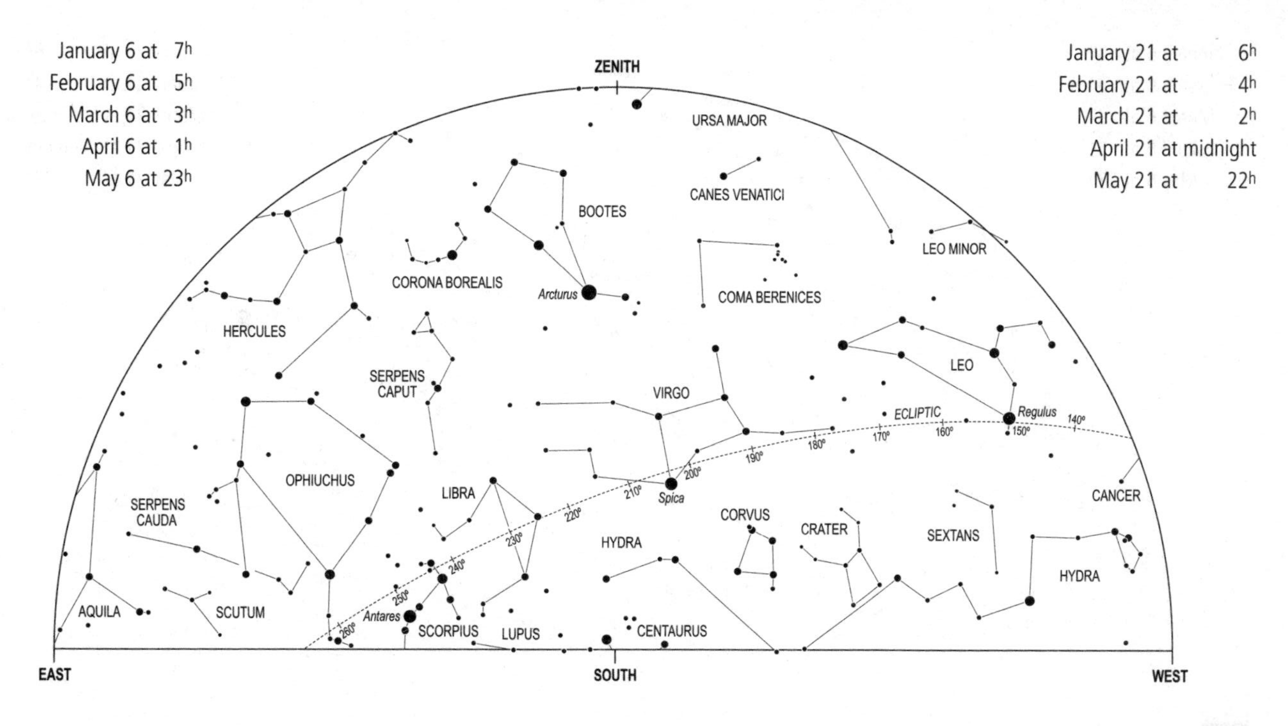
January 6 at 7h
February 6 at 5h
March 6 at 3h
April 6 at 1h
May 6 at 23h
January 21 at 6h
February 21 at 4h
March 21 at 2h
April 21 at midnight
May 21 at 22h
ZENITH
URSA MAJOR
CANES VENATICI
BOOTES
LEO MINOR
CORONA BOREALIS
Arcturus
COMA BERENICES
HERCULES
LEO
SERPENS CAPUT
VIRGO
ECLIPTIC
Regulus
140°
150°
160°
170°
180°
190°
200°
210°
220°
230°
240°
250°
260°
OPHIUCHUS
LIBRA
Spica
CANCER
SERPENS CAUDA
CORVUS
CRATER
SEXTANS
HYDRA
HYDRA
AQUILA
SCUTUM
Antares
SCORPIUS
LUPUS
CENTAURUS
EAST
SOUTH
WEST

6N

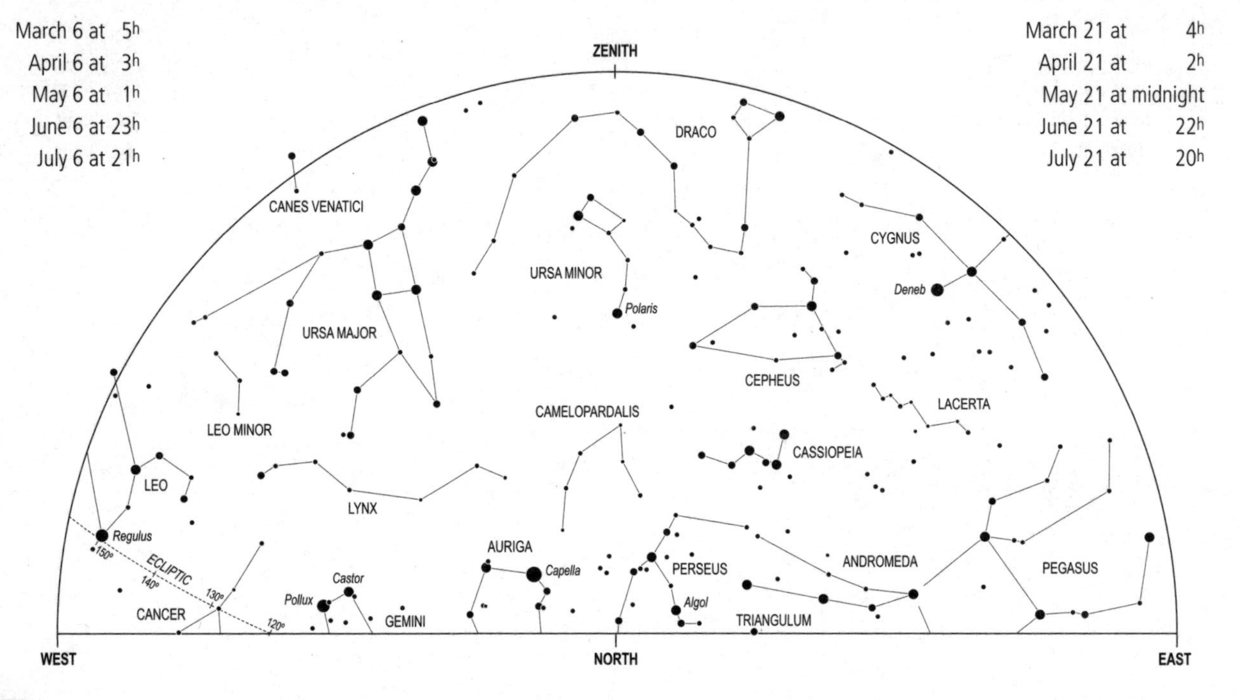
March 6 at 5h
April 6 at 3h
May 6 at 1h
June 6 at 23h
July 6 at 21h
March 21 at 4h
April 21 at 2h
May 21 at midnight
June 21 at 22h
July 21 at 20h
ZENITH
WEST
NORTH
EAST
CANES VENATICI
DRACO
URSA MINOR
Polaris
CYGNUS
Deneb
URSA MAJOR
CEPHEUS
LACERTA
LEO MINOR
CAMELOPARDALIS
CASSIOPEIA
LEO
LYNX
Regulus
ECLIPTIC
150°
140°
130°
120°
AURIGA
Capella
PERSEUS
ANDROMEDA
PEGASUS
Castor
Pollux
CANCER
GEMINI
Algol
TRIANGULUM

March 6 at 5^h
April 6 at 3^h
May 6 at 1^h
June 6 at 23^h
July 6 at 21^h

March 21 at 4^h
April 21 at 2^h
May 21 at midnight
June 21 at 22^h
July 21 at 20^h

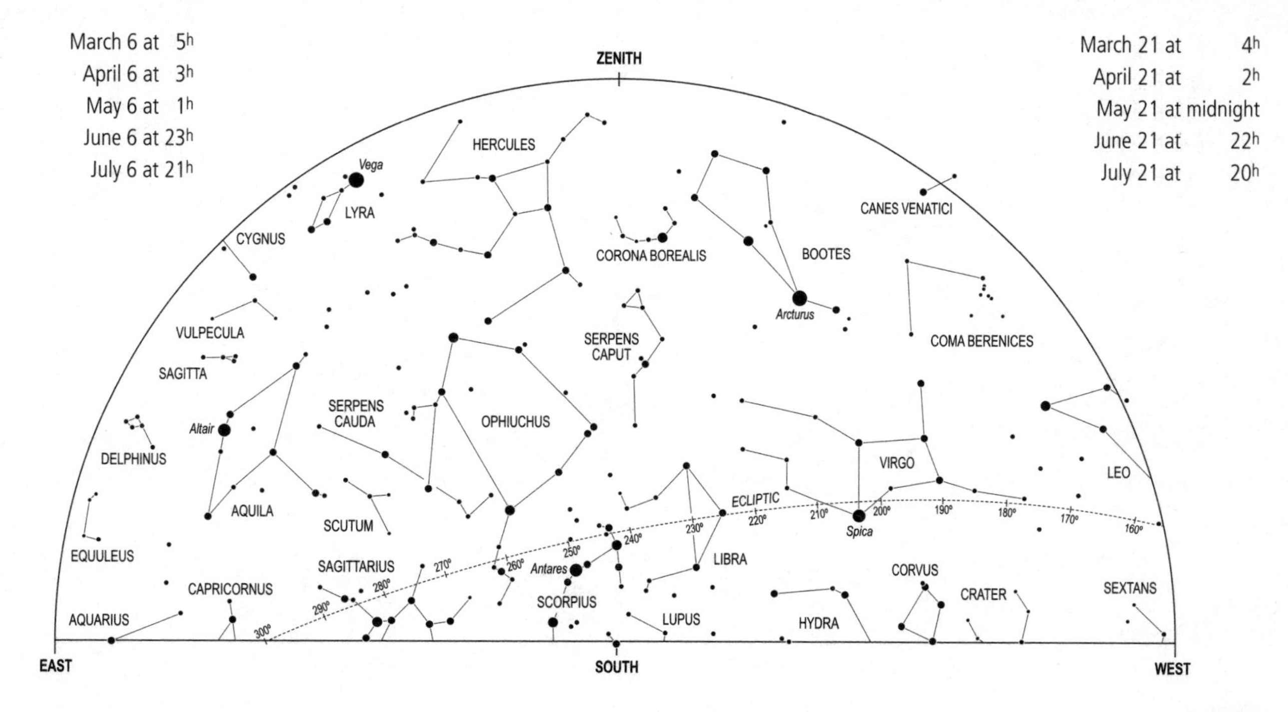

7N

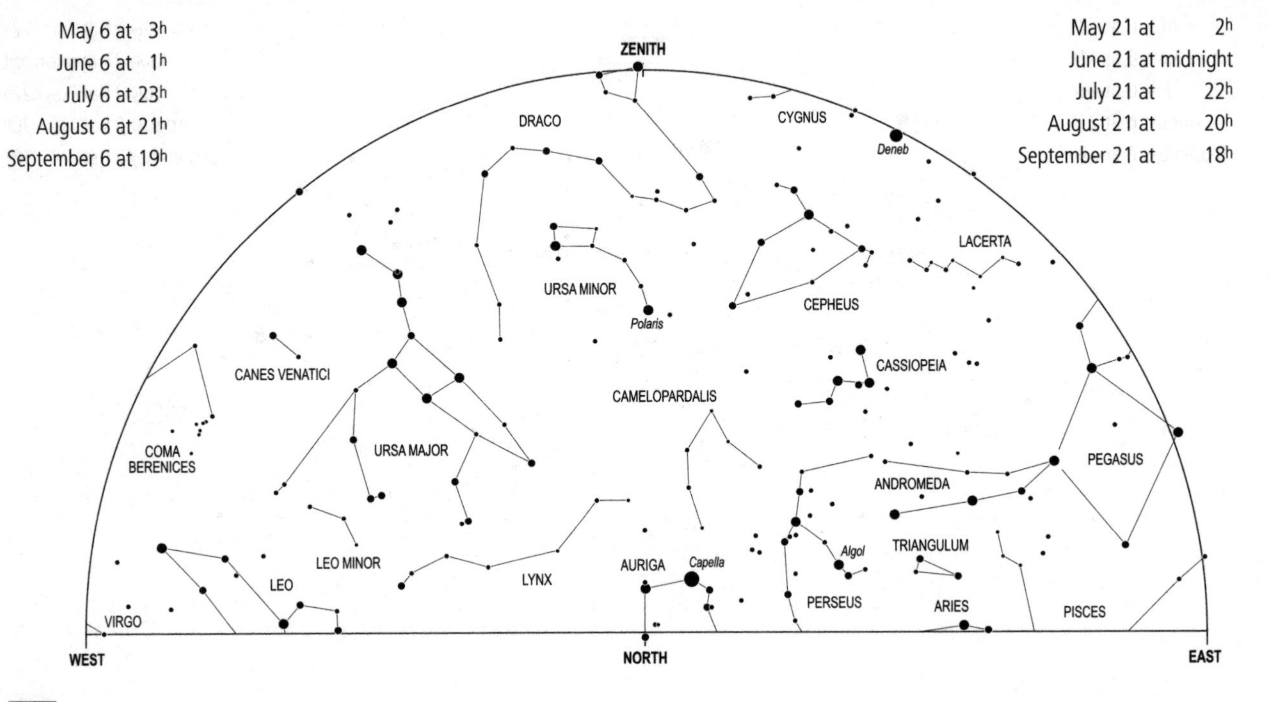
May 6 at 3h
June 6 at 1h
July 6 at 23h
August 6 at 21h
September 6 at 19h
May 21 at 2h
June 21 at midnight
July 21 at 22h
August 21 at 20h
September 21 at 18h
ZENITH
DRACO
CYGNUS
Deneb
LACERTA
URSA MINOR
Polaris
CEPHEUS
CANES VENATICI
CASSIOPEIA
CAMELOPARDALIS
COMA
BERENICES
URSA MAJOR
PEGASUS
ANDROMEDA
LEO MINOR
LYNX
AURIGA
Capella
Algol
TRIANGULUM
LEO
VIRGO
PERSEUS
ARIES
PISCES
WEST
NORTH
EAST

7S

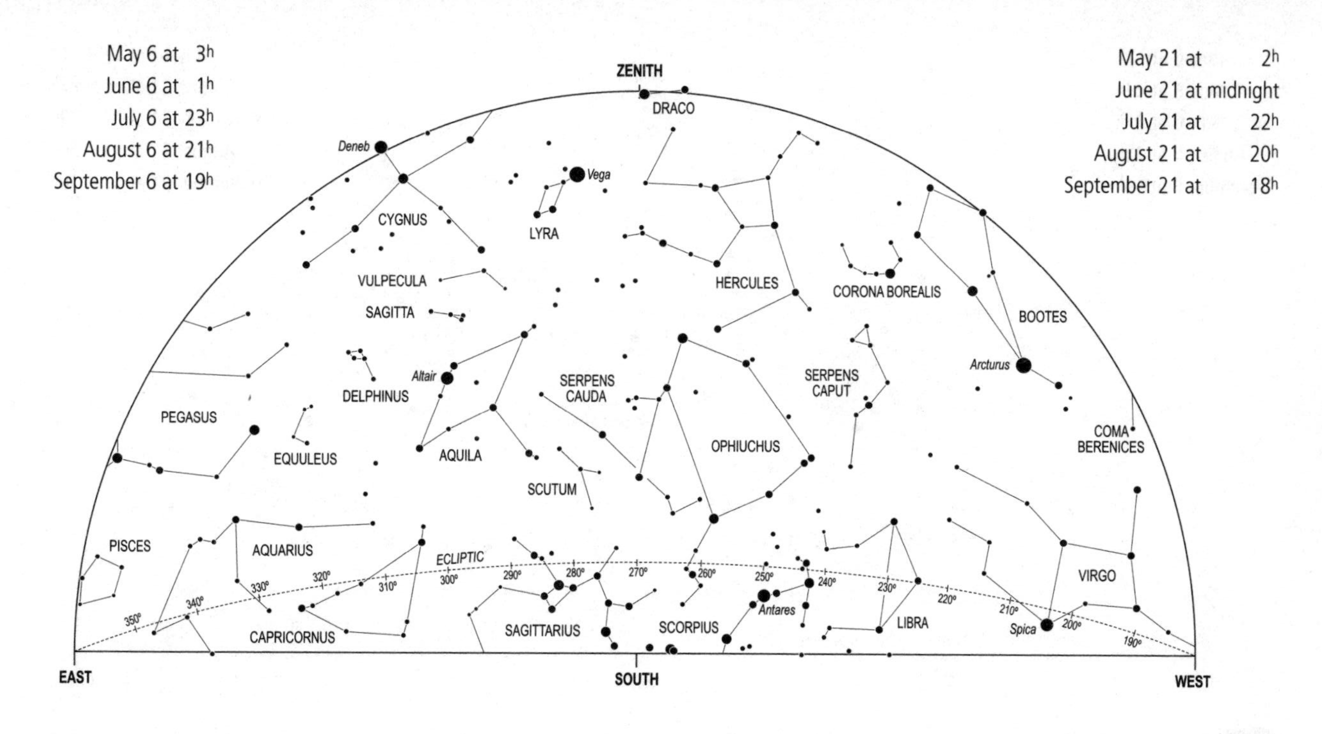
May 6 at 3h
June 6 at 1h
July 6 at 23h
August 6 at 21h
September 6 at 19h
May 21 at 2h
June 21 at midnight
July 21 at 22h
August 21 at 20h
September 21 at 18h
ZENITH
DRACO
Deneb
Vega
CYGNUS
LYRA
VULPECULA
HERCULES
CORONA BOREALIS
SAGITTA
BOOTES
Altair
DELPHINUS
SERPENS CAUDA
SERPENS CAPUT
Arcturus
PEGASUS
EQUULEUS
AQUILA
OPHIUCHUS
COMA BERENICES
SCUTUM
PISCES
AQUARIUS
ECLIPTIC
VIRGO
350°
340°
330°
320°
310°
300°
290°
280°
270°
260°
250°
240°
230°
220°
210°
200°
190°
Antares
LIBRA
Spica
CAPRICORNUS
SAGITTARIUS
SCORPIUS
EAST
SOUTH
WEST

8N

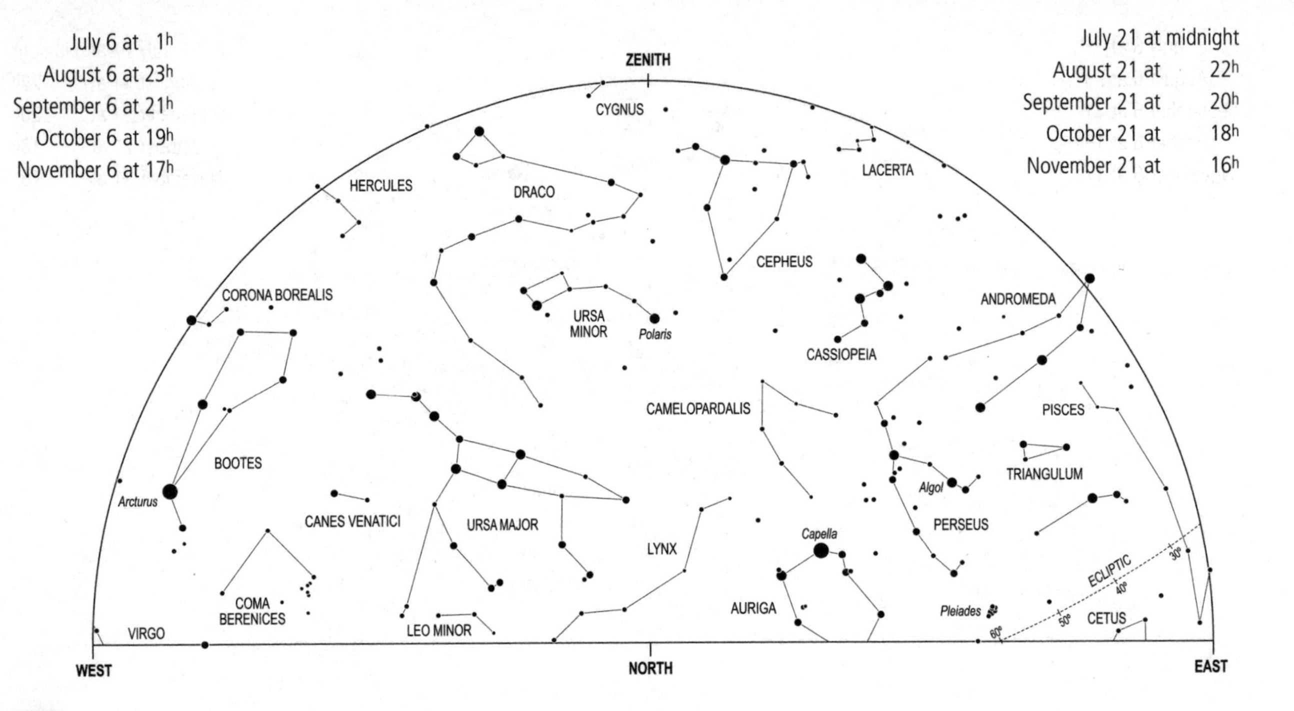

July 6 at 1h
August 6 at 23h
September 6 at 21h
October 6 at 19h
November 6 at 17h
July 21 at midnight
August 21 at 22h
September 21 at 20h
October 21 at 18h
November 21 at 16h
ZENITH
CYGNUS
HERCULES
DRACO
LACERTA
CEPHEUS
CORONA BOREALIS
URSA MINOR
Polaris
CASSIOPEIA
ANDROMEDA
CAMELOPARDALIS
PISCES
BOOTES
TRIANGULUM
Algol
Arcturus
CANES VENATICI
URSA MAJOR
PERSEUS
Capella
LYNX
ECLIPTIC
30°
40°
50°
60°
COMA BERENICES
AURIGA
Pleiades
CETUS
VIRGO
LEO MINOR
WEST
NORTH
EAST

S8

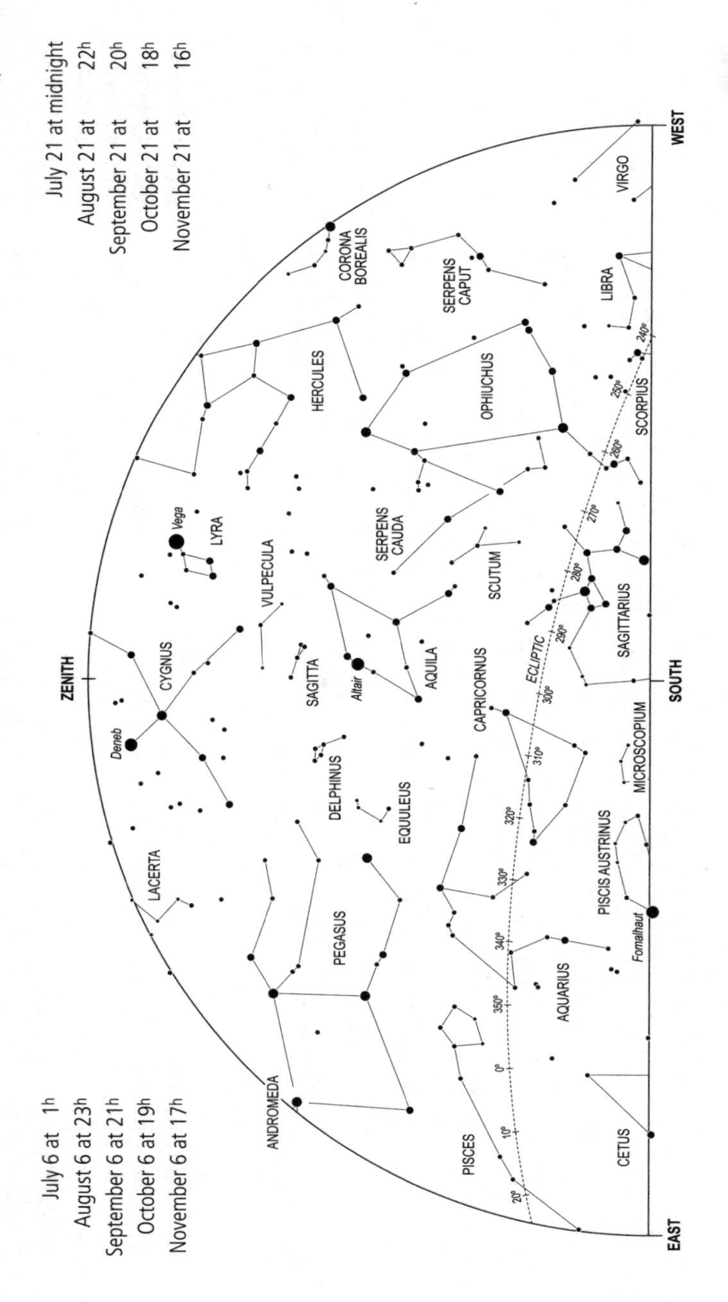
July 21 at midnight
August 21 at 22h
September 21 at 20h
October 21 at 18h
November 21 at 16h
July 6 at 1h
August 6 at 23h
September 6 at 21h
October 6 at 19h
November 6 at 17h
ZENITH
WEST
SOUTH
EAST
VIRGO
CORONA BOREALIS
SERPENS CAPUT
LIBRA
HERCULES
OPHIUCHUS
SCORPIUS
SERPENS CAUDA
LYRA
Vega
VULPECULA
SCUTUM
SAGITTARIUS
CYGNUS
SAGITTA
Altair
AQUILA
CAPRICORNUS
ECLIPTIC
Deneb
DELPHINUS
EQUULEUS
MICROSCOPIUM
PISCIS AUSTRINUS
LACERTA
PEGASUS
Fomalhaut
AQUARIUS
ANDROMEDA
PISCES
CETUS
240°
250°
260°
270°
280°
290°
300°
310°
320°
330°
340°
350°
0°
10°
20°

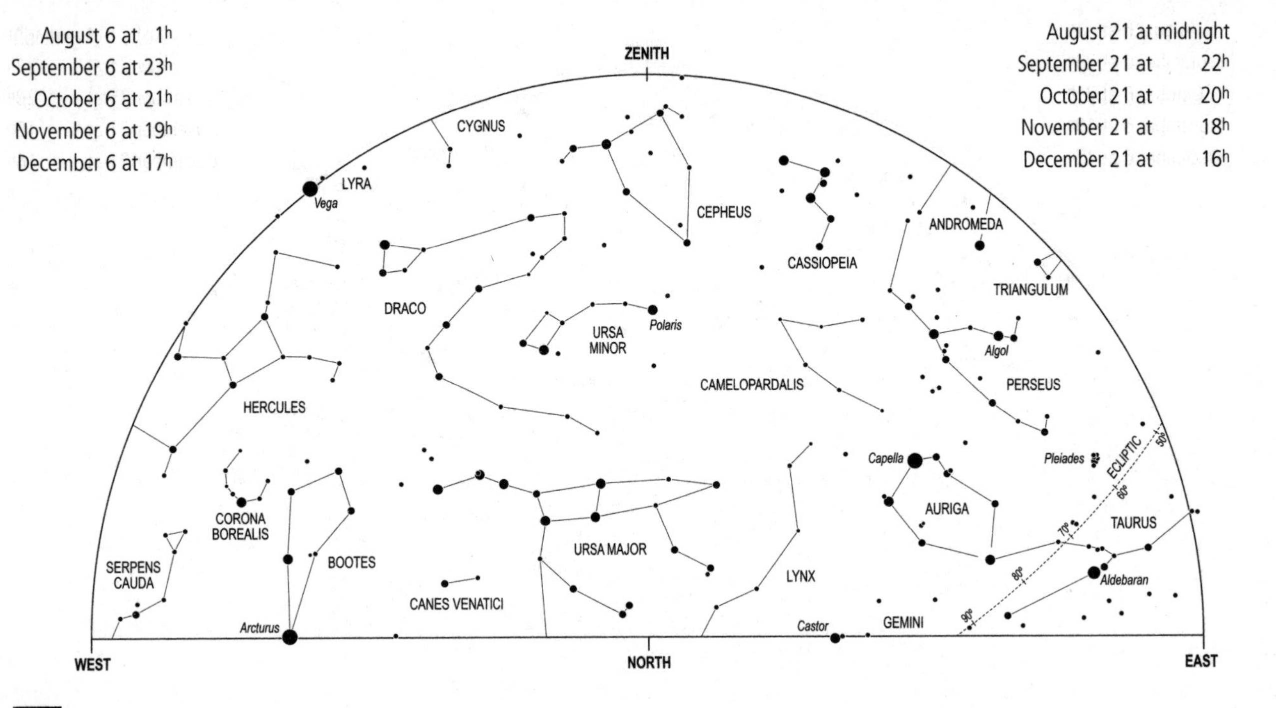
August 6 at 1h
September 6 at 23h
October 6 at 21h
November 6 at 19h
December 6 at 17h
August 21 at midnight
September 21 at 22h
October 21 at 20h
November 21 at 18h
December 21 at 16h
ZENITH
CYGNUS
LYRA
Vega
CEPHEUS
CASSIOPEIA
ANDROMEDA
TRIANGULUM
DRACO
URSA MINOR
Polaris
Algol
HERCULES
CAMELOPARDALIS
PERSEUS
Capella
Pleiades
ECLIPTIC
50°
60°
70°
80°
90°
AURIGA
TAURUS
CORONA BOREALIS
BOOTES
URSA MAJOR
SERPENS CAUDA
LYNX
Aldebaran
CANES VENATICI
Arcturus
Castor
GEMINI
WEST
NORTH
EAST

9N

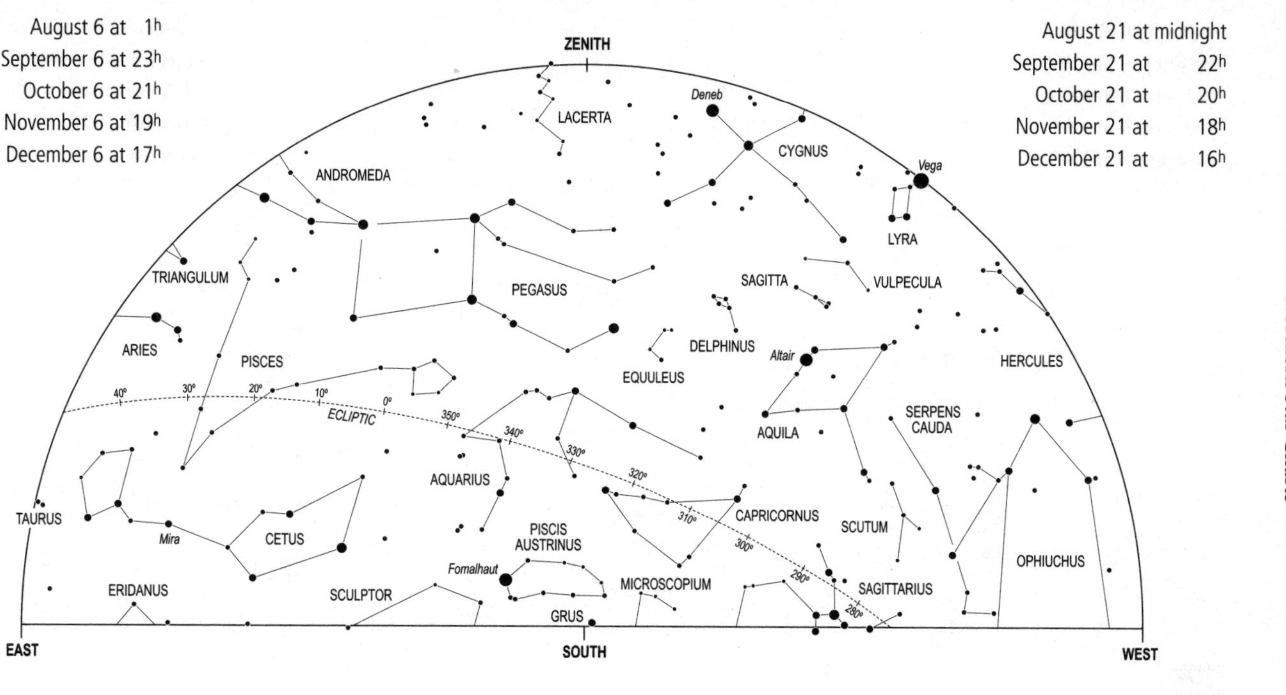
August 6 at 1h
September 6 at 23h
October 6 at 21h
November 6 at 19h
December 6 at 17h
August 21 at midnight
September 21 at 22h
October 21 at 20h
November 21 at 18h
December 21 at 16h
ZENITH
LACERTA
Deneb
CYGNUS
Vega
LYRA
ANDROMEDA
TRIANGULUM
PEGASUS
SAGITTA
VULPECULA
ARIES
PISCES
DELPHINUS
Altair
HERCULES
EQUULEUS
40°
30°
20°
10°
0°
350°
340°
330°
320°
310°
300°
290°
280°
ECLIPTIC
AQUILA
SERPENS CAUDA
AQUARIUS
TAURUS
CAPRICORNUS
SCUTUM
PISCIS AUSTRINUS
Mira
CETUS
OPHIUCHUS
Fomalhaut
ERIDANUS
SCULPTOR
MICROSCOPIUM
SAGITTARIUS
GRUS
EAST
SOUTH
WEST
9S

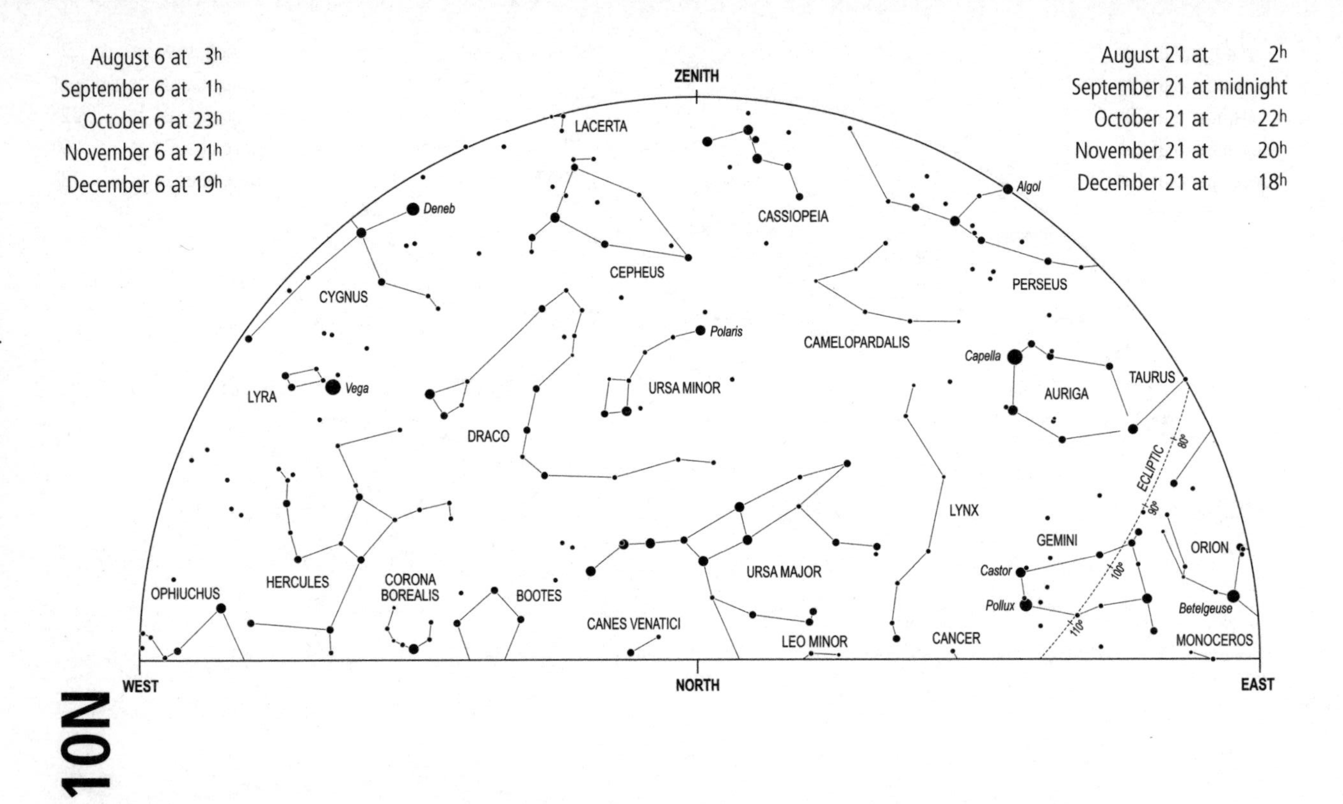
10N
August 6 at 3h
September 6 at 1h
October 6 at 23h
November 6 at 21h
December 6 at 19h
August 21 at 2h
September 21 at midnight
October 21 at 22h
November 21 at 20h
December 21 at 18h
ZENITH
WEST
NORTH
EAST
LACERTA
CASSIOPEIA
Algol
Deneb
CEPHEUS
PERSEUS
CYGNUS
Polaris
CAMELOPARDALIS
Capella
LYRA
Vega
URSA MINOR
AURIGA
TAURUS
DRACO
ECLIPTIC
80°
90°
100°
110°
LYNX
GEMINI
ORION
URSA MAJOR
Castor
Pollux
Betelgeuse
OPHIUCHUS
HERCULES
CORONA BOREALIS
BOOTES
CANES VENATICI
LEO MINOR
CANCER
MONOCEROS

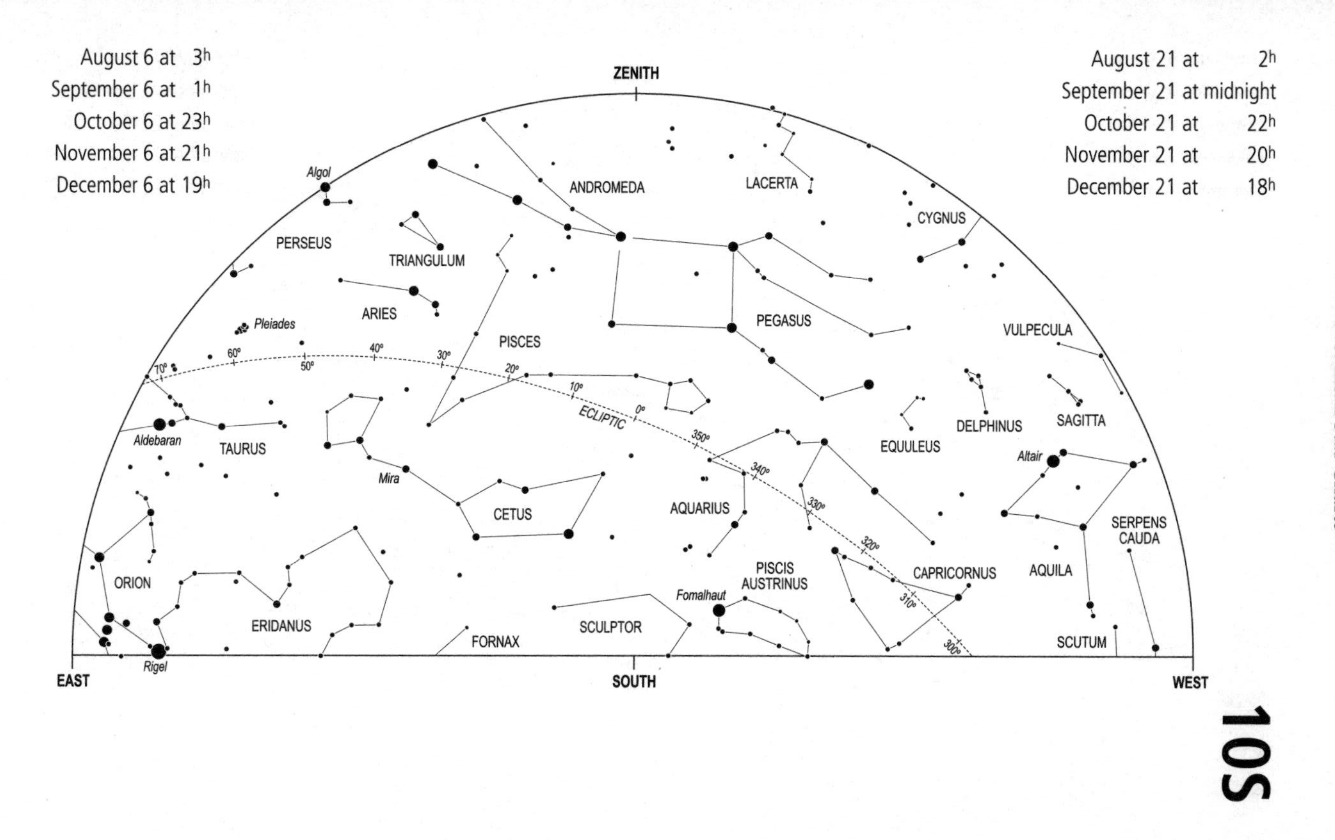
10S
August 6 at 3h
September 6 at 1h
October 6 at 23h
November 6 at 21h
December 6 at 19h
August 21 at 2h
September 21 at midnight
October 21 at 22h
November 21 at 20h
December 21 at 18h
ZENITH
EAST
SOUTH
WEST
ECLIPTIC
70°
60°
50°
40°
30°
20°
10°
0°
350°
340°
330°
320°
310°
300°
Algol
PERSEUS
TRIANGULUM
ANDROMEDA
LACERTA
CYGNUS
Pleiades
ARIES
PISCES
PEGASUS
VULPECULA
Aldebaran
TAURUS
Mira
CETUS
AQUARIUS
EQUULEUS
DELPHINUS
SAGITTA
Altair
ORION
ERIDANUS
FORNAX
SCULPTOR
Fomalhaut
PISCIS
AUSTRINUS
CAPRICORNUS
AQUILA
SERPENS
CAUDA
SCUTUM
Rigel

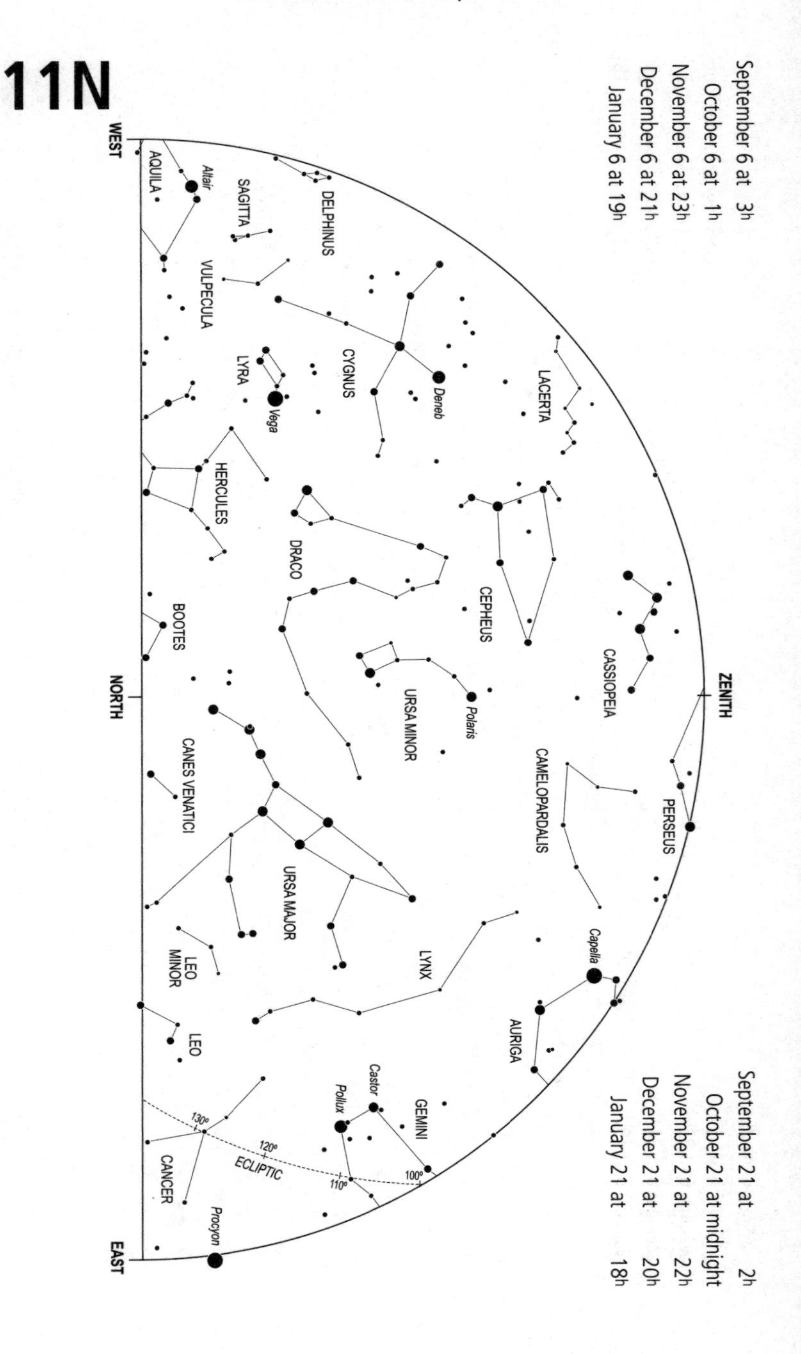
11N
September 6 at 3h
October 6 at 1h
November 6 at 23h
December 6 at 21h
January 6 at 19h
September 21 at 2h
October 21 at midnight
November 21 at 22h
December 21 at 20h
January 21 at 18h
WEST
NORTH
EAST
ZENITH
AQUILA
Altair
SAGITTA
DELPHINUS
VULPECULA
LYRA
Vega
CYGNUS
Deneb
LACERTA
HERCULES
DRACO
CEPHEUS
CASSIOPEIA
BOOTES
URSA MINOR
Polaris
CAMELOPARDALIS
PERSEUS
CANES VENATICI
URSA MAJOR
LYNX
Capella
AURIGA
LEO MINOR
LEO
GEMINI
Castor
Pollux
CANCER
Procyon
ECLIPTIC
130°
120°
110°
100°

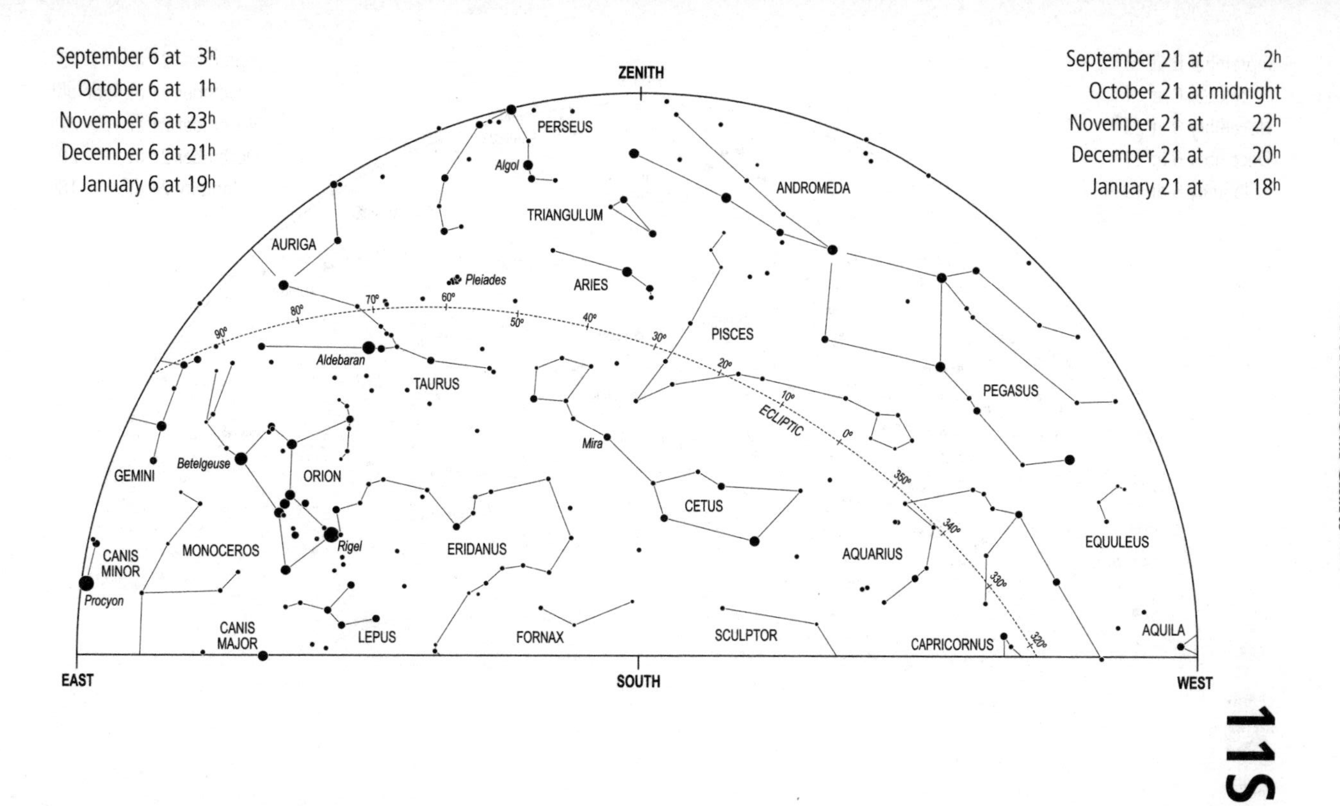

11S
September 6 at 3h
October 6 at 1h
November 6 at 23h
December 6 at 21h
January 6 at 19h
September 21 at 2h
October 21 at midnight
November 21 at 22h
December 21 at 20h
January 21 at 18h
ZENITH
EAST
SOUTH
WEST
ECLIPTIC
PERSEUS
Algol
TRIANGULUM
ANDROMEDA
AURIGA
Pleiades
ARIES
PISCES
PEGASUS
Aldebaran
TAURUS
Mira
GEMINI
Betelgeuse
ORION
CETUS
AQUARIUS
EQUULEUS
CANIS MINOR
Procyon
MONOCEROS
Rigel
ERIDANUS
CANIS MAJOR
LEPUS
FORNAX
SCULPTOR
CAPRICORNUS
AQUILA

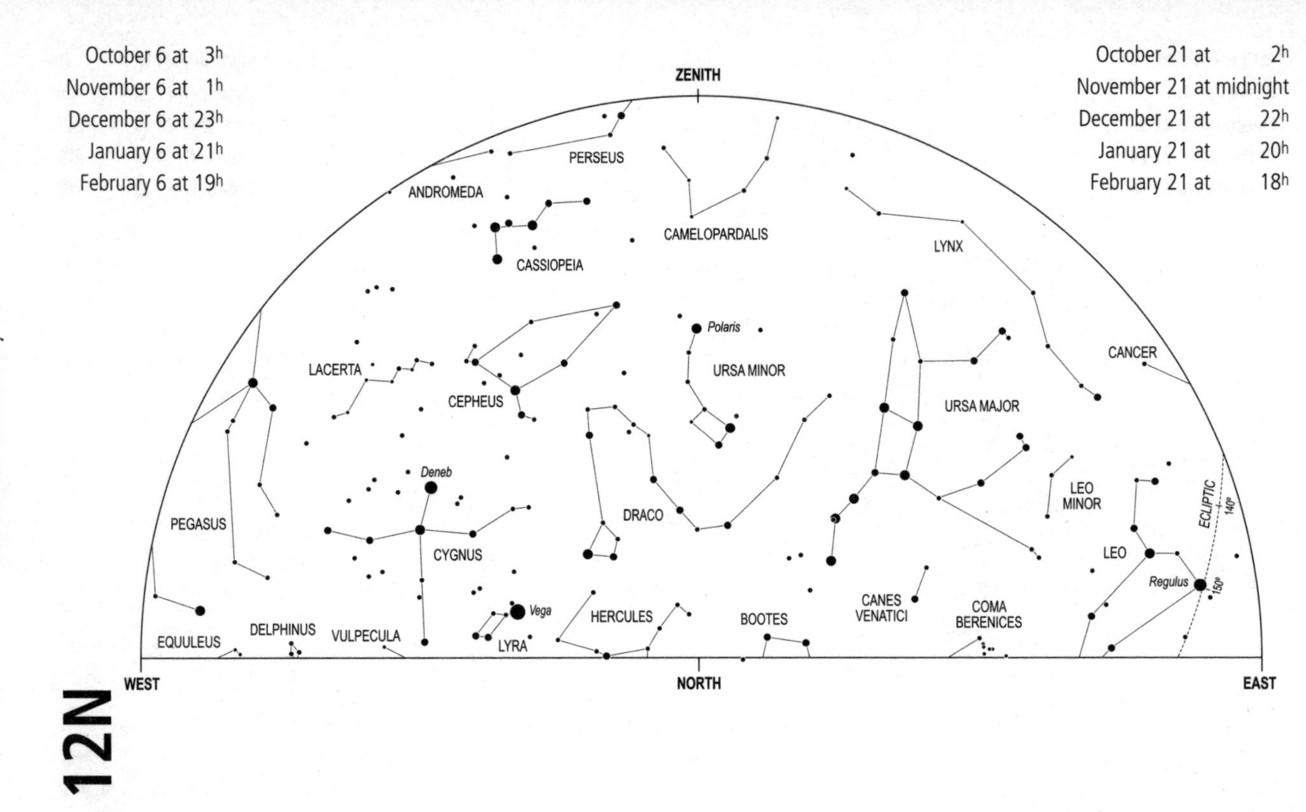

12N
October 6 at 3h
November 6 at 1h
December 6 at 23h
January 6 at 21h
February 6 at 19h
October 21 at 2h
November 21 at midnight
December 21 at 22h
January 21 at 20h
February 21 at 18h
ZENITH
WEST
NORTH
EAST
PERSEUS
ANDROMEDA
CASSIOPEIA
CAMELOPARDALIS
LYNX
Polaris
URSA MINOR
CANCER
LACERTA
CEPHEUS
URSA MAJOR
Deneb
LEO MINOR
PEGASUS
DRACO
CYGNUS
LEO
ECLIPTIC
140°
Regulus
150°
Vega
HERCULES
BOOTES
CANES VENATICI
COMA BERENICES
EQUULEUS
DELPHINUS
VULPECULA
LYRA

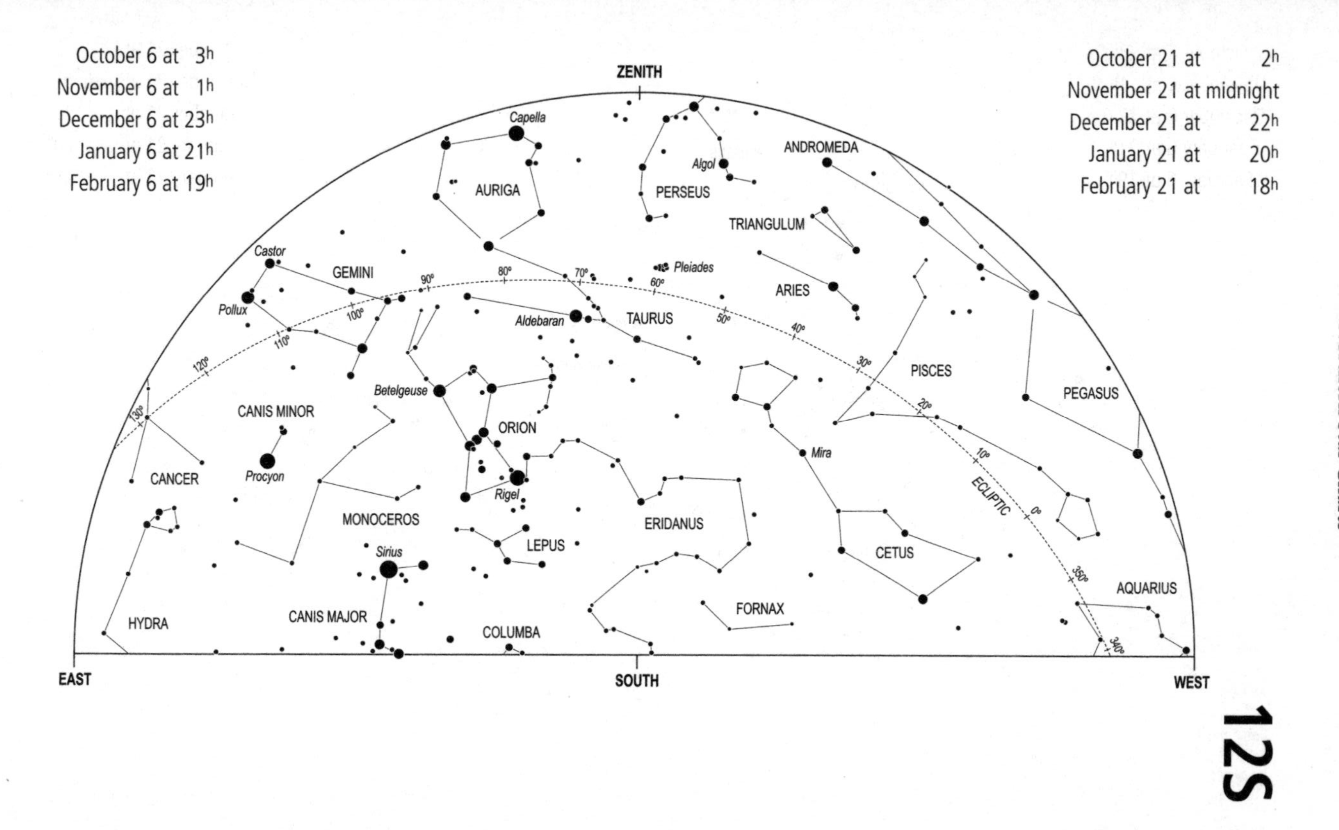

October 6 at 3h
November 6 at 1h
December 6 at 23h
January 6 at 21h
February 6 at 19h
October 21 at 2h
November 21 at midnight
December 21 at 22h
January 21 at 20h
February 21 at 18h
ZENITH
EAST
SOUTH
WEST
12S
Capella
AURIGA
PERSEUS
Algol
ANDROMEDA
TRIANGULUM
Pleiades
ARIES
Castor
GEMINI
Pollux
Aldebaran
TAURUS
PISCES
PEGASUS
Betelgeuse
CANIS MINOR
ORION
Procyon
CANCER
Mira
Rigel
MONOCEROS
ERIDANUS
ECLIPTIC
LEPUS
CETUS
Sirius
CANIS MAJOR
COLUMBA
FORNAX
AQUARIUS
HYDRA
130°
120°
110°
100°
90°
80°
70°
60°
50°
40°
30°
20°
10°
0°
350°
340°

Southern Star Charts

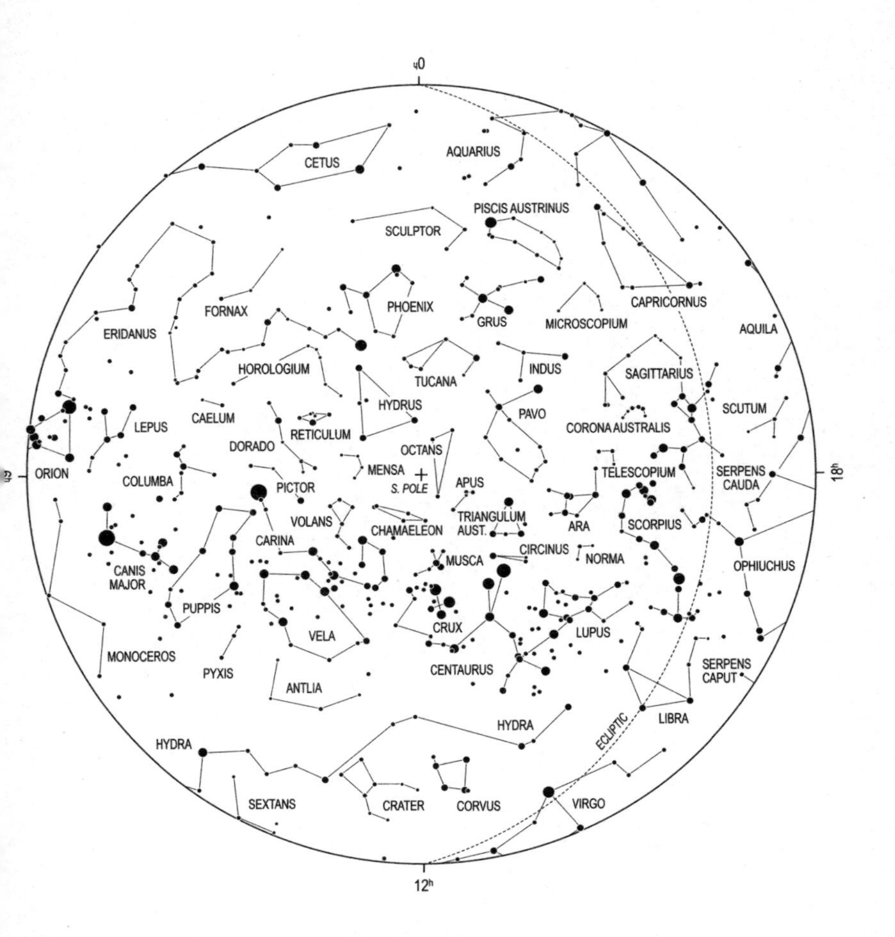

Southern Hemisphere

Note that the markers at 0^h, 6^h, 12^h and 18^h indicate hours of Right Ascension.

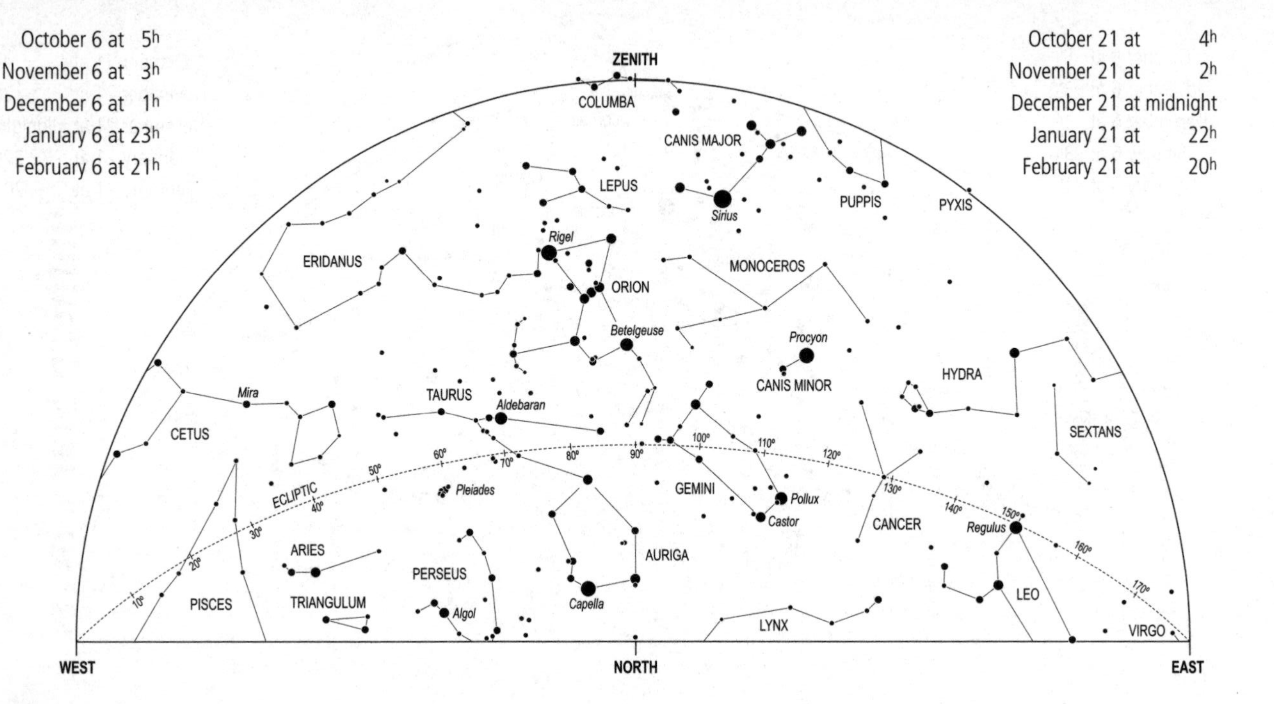
October 6 at 5h
November 6 at 3h
December 6 at 1h
January 6 at 23h
February 6 at 21h
October 21 at 4h
November 21 at 2h
December 21 at midnight
January 21 at 22h
February 21 at 20h
ZENITH
COLUMBA
CANIS MAJOR
LEPUS
Sirius
PUPPIS
PYXIS
ERIDANUS
Rigel
ORION
MONOCEROS
Betelgeuse
Procyon
CANIS MINOR
HYDRA
Mira
TAURUS
Aldebaran
CETUS
SEXTANS
ECLIPTIC
10°
20°
30°
40°
50°
60°
70°
80°
90°
100°
110°
120°
130°
140°
150°
160°
170°
Pleiades
GEMINI
Pollux
Castor
CANCER
Regulus
ARIES
PERSEUS
AURIGA
PISCES
TRIANGULUM
Algol
Capella
LYNX
LEO
VIRGO
WEST
NORTH
EAST

1N

1S

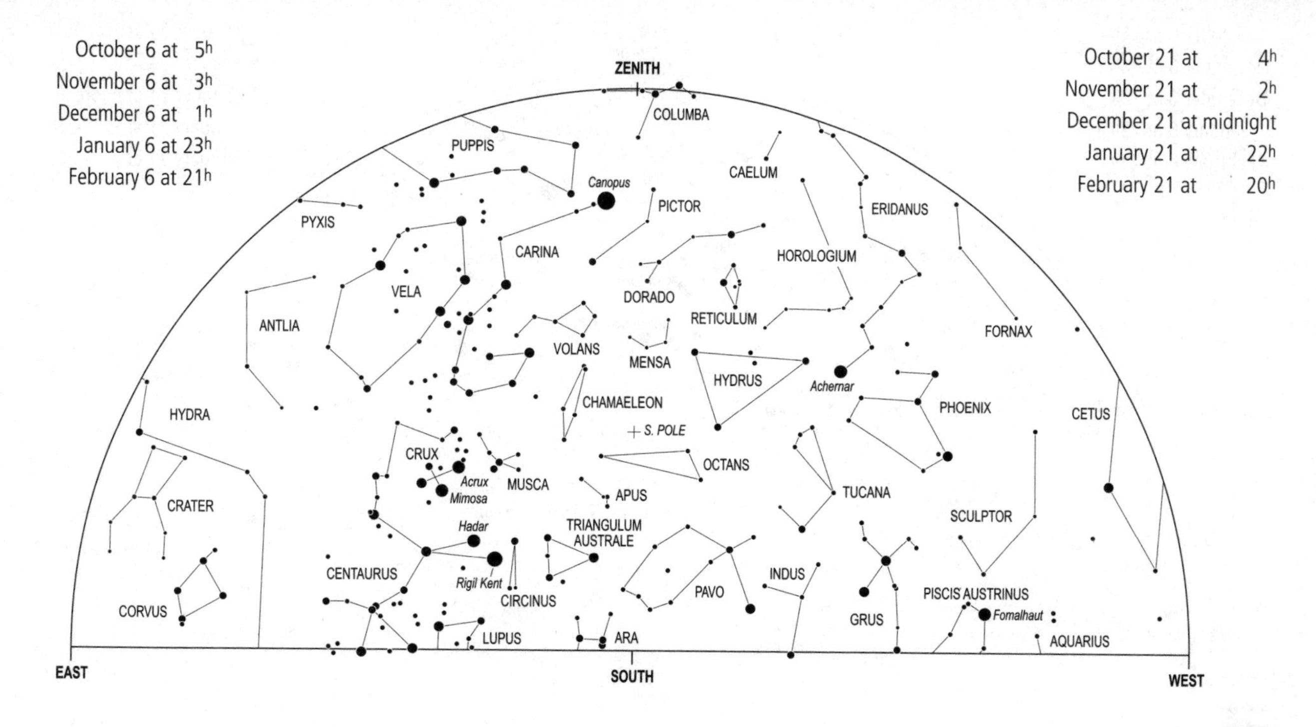
October 6 at 5h
November 6 at 3h
December 6 at 1h
January 6 at 23h
February 6 at 21h
October 21 at 4h
November 21 at 2h
December 21 at midnight
January 21 at 22h
February 21 at 20h
ZENITH
COLUMBA
PUPPIS
CAELUM
Canopus
PICTOR
ERIDANUS
PYXIS
CARINA
HOROLOGIUM
VELA
DORADO
RETICULUM
FORNAX
ANTLIA
VOLANS
MENSA
HYDRUS
Achernar
CHAMAELEON
PHOENIX
HYDRA
+ S. POLE
CETUS
CRUX
OCTANS
Acrux
MUSCA
Mimosa
APUS
TUCANA
CRATER
SCULPTOR
Hadar
TRIANGULUM AUSTRALE
CENTAURUS
INDUS
Rigil Kent
CIRCINUS
PAVO
PISCIS AUSTRINUS
CORVUS
Fomalhaut
GRUS
LUPUS
ARA
AQUARIUS
EAST
SOUTH
WEST

2N

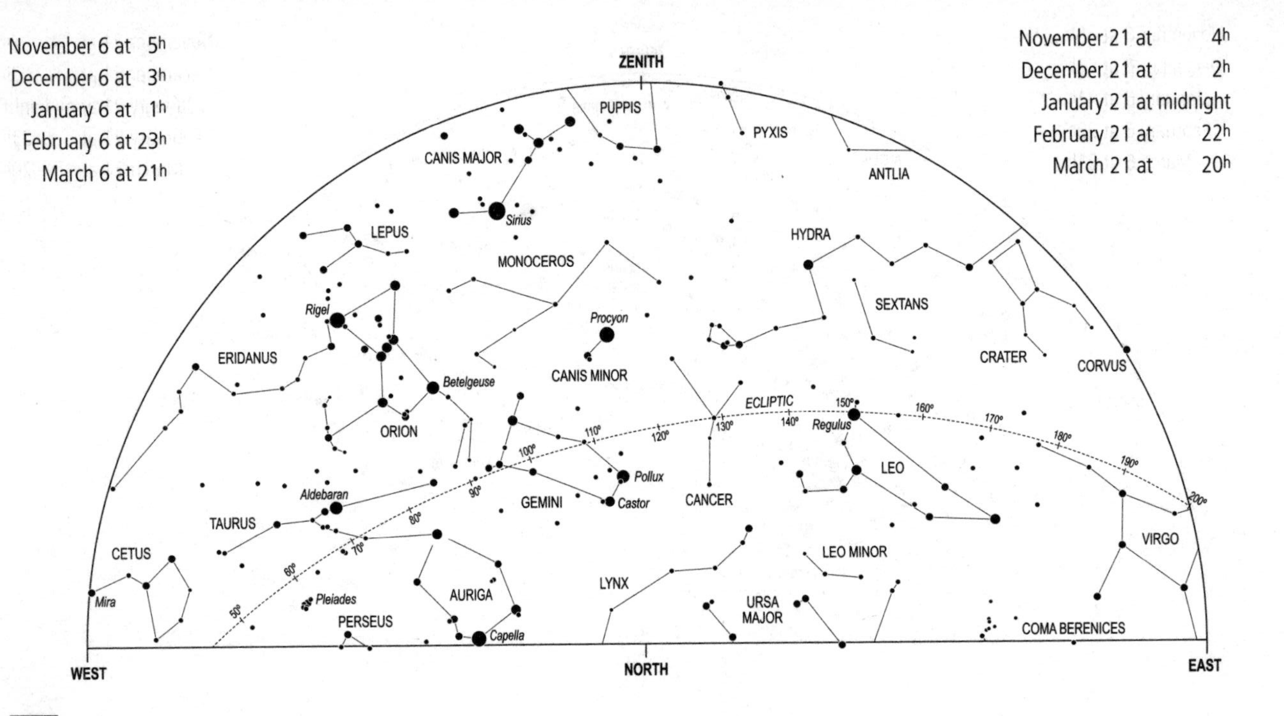
November 6 at 5h
December 6 at 3h
January 6 at 1h
February 6 at 23h
March 6 at 21h
November 21 at 4h
December 21 at 2h
January 21 at midnight
February 21 at 22h
March 21 at 20h
ZENITH
WEST
NORTH
EAST
ECLIPTIC
50°
60°
70°
80°
90°
100°
110°
120°
130°
140°
150°
160°
170°
180°
190°
200°
PUPPIS
PYXIS
CANIS MAJOR
ANTLIA
Sirius
LEPUS
HYDRA
MONOCEROS
SEXTANS
Rigel
Procyon
ERIDANUS
CRATER
CORVUS
CANIS MINOR
Betelgeuse
ORION
Regulus
LEO
Pollux
Aldebaran
GEMINI
Castor
CANCER
TAURUS
VIRGO
CETUS
LEO MINOR
AURIGA
LYNX
Mira
Pleiades
URSA MAJOR
PERSEUS
Capella
COMA BERENICES

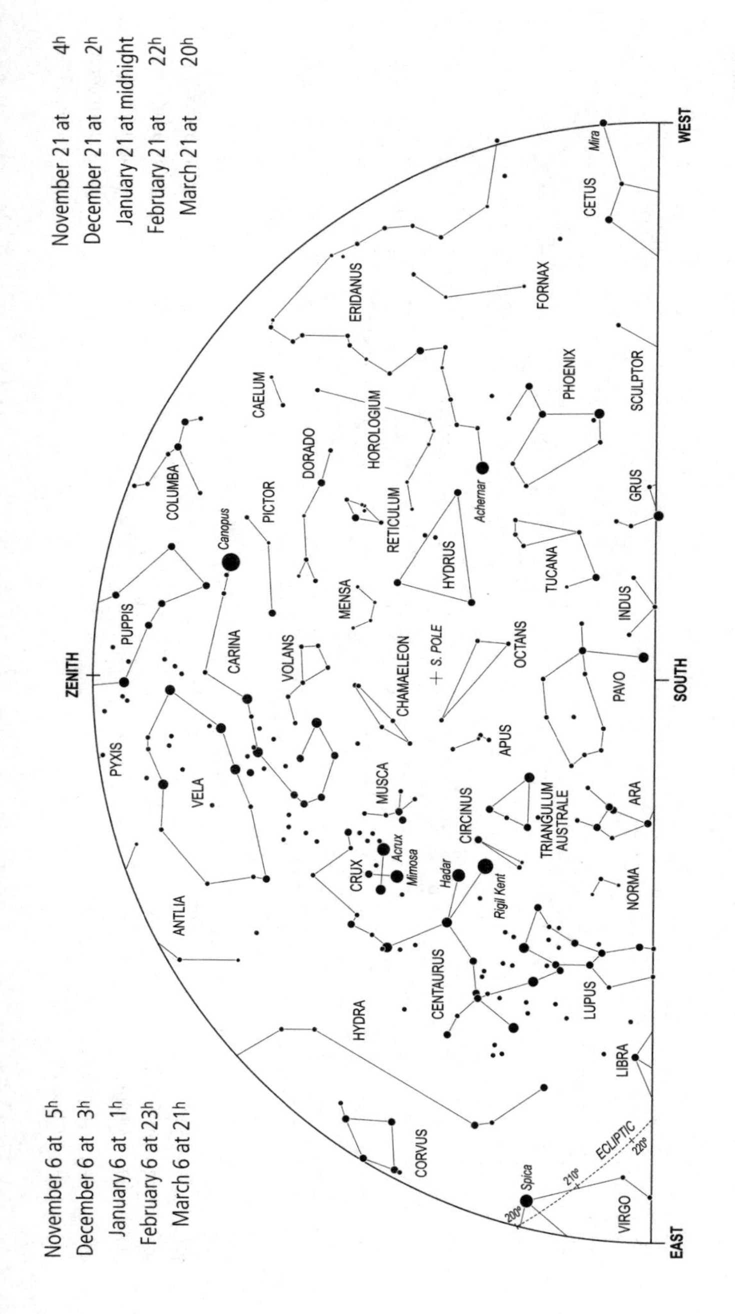
November 21 at 4h
December 21 at 2h
January 21 at midnight
February 21 at 22h
March 21 at 20h
November 6 at 5h
December 6 at 3h
January 6 at 1h
February 6 at 23h
March 6 at 21h
ZENITH
WEST
SOUTH
EAST
Mira
CETUS
FORNAX
ERIDANUS
SCULPTOR
PHOENIX
CAELUM
HOROLOGIUM
DORADO
COLUMBA
PICTOR
Achernar
GRUS
RETICULUM
Canopus
TUCANA
HYDRUS
PUPPIS
INDUS
MENSA
CARINA
VOLANS
S. POLE
OCTANS
CHAMAELEON
PAVO
PYXIS
APUS
VELA
MUSCA
ARA
CIRCINUS
TRIANGULUM AUSTRALE
Acrux
Mimosa
CRUX
Hadar
Rigil Kent
NORMA
ANTLIA
CENTAURUS
LUPUS
HYDRA
LIBRA
CORVUS
ECLIPTIC
220°
210°
200°
Spica
VIRGO

2S

3N

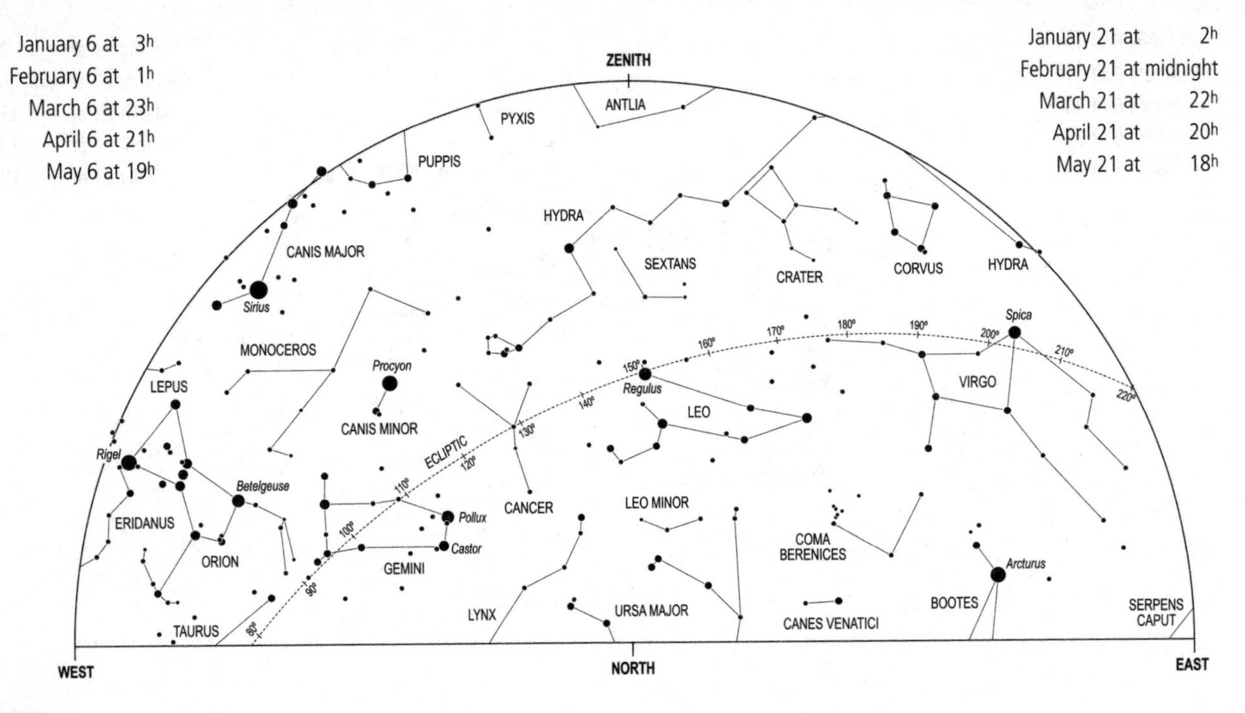

January 6 at 3h
February 6 at 1h
March 6 at 23h
April 6 at 21h
May 6 at 19h
January 21 at 2h
February 21 at midnight
March 21 at 22h
April 21 at 20h
May 21 at 18h
ZENITH
WEST
NORTH
EAST
ECLIPTIC
PYXIS
ANTLIA
PUPPIS
HYDRA
CANIS MAJOR
Sirius
SEXTANS
CRATER
CORVUS
HYDRA
Spica
MONOCEROS
Procyon
LEPUS
CANIS MINOR
Regulus
LEO
VIRGO
Rigel
Betelgeuse
Pollux
Castor
CANCER
LEO MINOR
ERIDANUS
ORION
GEMINI
COMA BERENICES
Arcturus
TAURUS
LYNX
URSA MAJOR
CANES VENATICI
BOOTES
SERPENS CAPUT
80°
90°
100°
110°
120°
130°
140°
150°
160°
170°
180°
190°
200°
210°
220°

3S

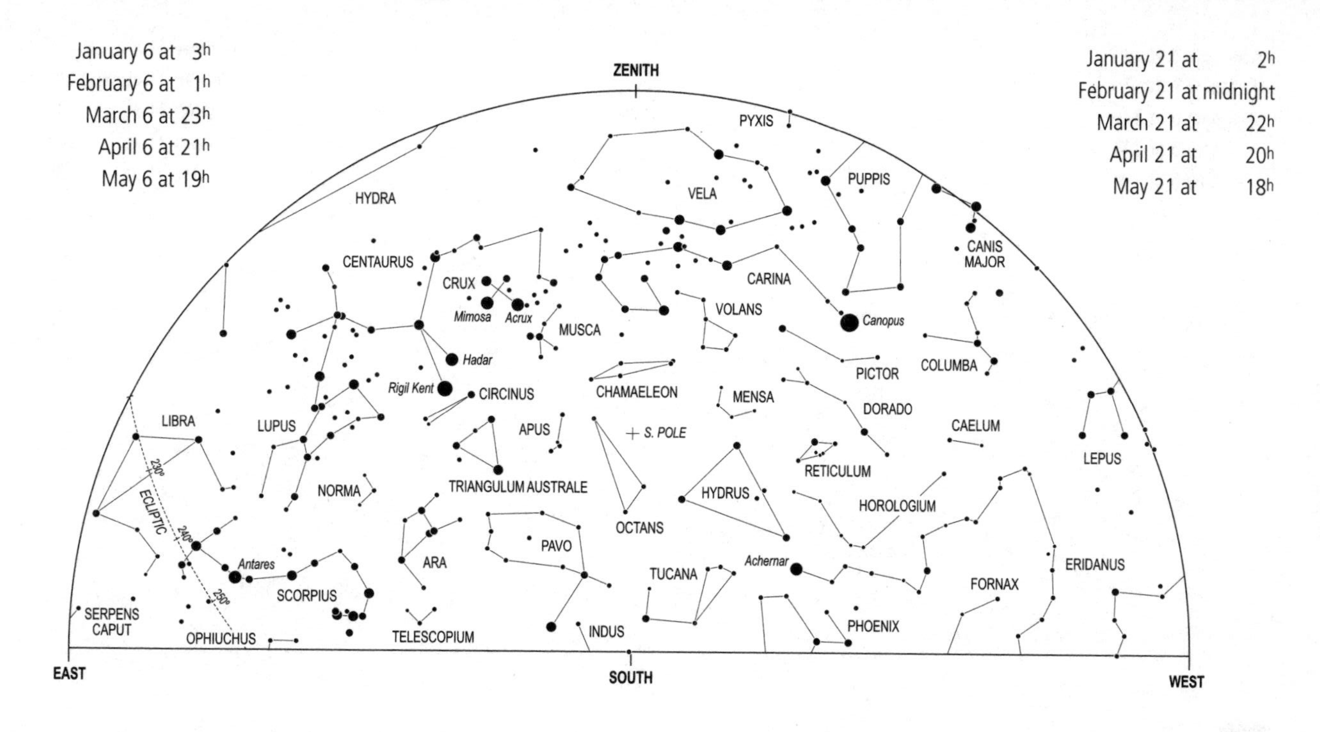
January 6 at 3h
February 6 at 1h
March 6 at 23h
April 6 at 21h
May 6 at 19h
ZENITH
January 21 at 2h
February 21 at midnight
March 21 at 22h
April 21 at 20h
May 21 at 18h
PYXIS
HYDRA
VELA
PUPPIS
CANIS MAJOR
CENTAURUS
CRUX
Mimosa
Acrux
MUSCA
CARINA
VOLANS
Canopus
Hadar
Rigil Kent
CIRCINUS
CHAMAELEON
MENSA
PICTOR
COLUMBA
DORADO
CAELUM
LEPUS
LIBRA
LUPUS
APUS
S. POLE
RETICULUM
NORMA
TRIANGULUM AUSTRALE
HYDRUS
HOROLOGIUM
ECLIPTIC
230°
240°
250°
OCTANS
PAVO
Antares
ARA
Achernar
TUCANA
ERIDANUS
FORNAX
SCORPIUS
SERPENS CAPUT
OPHIUCHUS
TELESCOPIUM
INDUS
PHOENIX
EAST
SOUTH
WEST

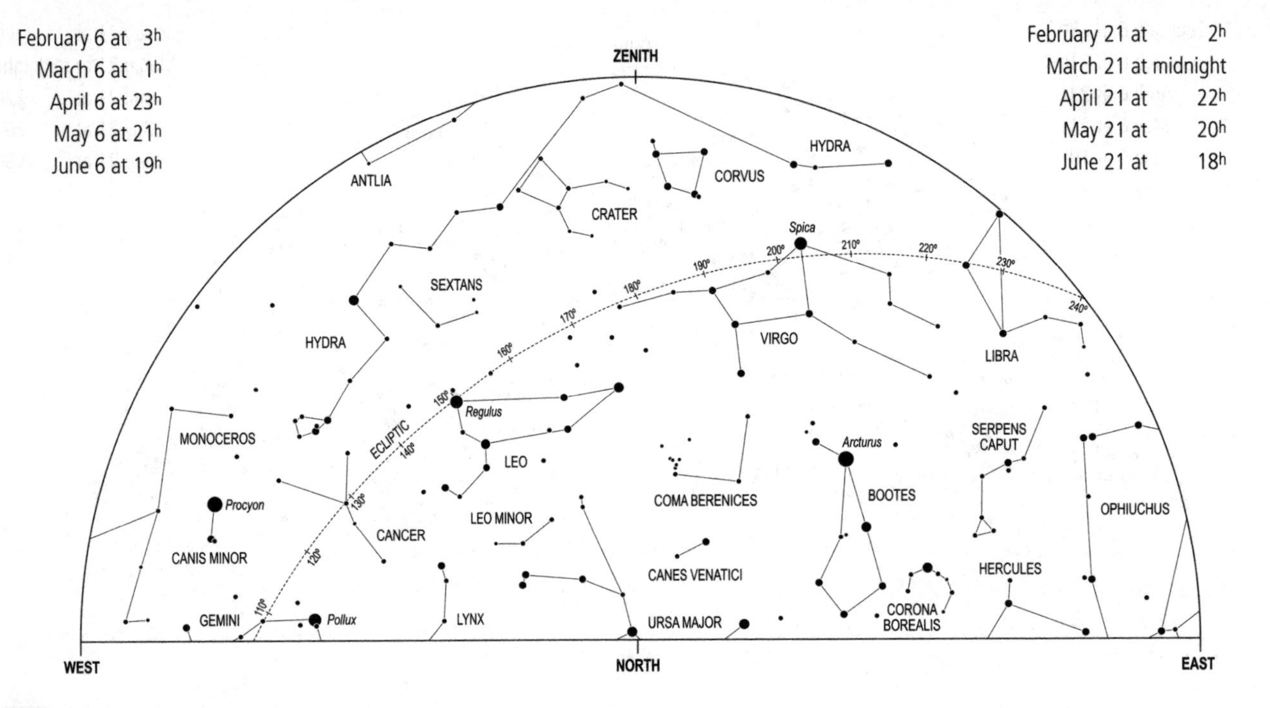
4N
February 6 at 3h
March 6 at 1h
April 6 at 23h
May 6 at 21h
June 6 at 19h
February 21 at 2h
March 21 at midnight
April 21 at 22h
May 21 at 20h
June 21 at 18h
ZENITH
WEST
NORTH
EAST
ECLIPTIC
110°
120°
130°
140°
150°
160°
170°
180°
190°
200°
210°
220°
230°
240°
ANTLIA
HYDRA
CORVUS
CRATER
Spica
SEXTANS
VIRGO
LIBRA
MONOCEROS
Regulus
LEO
Arcturus
SERPENS CAPUT
Procyon
CANIS MINOR
CANCER
LEO MINOR
COMA BERENICES
BOOTES
OPHIUCHUS
CANES VENATICI
HERCULES
GEMINI
Pollux
LYNX
URSA MAJOR
CORONA BOREALIS

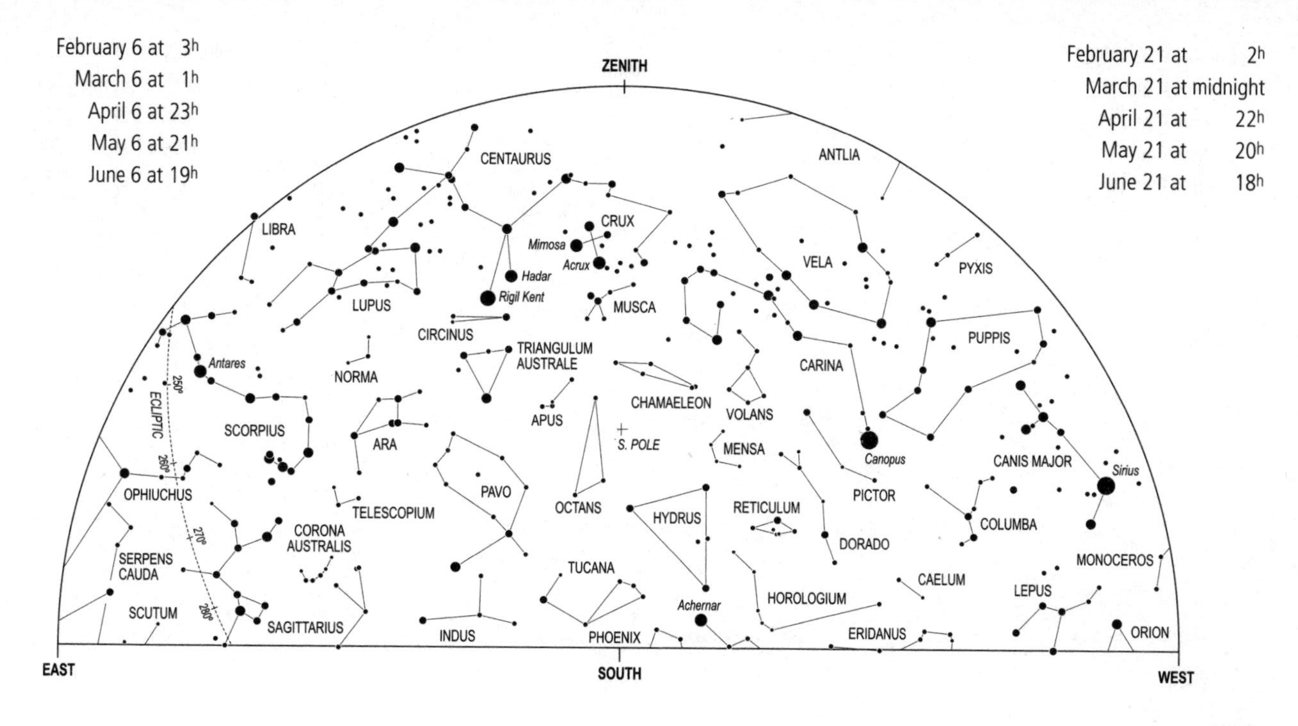
February 6 at 3h
March 6 at 1h
April 6 at 23h
May 6 at 21h
June 6 at 19h
February 21 at 2h
March 21 at midnight
April 21 at 22h
May 21 at 20h
June 21 at 18h
ZENITH
CENTAURUS
ANTLIA
LIBRA
CRUX
Mimosa
Acrux
Hadar
Rigil Kent
VELA
PYXIS
LUPUS
MUSCA
CIRCINUS
PUPPIS
Antares
TRIANGULUM AUSTRALE
NORMA
CARINA
CHAMAELEON
VOLANS
APUS
250°
ECLIPTIC
SCORPIUS
ARA
S. POLE
MENSA
Canopus
CANIS MAJOR
Sirius
260°
PAVO
OPHIUCHUS
OCTANS
PICTOR
TELESCOPIUM
HYDRUS
RETICULUM
COLUMBA
270°
CORONA AUSTRALIS
DORADO
SERPENS CAUDA
MONOCEROS
TUCANA
CAELUM
HOROLOGIUM
LEPUS
Achernar
SCUTUM
280°
SAGITTARIUS
INDUS
PHOENIX
ERIDANUS
ORION
EAST
SOUTH
WEST
4S

5N

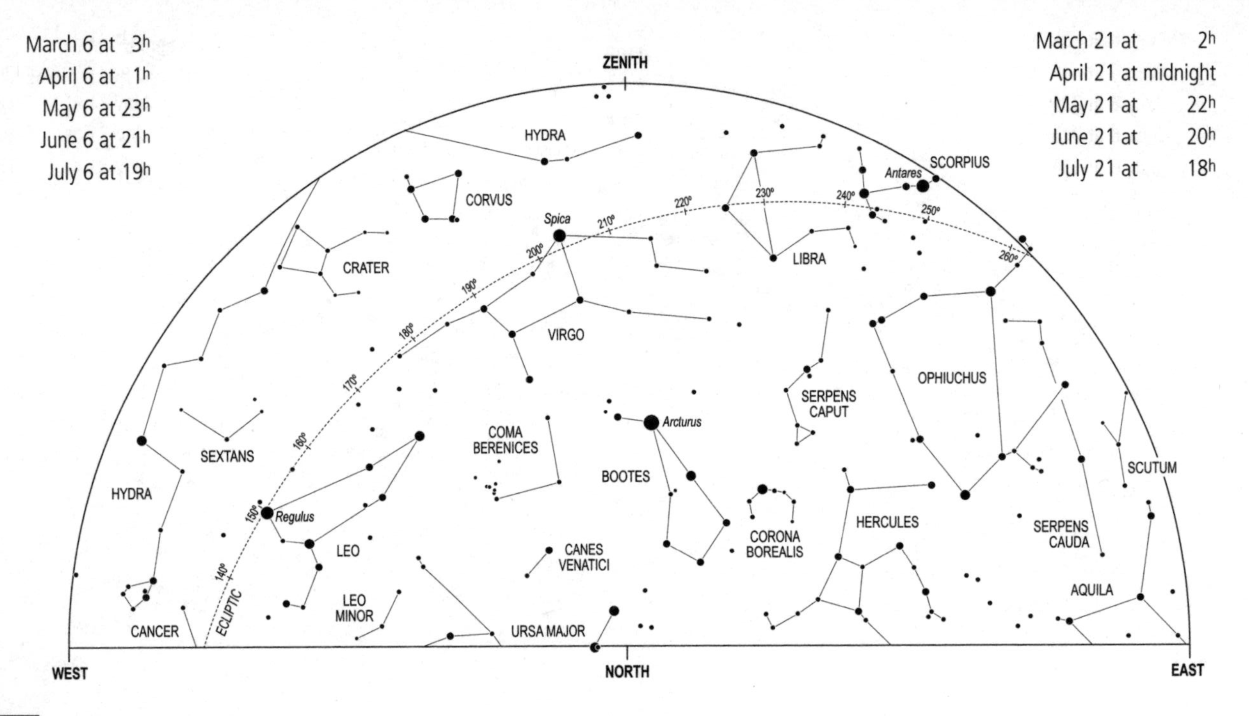
March 6 at 3h
April 6 at 1h
May 6 at 23h
June 6 at 21h
July 6 at 19h
March 21 at 2h
April 21 at midnight
May 21 at 22h
June 21 at 20h
July 21 at 18h
ZENITH
WEST
NORTH
EAST
ECLIPTIC
140°
150°
160°
170°
180°
190°
200°
210°
220°
230°
240°
250°
260°
HYDRA
CORVUS
Spica
LIBRA
Antares
SCORPIUS
CRATER
VIRGO
SEXTANS
COMA BERENICES
Arcturus
SERPENS CAPUT
OPHIUCHUS
BOOTES
Regulus
LEO
CANES VENATICI
CORONA BOREALIS
HERCULES
SERPENS CAUDA
SCUTUM
LEO MINOR
CANCER
URSA MAJOR
AQUILA

5S

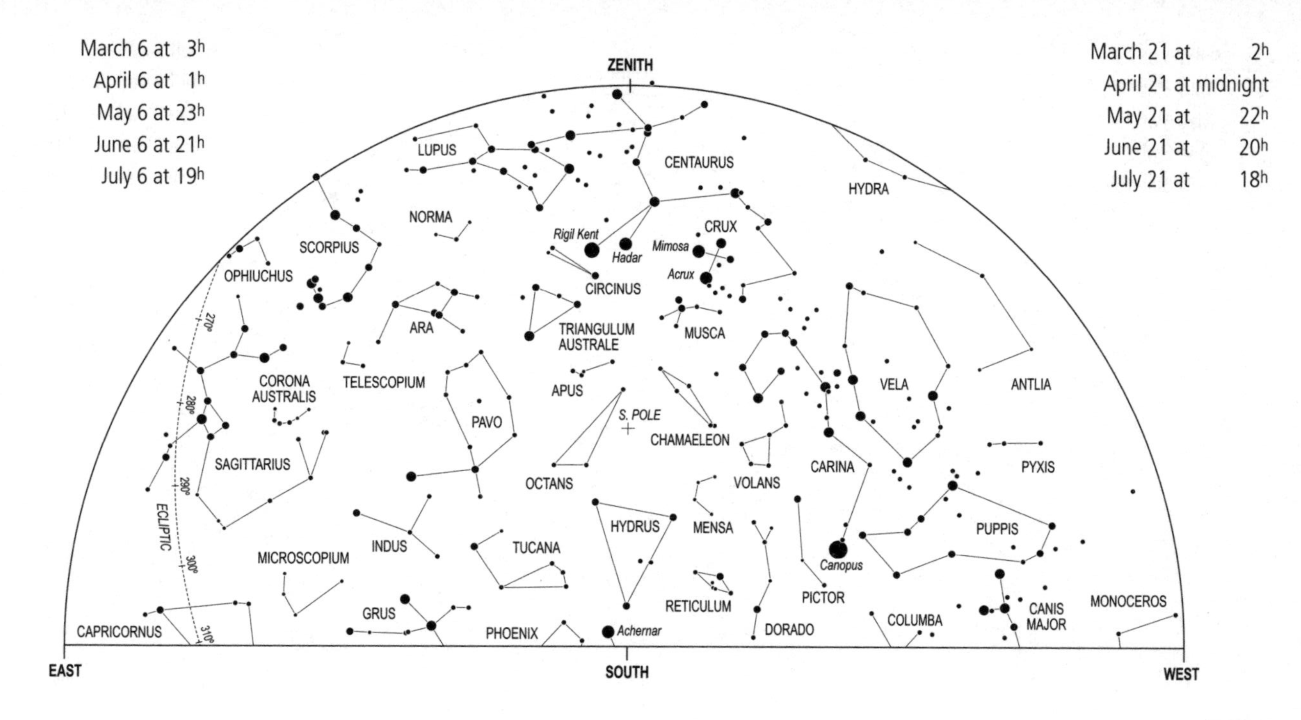
March 6 at 3h
April 6 at 1h
May 6 at 23h
June 6 at 21h
July 6 at 19h
March 21 at 2h
April 21 at midnight
May 21 at 22h
June 21 at 20h
July 21 at 18h
ZENITH
LUPUS
CENTAURUS
HYDRA
NORMA
SCORPIUS
OPHIUCHUS
Rigil Kent
Hadar
Mimosa
CRUX
Acrux
CIRCINUS
ARA
TRIANGULUM AUSTRALE
MUSCA
VELA
ANTLIA
CORONA AUSTRALIS
TELESCOPIUM
APUS
S. POLE
PAVO
CHAMAELEON
SAGITTARIUS
OCTANS
VOLANS
CARINA
PYXIS
HYDRUS
MENSA
PUPPIS
INDUS
TUCANA
MICROSCOPIUM
Canopus
RETICULUM
PICTOR
GRUS
CANIS MAJOR
MONOCEROS
COLUMBA
CAPRICORNUS
PHOENIX
Achernar
DORADO
ECLIPTIC
270°
280°
290°
300°
310°
EAST
SOUTH
WEST

6N

March 6 at 5h
April 6 at 3h
May 6 at 1h
June 6 at 23h
July 6 at 21h

March 21 at 4h
April 21 at 2h
May 21 at midnight
June 21 at 22h
July 21 at 20h

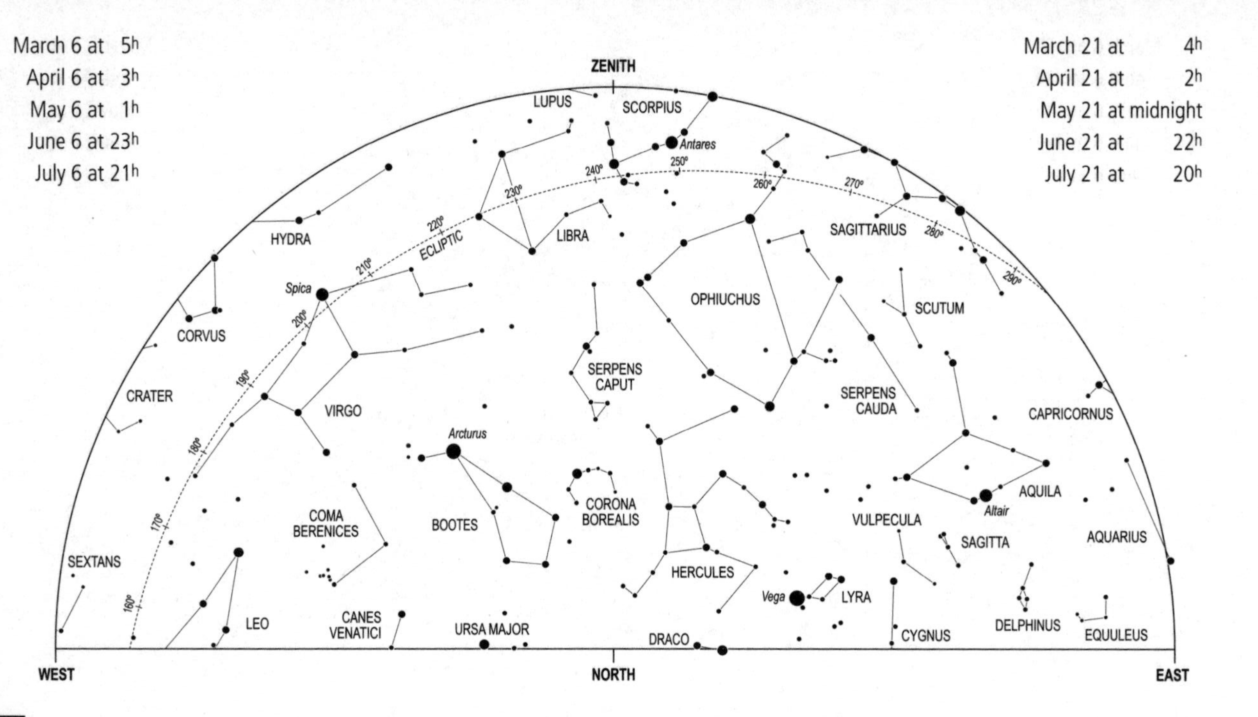

6S

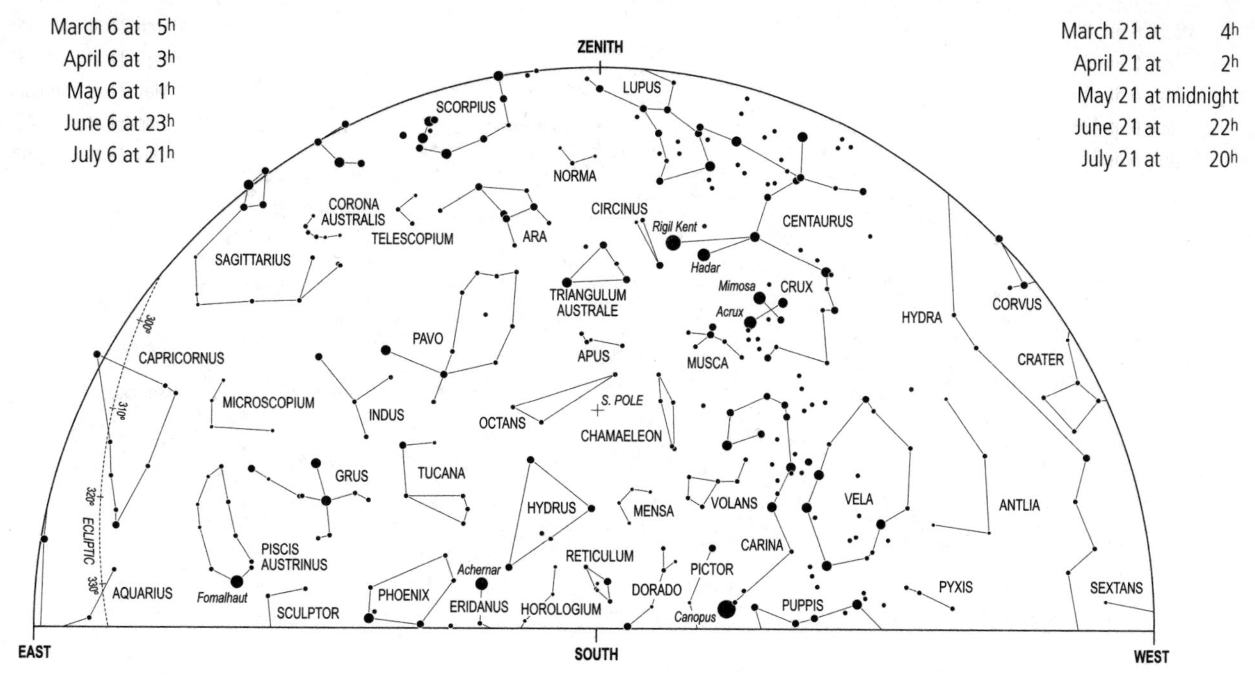
March 6 at 5h
April 6 at 3h
May 6 at 1h
June 6 at 23h
July 6 at 21h
March 21 at 4h
April 21 at 2h
May 21 at midnight
June 21 at 22h
July 21 at 20h
ZENITH
EAST
SOUTH
WEST
SCORPIUS
LUPUS
NORMA
CORONA AUSTRALIS
TELESCOPIUM
ARA
CIRCINUS
Rigil Kent
CENTAURUS
Hadar
SAGITTARIUS
TRIANGULUM AUSTRALE
Mimosa
CRUX
Acrux
HYDRA
CORVUS
PAVO
APUS
MUSCA
CAPRICORNUS
CRATER
MICROSCOPIUM
INDUS
OCTANS
S. POLE
CHAMAELEON
GRUS
TUCANA
HYDRUS
MENSA
VOLANS
VELA
ANTLIA
PISCIS AUSTRINUS
RETICULUM
CARINA
Achernar
PICTOR
AQUARIUS
Fomalhaut
PHOENIX
DORADO
PYXIS
SEXTANS
SCULPTOR
ERIDANUS
HOROLOGIUM
Canopus
PUPPIS
300°
310°
320°
330°
ECLIPTIC

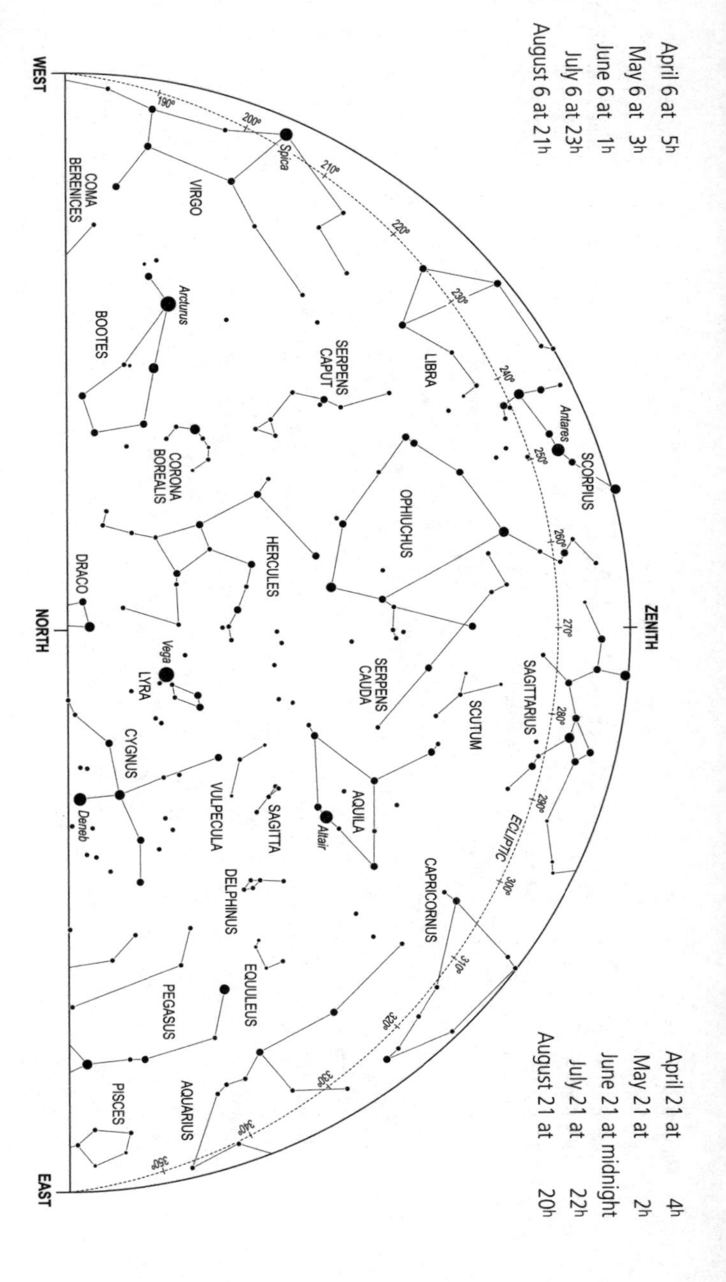
7N
April 6 at 5h
May 6 at 3h
June 6 at 1h
July 6 at 23h
August 6 at 21h
April 21 at 4h
May 21 at 2h
June 21 at midnight
July 21 at 22h
August 21 at 20h
WEST
NORTH
EAST
ZENITH
ECLIPTIC
COMA BERENICES
VIRGO
Spica
BOOTES
Arcturus
SERPENS CAPUT
LIBRA
SCORPIUS
Antares
CORONA BOREALIS
OPHIUCHUS
HERCULES
DRACO
Vega
LYRA
SERPENS CAUDA
SAGITTARIUS
SCUTUM
CYGNUS
Deneb
VULPECULA
SAGITTA
AQUILA
Altair
CAPRICORNUS
DELPHINUS
EQUULEUS
PEGASUS
AQUARIUS
PISCES
190°
200°
210°
220°
230°
240°
250°
260°
270°
280°
290°
300°
310°
320°
330°
340°
350°

7S

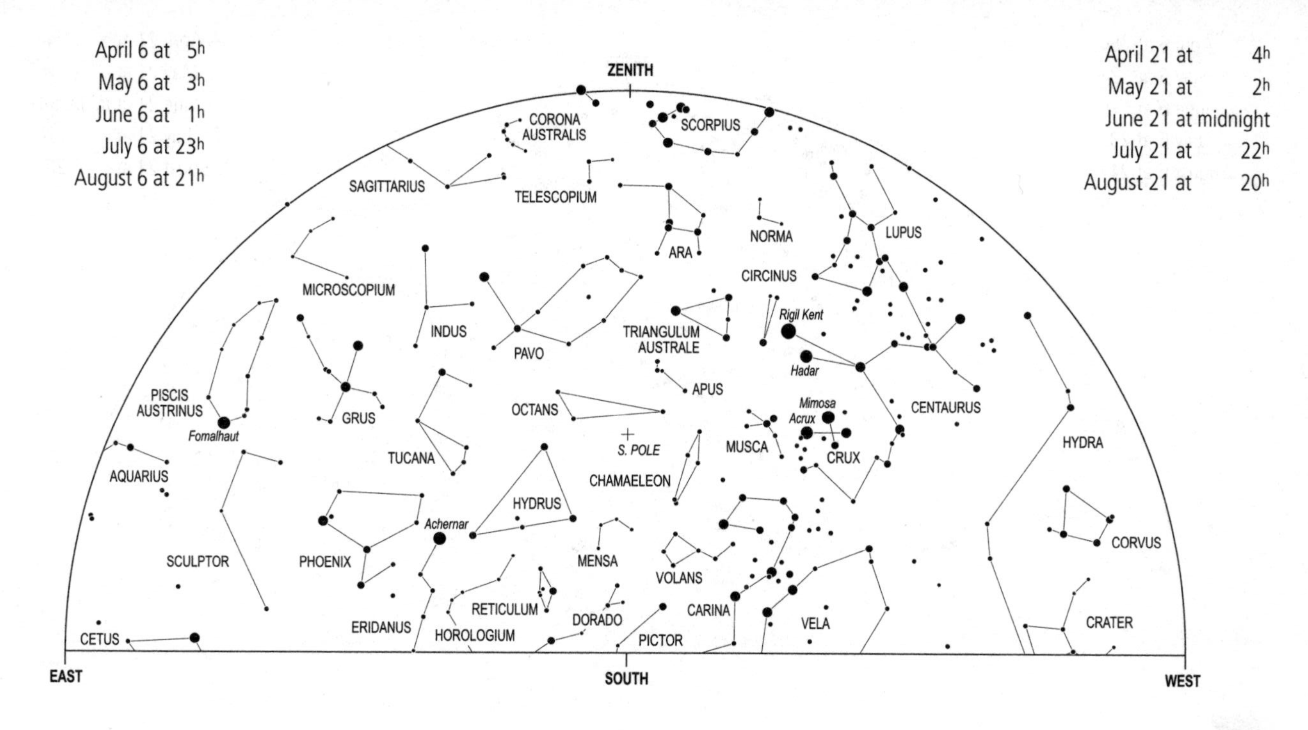
April 6 at 5^{h}
May 6 at 3^{h}
June 6 at 1^{h}
July 6 at 23^{h}
August 6 at 21^{h}
April 21 at 4^{h}
May 21 at 2^{h}
June 21 at midnight
July 21 at 22^{h}
August 21 at 20^{h}
ZENITH
CORONA AUSTRALIS
SCORPIUS
SAGITTARIUS
TELESCOPIUM
NORMA
LUPUS
ARA
CIRCINUS
MICROSCOPIUM
INDUS
PAVO
TRIANGULUM AUSTRALE
Rigil Kent
Hadar
APUS
CENTAURUS
PISCIS AUSTRINUS
Fomalhaut
GRUS
OCTANS
Mimosa
Acrux
S. POLE
MUSCA
CRUX
HYDRA
TUCANA
AQUARIUS
CHAMAELEON
HYDRUS
Achernar
CORVUS
SCULPTOR
PHOENIX
MENSA
VOLANS
RETICULUM
CARINA
DORADO
VELA
CRATER
ERIDANUS
HOROLOGIUM
PICTOR
CETUS
EAST
SOUTH
WEST

8N

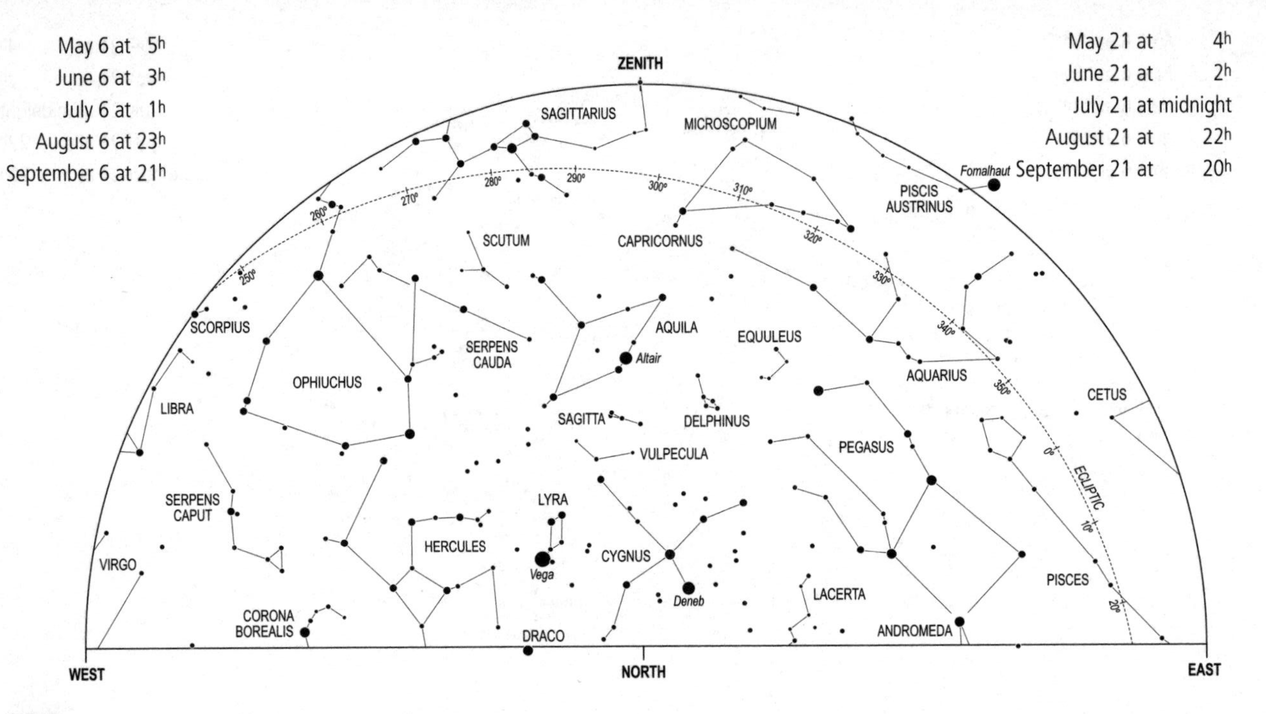
May 6 at 5h
June 6 at 3h
July 6 at 1h
August 6 at 23h
September 6 at 21h
May 21 at 4h
June 21 at 2h
July 21 at midnight
August 21 at 22h
September 21 at 20h
ZENITH
SAGITTARIUS
MICROSCOPIUM
Fomalhaut
PISCIS AUSTRINUS
SCUTUM
CAPRICORNUS
SCORPIUS
SERPENS CAUDA
AQUILA
EQUULEUS
Altair
OPHIUCHUS
AQUARIUS
CETUS
LIBRA
SAGITTA
DELPHINUS
VULPECULA
PEGASUS
ECLIPTIC
LYRA
SERPENS CAPUT
HERCULES
CYGNUS
VIRGO
Vega
Deneb
LACERTA
PISCES
CORONA BOREALIS
DRACO
ANDROMEDA
WEST
NORTH
EAST
250°
260°
270°
280°
290°
300°
310°
320°
330°
340°
350°
0°
10°
20°

8S

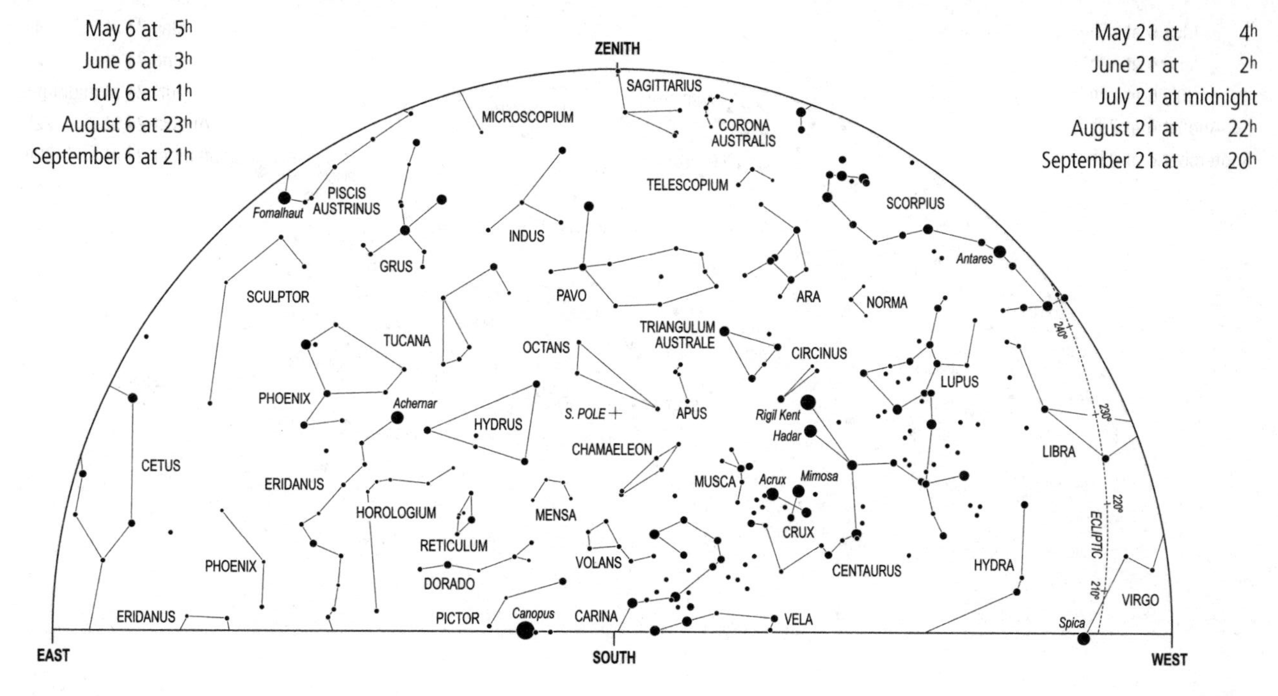
May 6 at 5h
June 6 at 3h
July 6 at 1h
August 6 at 23h
September 6 at 21h
May 21 at 4h
June 21 at 2h
July 21 at midnight
August 21 at 22h
September 21 at 20h
ZENITH
EAST
SOUTH
WEST
ECLIPTIC
210°
220°
230°
240°
S. POLE
SAGITTARIUS
MICROSCOPIUM
CORONA AUSTRALIS
TELESCOPIUM
SCORPIUS
Antares
PISCIS AUSTRINUS
Fomalhaut
GRUS
INDUS
PAVO
ARA
NORMA
SCULPTOR
TUCANA
OCTANS
TRIANGULUM AUSTRALE
CIRCINUS
LUPUS
PHOENIX
Achernar
HYDRUS
APUS
Rigil Kent
Hadar
LIBRA
CETUS
ERIDANUS
CHAMAELEON
MUSCA
Acrux
Mimosa
HOROLOGIUM
MENSA
CRUX
RETICULUM
VOLANS
CENTAURUS
HYDRA
DORADO
PICTOR
Canopus
CARINA
VELA
VIRGO
Spica

9N

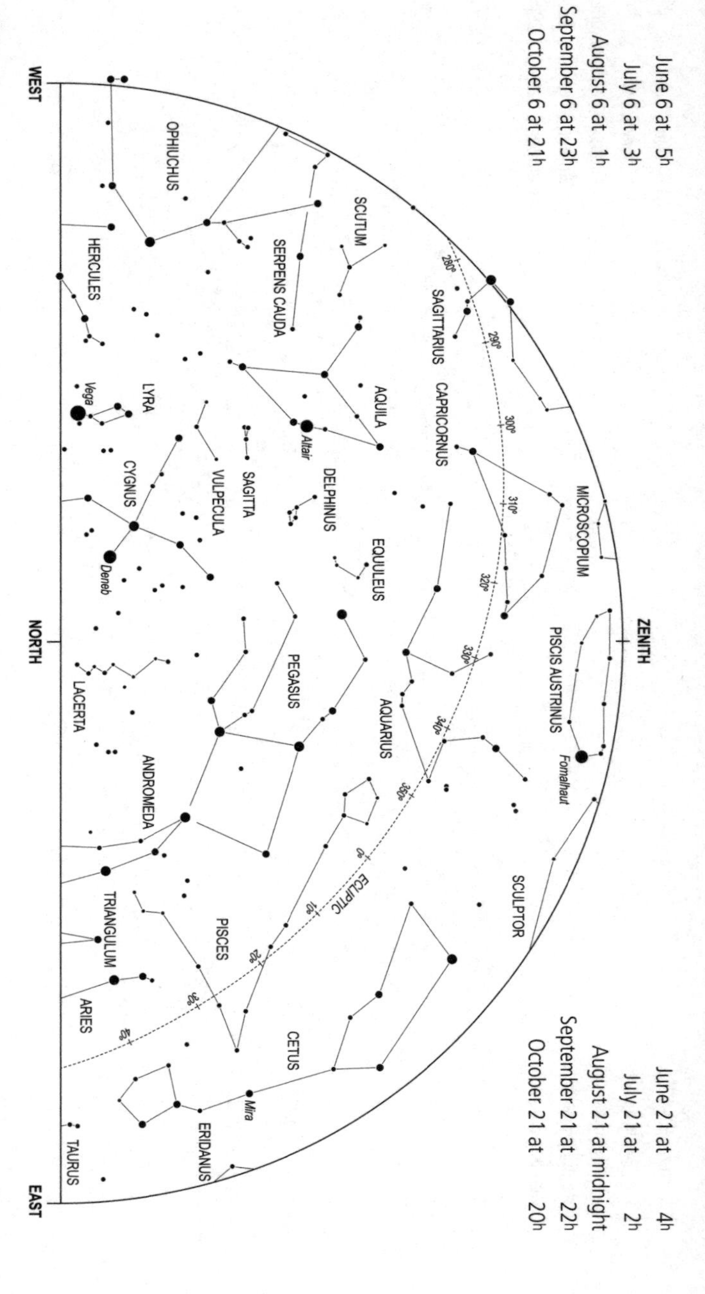

June 6 at 5h
July 6 at 3h
August 6 at 1h
September 6 at 23h
October 6 at 21h
June 21 at 4h
July 21 at 2h
August 21 at midnight
September 21 at 22h
October 21 at 20h
WEST
NORTH
EAST
ZENITH
OPHIUCHUS
HERCULES
SERPENS CAUDA
SCUTUM
SAGITTARIUS
AQUILA
Altair
LYRA
Vega
CYGNUS
Deneb
VULPECULA
SAGITTA
DELPHINUS
CAPRICORNUS
EQUULEUS
MICROSCOPIUM
PISCIS AUSTRINUS
Fomalhaut
LACERTA
PEGASUS
AQUARIUS
ANDROMEDA
SCULPTOR
TRIANGULUM
PISCES
ECLIPTIC
ARIES
CETUS
Mira
ERIDANUS
TAURUS
280°
290°
300°
310°
320°
330°
340°
350°
0°
10°
20°
30°
40°

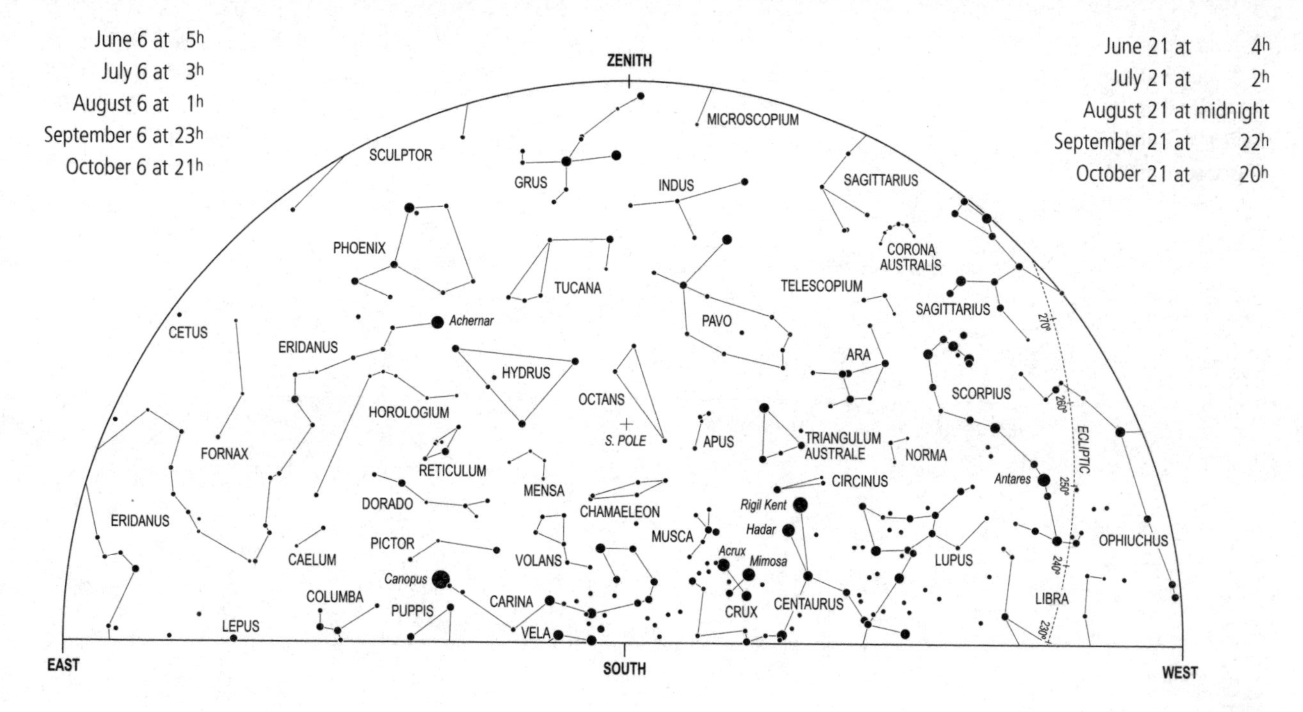
June 6 at 5h
July 6 at 3h
August 6 at 1h
September 6 at 23h
October 6 at 21h
June 21 at 4h
July 21 at 2h
August 21 at midnight
September 21 at 22h
October 21 at 20h
ZENITH
EAST
SOUTH
WEST
ECLIPTIC
270°
260°
250°
240°
230°
MICROSCOPIUM
SCULPTOR
GRUS
INDUS
SAGITTARIUS
PHOENIX
CORONA AUSTRALIS
TUCANA
TELESCOPIUM
SAGITTARIUS
CETUS
Achernar
PAVO
ERIDANUS
ARA
HYDRUS
SCORPIUS
OCTANS
HOROLOGIUM
S. POLE
APUS
TRIANGULUM AUSTRALE
NORMA
FORNAX
RETICULUM
Antares
MENSA
CIRCINUS
DORADO
CHAMAELEON
Rigil Kent
ERIDANUS
MUSCA
Hadar
OPHIUCHUS
PICTOR
CAELUM
VOLANS
Acrux
Mimosa
LUPUS
Canopus
LIBRA
COLUMBA
CARINA
CENTAURUS
PUPPIS
CRUX
LEPUS
VELA

9S

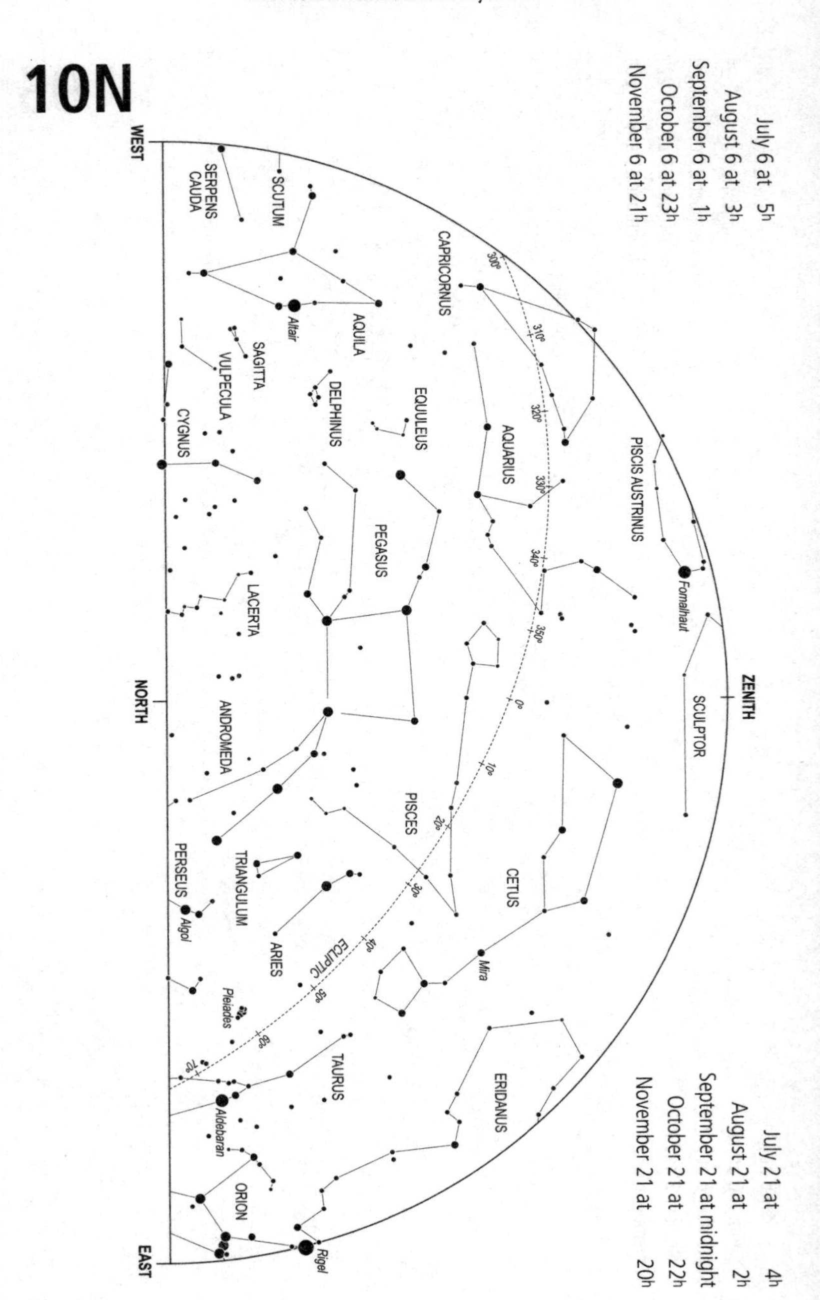
10N
July 6 at 5h
August 6 at 3h
September 6 at 1h
October 6 at 23h
November 6 at 21h
July 21 at 4h
August 21 at 2h
September 21 at midnight
October 21 at 22h
November 21 at 20h
WEST
NORTH
EAST
ZENITH
SERPENS CAUDA
SCUTUM
CAPRICORNUS
AQUILA
Altair
SAGITTA
VULPECULA
DELPHINUS
EQUULEUS
CYGNUS
AQUARIUS
PISCIS AUSTRINUS
Fomalhaut
PEGASUS
LACERTA
SCULPTOR
ANDROMEDA
PISCES
CETUS
Mira
TRIANGULUM
PERSEUS
Algol
ARIES
ECLIPTIC
Pleiades
TAURUS
Aldebaran
ERIDANUS
ORION
Rigel
300°
310°
320°
330°
340°
350°
0°
10°
20°
30°
40°
50°
60°
70°

10S

July 6 at 5h
August 6 at 3h
September 6 at 1h
October 6 at 23h
November 6 at 21h

July 21 at 4h
August 21 at 2h
September 21 at midnight
October 21 at 22h
November 21 at 20h

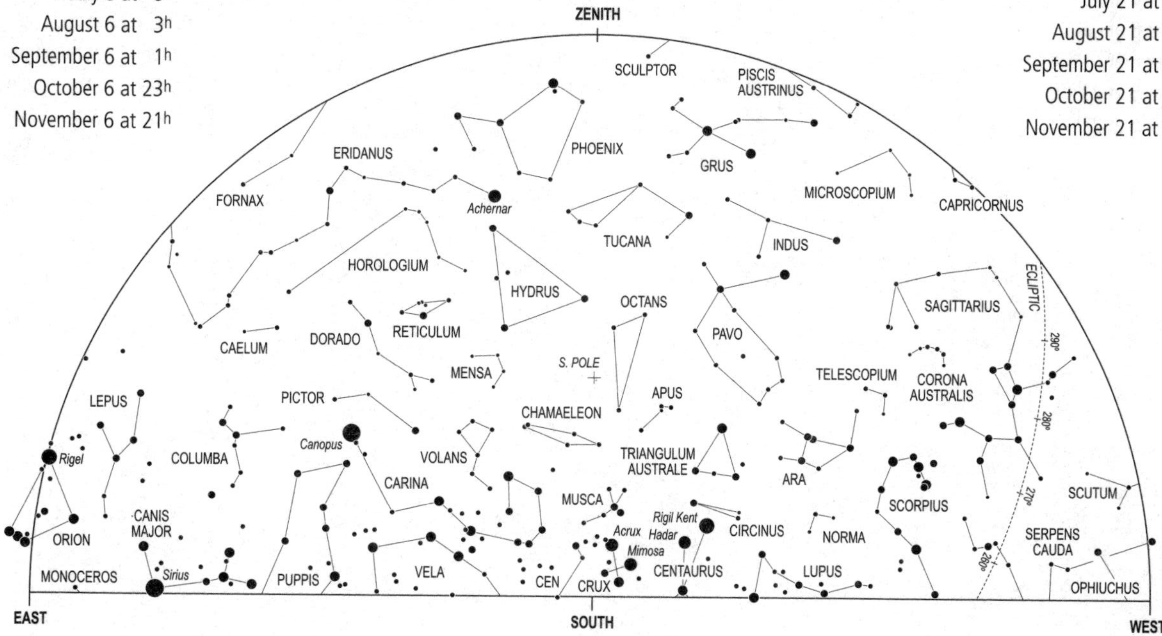
ZENITH
EAST
SOUTH
WEST
SCULPTOR
PISCIS AUSTRINUS
PHOENIX
GRUS
ERIDANUS
MICROSCOPIUM
CAPRICORNUS
FORNAX
Achernar
TUCANA
INDUS
HOROLOGIUM
ECLIPTIC
HYDRUS
OCTANS
SAGITTARIUS
RETICULUM
PAVO
CAELUM
DORADO
290°
S. POLE
MENSA
TELESCOPIUM
CORONA AUSTRALIS
LEPUS
PICTOR
APUS
CHAMAELEON
280°
Canopus
Rigel
COLUMBA
VOLANS
TRIANGULUM AUSTRALE
ARA
CARINA
SCUTUM
270°
SCORPIUS
MUSCA
Rigil Kent
CANIS MAJOR
Acrux
Hadar
CIRCINUS
NORMA
ORION
SERPENS CAUDA
Mimosa
260°
Sirius
VELA
CENTAURUS
LUPUS
MONOCEROS
PUPPIS
CEN
CRUX
OPHIUCHUS

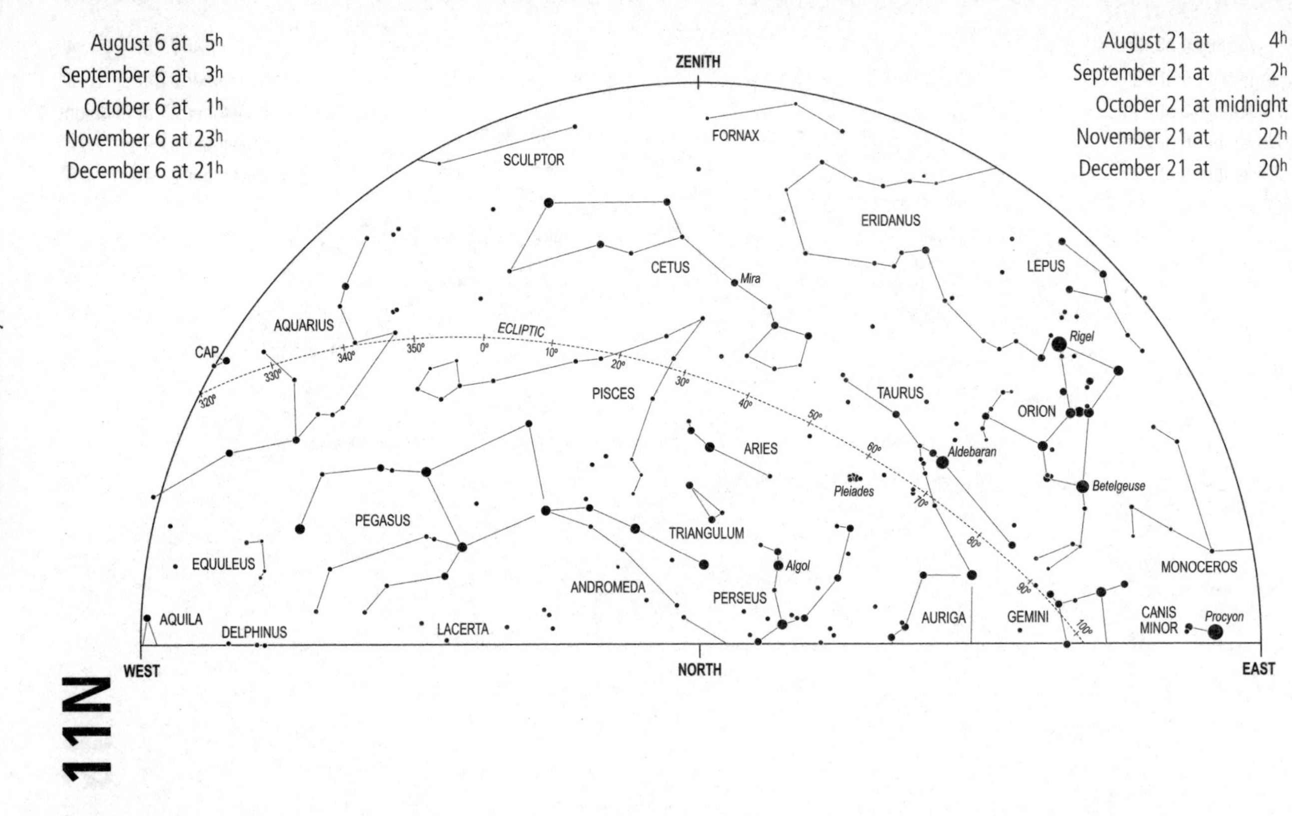
11N
August 6 at 5h
September 6 at 3h
October 6 at 1h
November 6 at 23h
December 6 at 21h
August 21 at 4h
September 21 at 2h
October 21 at midnight
November 21 at 22h
December 21 at 20h
ZENITH
WEST
NORTH
EAST
SCULPTOR
FORNAX
ERIDANUS
CETUS
Mira
LEPUS
AQUARIUS
ECLIPTIC
CAP
Rigel
PISCES
TAURUS
ORION
ARIES
Aldebaran
Pleiades
Betelgeuse
PEGASUS
TRIANGULUM
EQUULEUS
Algol
MONOCEROS
ANDROMEDA
PERSEUS
AQUILA
DELPHINUS
LACERTA
AURIGA
GEMINI
CANIS MINOR
Procyon
320°
330°
340°
350°
0°
10°
20°
30°
40°
50°
60°
70°
80°
90°
100°

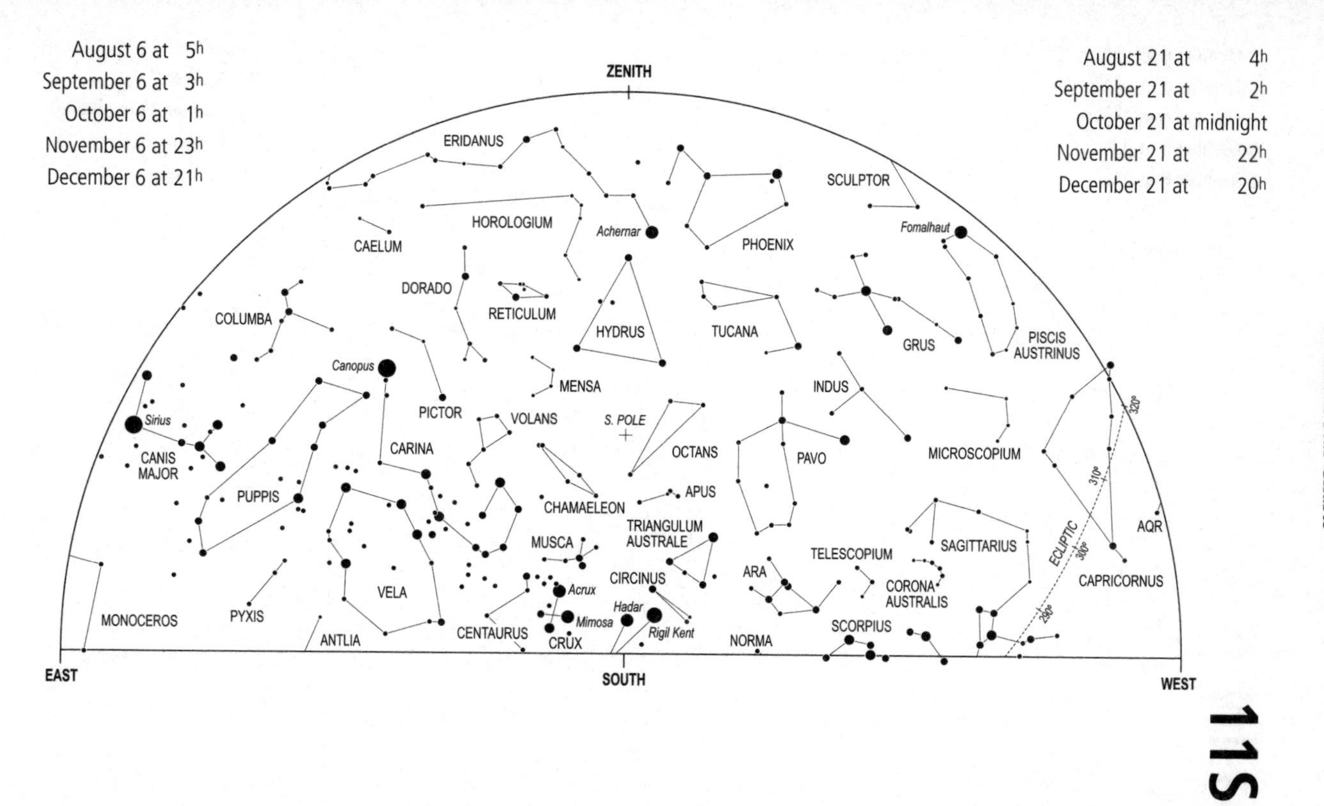
August 6 at 5h
September 6 at 3h
October 6 at 1h
November 6 at 23h
December 6 at 21h
August 21 at 4h
September 21 at 2h
October 21 at midnight
November 21 at 22h
December 21 at 20h
ZENITH
ERIDANUS
SCULPTOR
HOROLOGIUM
Achernar
PHOENIX
Fomalhaut
CAELUM
DORADO
RETICULUM
HYDRUS
TUCANA
GRUS
PISCIS AUSTRINUS
COLUMBA
Canopus
MENSA
INDUS
PICTOR
VOLANS
S. POLE
Sirius
CANIS MAJOR
CARINA
OCTANS
PAVO
MICROSCOPIUM
320°
310°
AQR
PUPPIS
CHAMAELEON
APUS
TRIANGULUM AUSTRALE
MUSCA
TELESCOPIUM
SAGITTARIUS
ECLIPTIC
300°
CAPRICORNUS
CIRCINUS
ARA
Acrux
VELA
CORONA AUSTRALIS
Hadar
290°
MONOCEROS
PYXIS
CENTAURUS
Mimosa
Rigil Kent
SCORPIUS
ANTLIA
CRUX
NORMA
EAST
SOUTH
WEST
11S

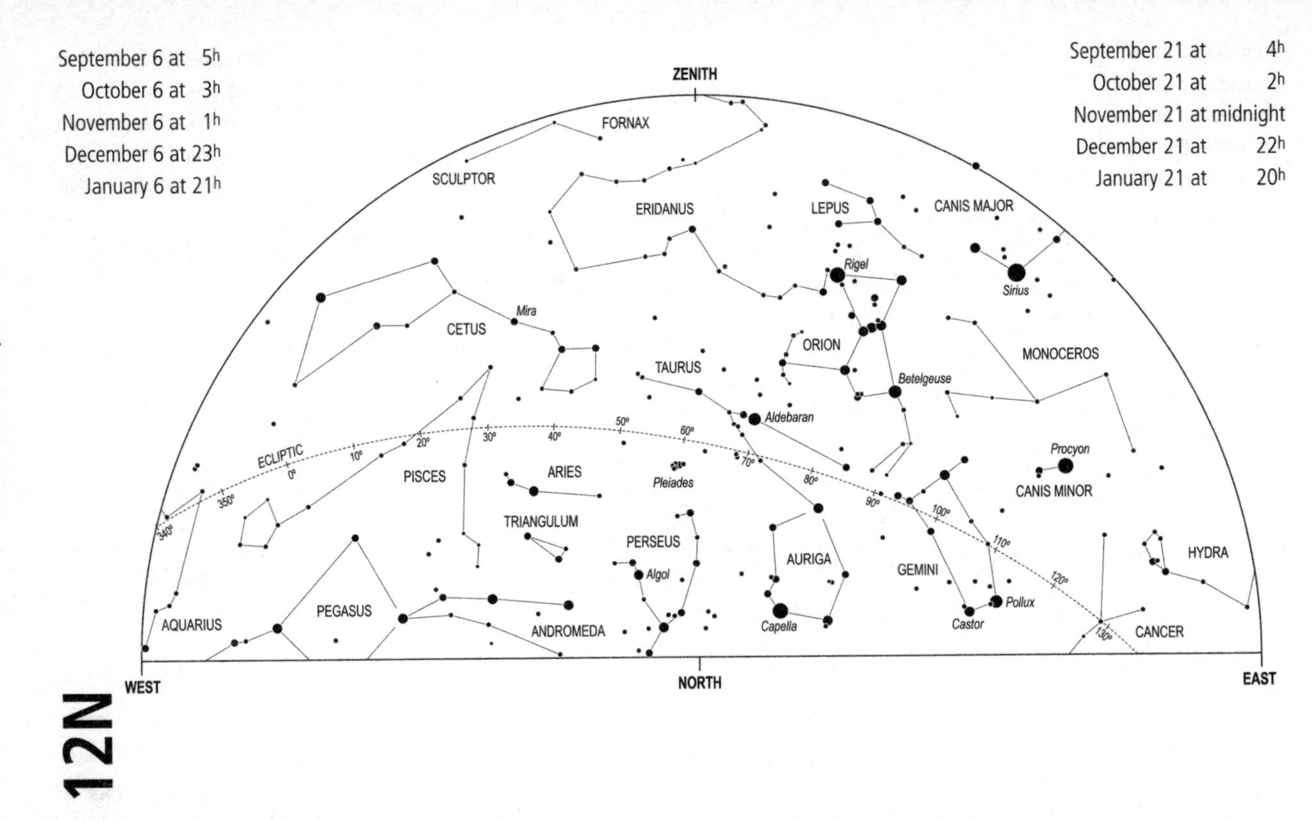
12N
September 6 at 5h
October 6 at 3h
November 6 at 1h
December 6 at 23h
January 6 at 21h
September 21 at 4h
October 21 at 2h
November 21 at midnight
December 21 at 22h
January 21 at 20h
ZENITH
WEST
NORTH
EAST
FORNAX
SCULPTOR
ERIDANUS
LEPUS
CANIS MAJOR
Rigel
Sirius
CETUS
Mira
ORION
TAURUS
MONOCEROS
Betelgeuse
Aldebaran
Procyon
ECLIPTIC
PISCES
ARIES
Pleiades
CANIS MINOR
TRIANGULUM
PERSEUS
AURIGA
GEMINI
HYDRA
Algol
PEGASUS
AQUARIUS
ANDROMEDA
Capella
Castor
Pollux
CANCER
340°
350°
0°
10°
20°
30°
40°
50°
60°
70°
80°
90°
100°
110°
120°
130°

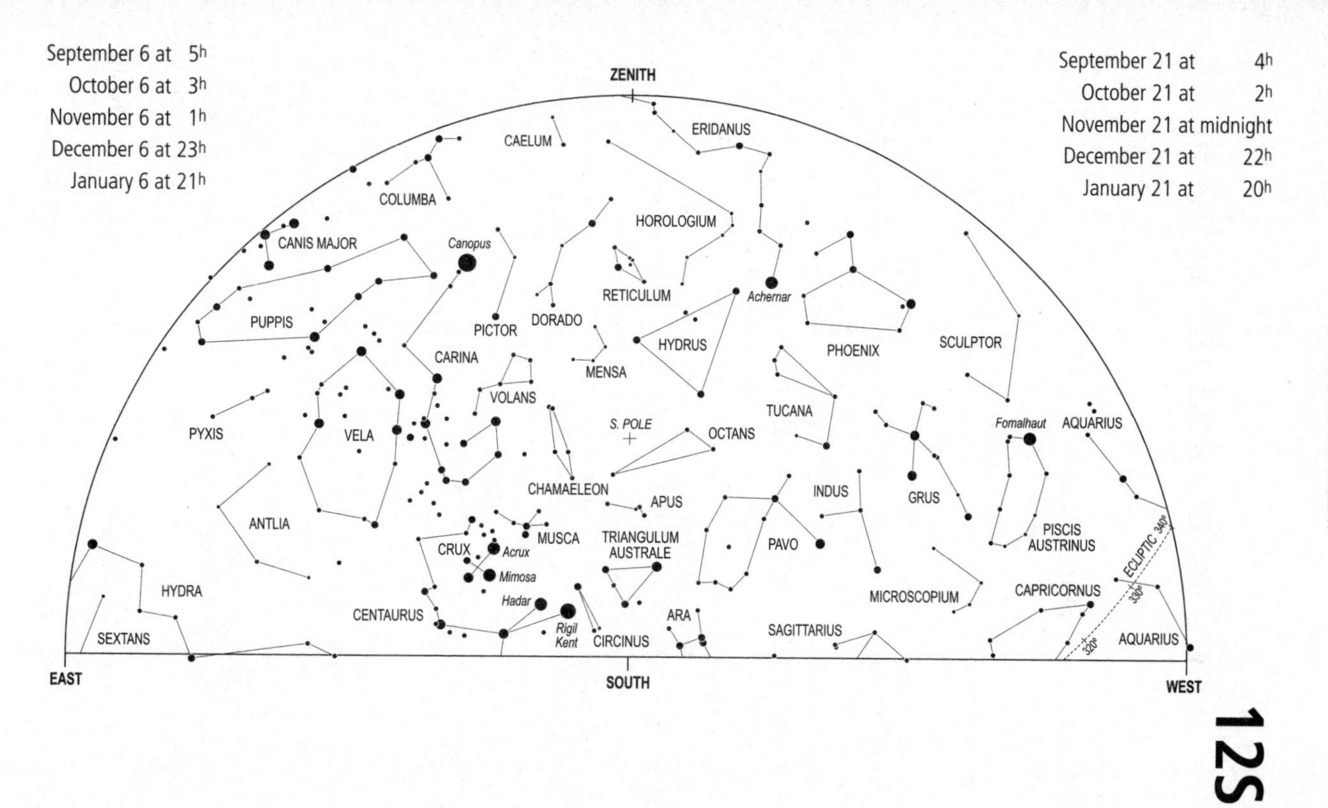
September 6 at 5h
October 6 at 3h
November 6 at 1h
December 6 at 23h
January 6 at 21h
September 21 at 4h
October 21 at 2h
November 21 at midnight
December 21 at 22h
January 21 at 20h
12S
ZENITH
EAST
SOUTH
WEST
CAELUM
ERIDANUS
COLUMBA
HOROLOGIUM
CANIS MAJOR
Canopus
RETICULUM
Achernar
PUPPIS
PICTOR
DORADO
HYDRUS
PHOENIX
SCULPTOR
CARINA
MENSA
VOLANS
TUCANA
S. POLE
OCTANS
Fomalhaut
AQUARIUS
PYXIS
VELA
CHAMAELEON
APUS
INDUS
GRUS
ANTLIA
MUSCA
TRIANGULUM AUSTRALE
PAVO
PISCIS AUSTRINUS
CRUX
Acrux
Mimosa
Hadar
Rigil Kent
ECLIPTIC
340°
330°
320°
HYDRA
MICROSCOPIUM
CAPRICORNUS
CENTAURUS
ARA
SAGITTARIUS
SEXTANS
CIRCINUS
AQUARIUS

The Planets and the Ecliptic

The paths of the planets about the Sun all lie close to the plane of the ecliptic, which is marked for us in the sky by the apparent path of the Sun among the stars, and is shown on the star charts by a broken line. The Moon and naked-eye planets will always be found close to this line, never departing from it by more than about 7°. Thus the planets are most favourably placed for observation when the ecliptic is well displayed, and this means that it should be as high in the sky as possible. This avoids the difficulty of finding a clear horizon, and also overcomes the problem of atmospheric absorption, which greatly reduces the light of the stars. Thus a star at an altitude of 10° suffers a loss of 60 per cent of its light, which corresponds to a whole magnitude; at an altitude of only 4°, the loss may amount to two magnitudes.

The position of the ecliptic in the sky is therefore of great importance, and since it is tilted at about 23.5° to the Equator, it is only at certain times of the day or year that it is displayed to the best advantage. It will be realized that the Sun (and therefore the ecliptic) is at its highest in the sky at noon in midsummer, and at its lowest at noon in midwinter. Allowing for the daily motion of the sky, it follows that the ecliptic is highest at midnight in the winter, at sunset in the spring, at noon in the summer and at sunrise in the autumn. Hence these are the best times to see the planets. Thus, if Venus is an evening object in the western sky after sunset, it will be seen to best advantage if this occurs in the spring, when the ecliptic is high in the sky and slopes down steeply to the horizon. This means that the planet is not only higher in the sky, but also will remain for a much longer period above the horizon. For similar reasons, a morning object will be seen at its best on autumn mornings before sunrise, when the ecliptic is high in the east. The outer planets, which can come to opposition (i.e. opposite the Sun), are best seen when opposition occurs in the winter months, when the ecliptic is high in the sky at midnight.

The seasons are reversed in the Southern Hemisphere, spring beginning at the September equinox, when the Sun crosses the Equator on its way south, summer beginning at the December solstice, when the

Sun is highest in the southern sky, and so on. Thus, the times when the ecliptic is highest in the sky, and therefore best placed for observing the planets, may be summarized as follows:

	Midnight	**Sunrise**	**Noon**	**Sunset**
Northern latitudes	December	September	June	March
Southern latitudes	June	March	December	September

In addition to the daily rotation of the celestial sphere from east to west, the planets have a motion of their own among the stars. The apparent movement is generally *direct*, i.e. to the east, in the direction of increasing longitude, but for a certain period (which depends on the distance of the planet) this apparent motion is reversed. With the outer planets this *retrograde* motion occurs about the time of opposition. Owing to the different inclination of the orbits of these planets, the actual effect is to cause the apparent path to form a loop, or sometimes an S-shaped curve. The same effect is present in the motion of the inferior planets, Mercury and Venus, but it is not so obvious, since it always occurs at the time of inferior conjunction.

The *inferior planets*, Mercury and Venus, move in smaller orbits than that of the Earth, and so are always seen near the Sun. They are most obvious at the times of greatest angular distance from the Sun (greatest elongation), which may reach 28° for Mercury, and 47° for Venus. They are seen as evening objects in the western sky after sunset (at eastern elongations) or as morning objects in the eastern sky before sunrise (at western elongations). The succession of phenomena, conjunctions and elongations, always follows the same order, but the intervals between them are not equal. Thus, if either planet is moving round the far side of its orbit, its motion will be to the east, in the same direction in which the Sun appears to be moving. It therefore takes much longer for the planet to overtake the Sun – that is, to come to superior conjunction – than it does when moving round to inferior conjunction, between Sun and Earth. The intervals given in the table at the top of p.60 are average values; they remain fairly constant in the case of Venus, which travels in an almost circular orbit. In the case of Mercury, however, conditions vary widely because of the great eccentricity and inclination of the planet's orbit.

		Mercury	Venus
Inferior Conjunction	to Elongation West	22 days	72 days
Elongation West	to Superior Conjunction	36 days	220 days
Superior Conjunction	to Elongation East	35 days	220 days
Elongation East	to Inferior Conjunction	22 days	72 days

The greatest brilliancy of Venus always occurs about thirty-six days before or after inferior conjunction. This will be about a month after greatest eastern elongation (as an evening object), or a month before greatest western elongation (as a morning object). No such rule can be given for Mercury, because its distances from the Earth and the Sun can vary over a wide range.

Mercury is not likely to be seen unless a clear horizon is available. It is seldom as much as 10° above the horizon in the twilight sky in northern temperate latitudes, but this figure is often exceeded in the Southern Hemisphere. This favourable condition arises because the maximum elongation of 28° can occur only when the planet is at aphelion (furthest from the Sun), and it then lies well south of the Equator. Northern observers must be content with smaller elongations, which may be as little as 18° at perihelion. In general, it may be said that the most favourable times for seeing Mercury as an evening object will be in spring, some days before greatest eastern elongation; in autumn, it may be seen as a morning object some days after greatest western elongation.

Venus is the brightest of the planets and may be seen on occasions in broad daylight. Like Mercury, it is alternately a morning and an evening object, and it will be highest in the sky when it is a morning object in autumn, or an evening object in spring. Venus is to be seen at its best as an evening object in northern latitudes when eastern elongation occurs in June. The planet is then well north of the Sun in the preceding spring months, and is a brilliant object in the evening sky over a long period. In the Southern Hemisphere a November elongation is best. For similar reasons, Venus gives a prolonged display as a morning object in the months following western elongation in October (in northern latitudes) or in June (in the Southern Hemisphere).

The *superior planets*, Mars, Jupiter, Saturn, Uranus and Neptune, which travel in orbits larger than that of the Earth, differ from Mercury and Venus in that they can be seen opposite the Sun in the sky. The superior planets are morning objects after conjunction with the Sun, rising earlier each day until they come to opposition. They will then be nearest to the Earth (and therefore at their brightest), and will be on the meridian at midnight, due south in northern latitudes, but due north in the Southern Hemisphere. After opposition they are evening objects, setting earlier each evening until they set in the west with the Sun at the next conjunction. The difference in brightness from one opposition to another is most noticeable in the case of Mars, whose distance from Earth can vary considerably and rapidly. The other superior planets are at such great distances that there is very little change in brightness from one opposition to the next. The effect of altitude is, however, of some importance, for at a December opposition in northern latitudes a planet will be among the stars of Taurus or Gemini, and can then be at an altitude of more than 60° from southern England. At a summer opposition, when a planet is in Sagittarius, it may only rise to about 15° above the southern horizon, and so makes a less impressive appearance. In the Southern Hemisphere the reverse conditions apply, a June opposition being the best, with the planet in Sagittarius at an altitude which can reach 80° above the northern horizon for observers in South Africa.

Mars, whose orbit is appreciably eccentric, comes nearest to the Earth at oppositions at the end of August. It may then be brighter even than Jupiter, but rather low in the sky in Aquarius for northern observers, though very well placed for those in southern latitudes. These favourable oppositions occur every fifteen or seventeen years (e.g. in 1988, 2003 and 2018). In the Northern Hemisphere the planet is probably better seen at oppositions in the autumn or winter months, when it is higher in the sky – such as in 2005 when opposition was in early November. Oppositions of Mars occur at an average interval of 780 days, and during this time the planet makes a complete circuit of the sky.

Jupiter is always a bright planet, and comes to opposition a month later each year, having moved, roughly speaking, from one Zodiacal constellation to the next.

Saturn moves much more slowly than Jupiter, taking just under 29.5 years to orbit the Sun, and may remain in the same constellation for several years. During each orbit, we see the rings of Saturn from different angles, the apparent brightness of the planet depending upon the aspects of its rings, as well as on its distance from the Earth and the Sun. The rings were last wide open, with Saturn at its brightest, in 2002 and 2003, when Saturn's southern hemisphere was tipped towards the Earth. The rings then began to close, displaying a narrower and narrower aspect, with Saturn becoming fainter at each successive opposition. The Earth last passed through the plane of Saturn's rings in 2009, when they appeared edge-on, and then they slowly began to open once again, with the northern hemisphere of Saturn now becoming increasingly angled towards the Earth. Saturn's rings will once again appear wide open in 2017.

Uranus and *Neptune* are both visible with binoculars or a small telescope, but you will need a finder chart to help you locate them (such as those reproduced in this Yearbook on pages 134 and 140). *Pluto* (now officially classified as a 'dwarf planet') is hardly likely to attract the attention of observers without adequate telescopes.

Phases of the Moon in 2016

NICK JAMES

New Moon				First Quarter				Full Moon				Last Quarter			
	d	h	m		d	h	m		d	h	m		d	h	m
												Jan	2	05	31
Jan	10	01	31	Jan	16	23	26	Jan	24	01	46	Feb	1	03	28
Feb	8	14	39	Feb	15	07	47	Feb	22	18	20	Mar	1	23	11
Mar	9	01	55	Mar	15	17	03	Mar	23	12	01	Mar	31	15	17
Apr	7	11	24	Apr	14	03	59	Apr	22	05	24	Apr	30	03	29
May	6	19	30	May	13	17	02	May	21	21	14	May	29	12	12
June	5	03	00	June	12	08	10	June	20	11	02	June	27	18	19
July	4	11	01	July	12	00	52	July	19	22	57	July	26	23	00
Aug	2	20	45	Aug	10	18	21	Aug	18	09	27	Aug	25	03	41
Sept	1	09	03	Sept	9	11	49	Sept	16	19	05	Sept	23	09	56
Oct	1	00	11	Oct	9	04	33	Oct	16	04	23	Oct	22	19	14
Oct	30	17	38	Nov	7	19	51	Nov	14	13	52	Nov	21	08	33
Nov	29	12	18	Dec	7	09	03	Dec	14	00	06	Dec	21	01	56
Dec	29	06	53												

All times are UTC (GMT)

Longitudes of the Sun, Moon and Planets in 2016

NICK JAMES

Date		Sun	Moon	Venus	Mars	Jupiter	Saturn	Uranus	Neptune
		°	°	°	°	°	°	°	°
Jan	6	285	236	248	211	173	252	17	338
	21	300	82	266	219	173	253	17	338
Feb	6	317	283	286	227	172	255	17	339
	21	332	132	305	234	170	255	18	339
Mar	6	346	304	322	240	169	256	19	340
	21	1	153	341	245	167	256	19	340
Apr	6	17	356	0	248	165	256	20	341
	21	31	198	19	249	164	256	21	341
May	6	46	34	37	247	163	255	22	342
	21	60	231	56	242	163	254	23	342
June	6	76	88	75	237	164	253	23	342
	21	90	276	94	234	166	252	24	342
July	6	104	124	112	233	168	251	24	342
	21	119	312	131	236	170	250	24	342
Aug	6	134	171	150	241	173	250	24	341
	21	148	3	169	248	176	250	24	341
Sept	6	164	216	188	257	179	250	24	341
	21	178	56	207	266	182	251	23	340
Oct	6	193	248	225	276	186	252	23	340
	21	208	95	243	286	189	253	22	339
Nov	6	224	293	263	298	192	255	22	339
	21	239	145	280	309	195	257	21	339
Dec	6	254	328	298	320	198	258	21	339
	21	270	179	315	331	200	260	21	340

Moon: Longitude of the ascending node: Jan 1: 176° Dec 31: 156°

Mercury moves so quickly among the stars that it is not possible to indicate its position on the star charts at convenient intervals. The monthly notes should be consulted for the best times at which the planet may be seen.

The positions of the Sun, Moon and planets other than Mercury are given in the table on p.64. These objects move along paths which remain close to the ecliptic and this list shows the apparent ecliptic longitude for each object on dates which correspond to those of the star charts. This information can be used to plot the position of the desired object on the selected chart.

EXAMPLES

Two planets are seen near the Moon in the south-west on the night of 11 August. What are they?

> The northern star chart 7S shows the south-western sky on 21 August at 20h. The ecliptic longitude visible to the south-west ranges from 250° to 230°. With reference to the table on p.64 it can be seen that two planets are in this range on 11 August. Mars is around 243° and Saturn is around 250°. The planets are therefore Saturn and Mars near the bright red star Antares in Scorpius.

The positions of the Sun and Moon can be plotted on the star maps in the same way as the planets. This is straightforward for the Sun since it always lies on the ecliptic and it moves on average at only 1° per day. The Moon is more difficult since it moves rapidly at an average of 13° per day and it moves up to 5° north or south of the ecliptic during the month. A rough indication of the Moon's position relative to the ecliptic may be obtained by considering its longitude relative to that of the ascending node. The longitude of the ascending node decreases by around 1.6° per month as will be seen from the values for the first and last day of the year given on p.64.

If d is the difference in longitude between the Moon and its ascending node, then the Moon is on the ecliptic when d = 0°, 180° or 360°. The Moon is 5° north of the ecliptic if d = 90°, and the Moon is 5° south of the ecliptic if d = 270°.

As an example, from the table on p.63, it can be seen that the Moon is at first quarter on the evening of 13 May. The table on p.64 shows that the Moon's longitude is 34° at 0h on 6 May. Moving forwards eight days at 13° per day we estimate the Moon's longitude at 0h on 14 May to be around 138°. At this time the longitude of the ascending node is found by interpolation to be around 167°. Thus d = (138° + 360°) – 167° = 331° and the Moon is south of the ecliptic moving north. Its position may be plotted on northern star chart 4S where it is found in the west between Cancer and Leo.

Some Events in 2016

Jan	2	Moon at Apogee (furthest from the Earth) (404,275 km)
	2	*Earth* at Perihelion (closest to the Sun)
	10	New Moon
	14	*Mercury* in Inferior Conjunction
	15	Moon at Perigee (closest to the Earth) (364,355 km)
	24	Full Moon
	30	Moon at Apogee (404,550 km)
Feb	7	*Mercury* at Greatest Western Elongation (26°)
	8	New Moon
	11	Moon at Perigee (364,355 km)
	22	Full Moon
	27	Moon at Apogee (405,380 km)
	28	*Neptune* in Conjunction with Sun
Mar	8	*Jupiter* at Opposition in Cancer
	9	New Moon
	9	Total Eclipse of the Sun
	10	Moon at Perigee (359,510 km)
	20	Equinox (Spring Equinox in Northern Hemisphere)
	23	Full Moon
	23	Penumbral Eclipse of the Moon
	23	*Mercury* in Superior Conjunction
	25	Moon at Apogee (406,125 km)
	27	Summer Time Begins in the UK
Apr	7	New Moon
	7	Moon at Perigee (357,165 km)
	9	*Uranus* in Conjunction with Sun
	18	*Mercury* at Greatest Eastern Elongation (20°)
	21	Moon at Apogee (406,350 km)
	22	Full Moon

May	6	Moon at Perigee (357,825 km)
	6	New Moon
	9	*Mercury* in Inferior Conjunction
	9	*Mercury* in Transit across the Sun
	18	Moon at Apogee (405,935 km)
	21	Full Moon
	22	*Mars* at Opposition in Scorpius
	30	*Mars* closest approach to Earth
June	3	*Saturn* at Opposition in Libra
	3	Moon at Perigee (361,140 km)
	5	New Moon
	5	*Mercury* at Greatest Western Elongation (24°)
	6	*Venus* in Superior Conjunction
	15	Moon at Apogee (405,020 km)
	20	Full Moon
	20	Solstice (Summer Solstice in Northern Hemisphere)
July	1	Moon at Perigee (365,980 km)
	4	New Moon
	4	*Earth* at Aphelion (furthest from the Sun)
	7	*Mercury* in Superior Conjunction
	7	*Pluto* at Opposition in Sagittarius
	13	Moon at Apogee (404,270 km)
	19	Full Moon
	27	Moon at Perigee (369,660 km)
Aug	2	New Moon
	10	Moon at Apogee (404,265 km)
	16	*Mercury* at Greatest Eastern Elongation (27°)
	18	Full Moon
	18	Penumbral Eclipse of the Moon
	22	Moon at Perigee (367,045 km)
Sept	1	New Moon
	1	Annular eclipse of the Sun
	2	*Neptune* at Opposition in Aquarius
	6	Moon at Apogee (405,055 km)
	13	*Mercury* in Inferior Conjunction

	16	Full Moon
	16	Penumbral Eclipse of the Moon
	18	Moon at Perigee (361,895 km)
	22	Equinox (Autumnal Equinox in Northern Hemisphere)
	26	*Jupiter* in Conjunction with Sun
	28	*Mercury* at Greatest Western Elongation (18°)
Oct	1	New Moon
	4	Moon at Apogee (406,100 km)
	15	*Uranus* at Opposition in Pisces
	16	Full Moon
	16	Moon at Perigee (357,860 km)
	27	*Mercury* in Superior Conjunction
	29	*Mars* at Perihelion
	30	Summer Time Ends in the UK
	30	New Moon
	31	Moon at Apogee (406,660 km)
Nov	14	Moon at Perigee (356,510 km)
	14	Full Moon
	27	Moon at Apogee (406,555 km)
	29	New Moon
Dec	10	*Saturn* in Conjunction with Sun
	11	*Mercury* at Greatest Eastern Elongation (21°)
	12	Moon at Perigee (358,460 km)
	14	Full Moon
	21	Solstice (Winter Solstice in Northern Hemisphere)
	25	Moon at Apogee (405,870 km)
	28	*Mercury* in Inferior Conjunction
	29	New Moon

Note: Dates are given for the longitude of the Greenwich Meridian

Monthly Notes 2016

January

New Moon: 10 January   *Full Moon:* 24 January

EARTH is at perihelion (nearest to the Sun) on 2 January at a distance of 147 million kilometres (91.4 million miles).

MERCURY passes through inferior conjunction on 14 January. For the first few days of the month it may be possible to glimpse the planet as a difficult evening object, low in the south-western sky at the end of evening civil twilight, about thirty-five minutes after sunset from northern temperate latitudes. The planet will be a slightly easier object for observers in equatorial latitudes at this time. The planet fades from magnitude –0.2 to +1.2 during the first week of January. In the last week of the month Mercury may possibly be seen by observers in the tropics and more southerly latitudes as a morning object in the south-eastern quadrant of the sky at the beginning of morning civil twilight. During this period its magnitude brightens from about +1.0 to +0.1. Despite the fact that Mercury is approaching greatest western elongation early next month it is not well placed for observation by those in the latitudes of the British Isles: since its declination is about –20 degrees, it will only be at a very low altitude above the horizon.

VENUS, magnitude –4.0, is a brilliant object, completely dominating the south-eastern sky before dawn, though the planet is best seen from the tropics and southern temperate latitudes from where it rises about two and a half hours before the Sun. For observers in the latitudes of the British Isles the duration of its period of visibility shortens noticeably during the month and because of its southerly declination it is never at any great altitude. The phase of the planet increases from 77 per cent to 85 per cent during the month. On 6 January, a very thin waning crescent Moon will make a nice pairing with Venus in the dawn twilight sky.

MARS is visible as a morning object, magnitude +1.2, rising at about

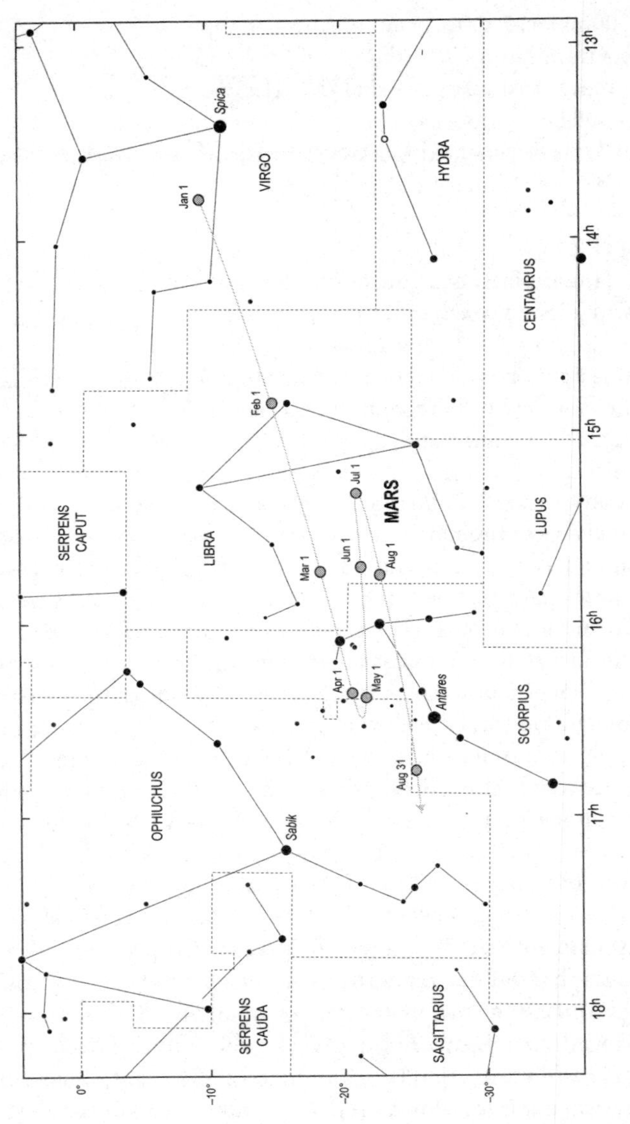

Figure 1. The path of Mars as it moves through the Zodiacal constellations of Virgo, Libra, Scorpius, Ophiuchus, back into Scorpius and Libra, then eastwards again through Scorpius and into Ophiuchus between 1 January and 31 August 2016. The planet is at opposition in Scorpius on 22 May.

02h 00m local time from northern temperate latitudes but rather earlier from latitudes further south. It begins the month in Virgo and then moves eastwards crossing into neighbouring Libra on 17 January. Mars will be at opposition in Scorpius on 22 May and the planet is visible at some time of the night throughout the whole of 2016. The path of Mars between January and August 2016 is shown in Figure 1.

JUPITER now rises in the mid-evening and reaches a stationary point on 8 January in Leo; thereafter its motion is retrograde (i.e., towards the west). The planet continues to increase in brightness, from magnitude –2.2 to –2.4 during the month, as it approaches opposition in early March. The waning gibbous Moon passes just south of Jupiter late on the evening of 27 January. The path of Jupiter during 2016 is shown in Figure 8, given with the notes for March.

SATURN was in conjunction with the Sun on 30 November last year and is visible in the south-eastern sky before dawn during January from northern temperate latitudes. By the end of the month, from such locations, the planet will be rising at about 04h 30m, but some two to three hours earlier from the tropics and locations further south. The planet, magnitude +0.6, is moving direct (i.e., towards the east) among the stars of Ophiuchus, north-east of the reddish Antares in Scorpius. (Ophiuchus intrudes into the Zodiac between the constellations of Scorpius and Sagittarius.) In the dawn twilight of 9 January, the brilliant Venus passes just five arc minutes (one sixth of a Moon diameter) north of Saturn and will be a useful guide to locating the very-much fainter and more distant planet. The path of Saturn in 2016 is shown in Figure 2 overleaf.

The Illusion of Size. Watch the full Moon rising in the early evening on 24 January; it will appear huge, as big as a plate. Six hours later, when the Moon is high up in the winter night sky, it will appear much smaller. Measurement will show that its apparent size has not altered. One can still cover the Moon with the tip of a little finger at arm's length when it is low down or high up. This is the so-called Moon illusion – an optical illusion which causes the Moon to appear much larger when it is seen near the horizon than it does when higher up in the sky. We are accustomed to judging the sizes of things by earthly standards and it seems that when we look up into the sky, with nothing to act as a

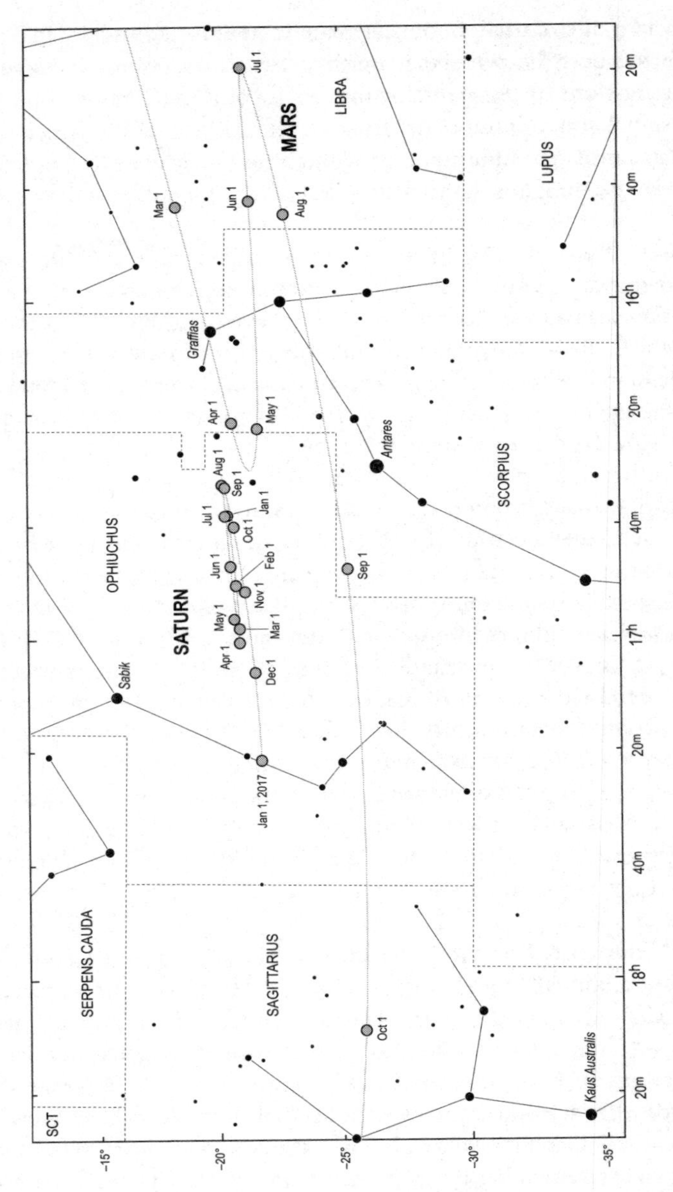

Figure 2. The path of Saturn against the background stars of Ophiuchus during 2016. The planet is at opposition on 3 June.

standard of comparison, our judgement seems to be affected. The illusion has been known since ancient times but, in spite of exhaustive investigations by psychologists over many years, an explanation of this illusion is still a matter of debate.

One explanation proposes that the Moon looks bigger when we see it near the horizon than when it is higher up because we think that objects near the horizon are further away from us. For example, when we see clouds or birds in the sky, those near the horizon are usually further away from us than those overhead. There may even be a perception that the sky itself is an upturned bowl but flattened at the top, in which objects moving near the horizon are perceived as being more distant and therefore as being larger than objects that are seen above us in the sky. Experiments by many researchers have shown that the horizon Moon is perceived to be at the end of a stretch of landscape receding into the distance, accompanied by distant trees, buildings and the like, all of which convince the brain that it must be a long way away, but such visual cues are absent when looking at the Moon high up in the sky.

Another explanation proposes that the perceived size of an object depends not only on the size of its image on the retina of the eye, but also on the size of objects in its immediate visual environment. In the case of the Moon illusion, objects on or near the horizon around where the Moon appears display fine detail that makes the Moon appear larger, while a Moon that is nearly overhead is surrounded by large expanses of empty sky that make it appear smaller. The effect may be simply illustrated by an example of the famous Ebbinghaus illusion shown in Figure 3. The lower central circle surrounded by small circles might represent the Moon near the horizon surrounded by objects of smaller visual extent, while the upper central circle represents the zenith Moon surrounded by expanses of empty sky of larger visual extent. Although both central circles are actually the same size, the lower one seems larger to many people.

This same optical illusion also occurs with the Sun and major constellations such as Orion and Ursa Major. Orion appears huge when it is low down in the winter night sky yet it can still be covered with a typical mobile phone held in an outstretched hand. Actually, this is an easy way to remember the shape of Orion: a rectangular mobile phone with a star at each corner for the shoulders and feet and three stars across the middle for his Belt. From Britain, Orion never appears particularly high in the sky, but the illusion of size is quite apparent with the Plough or Big Dipper, the principal stars of Ursa Major, the Great Bear. When seen low down in the northern sky early on a November evening, the Plough looks very large – like an enormous saucepan – but when it is overhead late on an April evening, it looks so much smaller and less impressive.

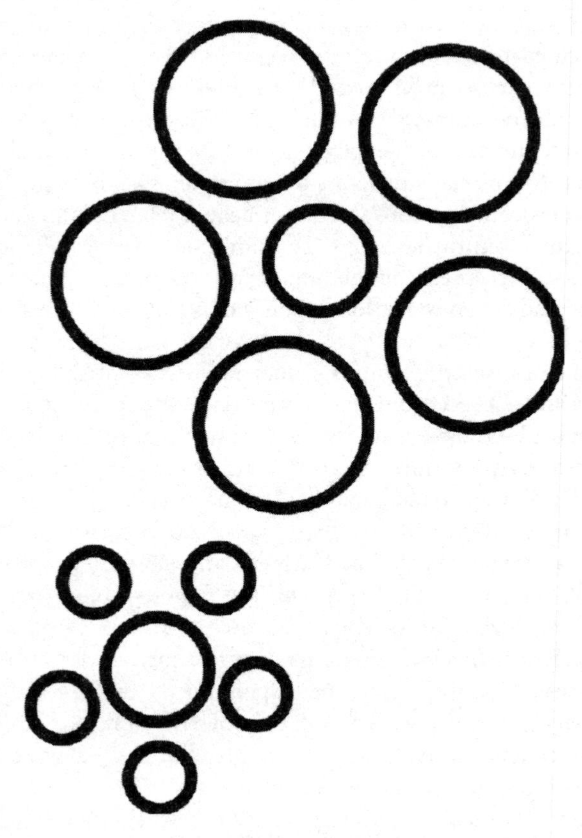

Figure 3. In this example of the Ebbinghaus illusion the lower central circle surrounded by small circles might represent the Moon near the horizon accompanied by objects of smaller visual extent, while the upper central circle represents the Moon high up in the sky surrounded by expanses of sky of larger visual extent. Although both central circles are the same size, the lower one will seem larger to many people. (Image courtesy of Wikimedia Commons.)

Forgotten Constellations. The first meteor shower of the year – which peaks on 4 January – has a strange-sounding name, not associated with any now-accepted constellation. The name the Quadrantids comes from the old group of Quadrans Muralis (the Mural Quadrant), which has long disappeared from the star maps. Quadrans is by no means the only rejected constellation; in fact, the list is quite extensive, but most have been long forgotten. From an historical point of view, it is interesting to look back at the various groups which have been proposed, but which have not survived.

Ptolemy, in his great work *The Almagest*, written in the second century AD, enumerated forty-eight constellations; twelve zodiacal; twenty-one in the Northern Hemisphere of the sky, and fifteen in the Southern. Not all the northern sky was covered, and Ptolemy naturally could know nothing about that part of the sky which never rose above the horizon from his home in Alexandria. However, all his constellations are still accepted, though in many cases the boundaries have been modified and the constellation Argo Navis was so big that it was broken up into three separate constellations – Carina, Puppis and Vela. The list includes almost all the famous groups visible from Europe and North America, such as Andromeda, Cassiopeia, Cygnus, Gemini, Hercules, Leo, Orion, Ursa Major and Ursa Minor. A few very small constellations also prove, surprisingly, to be 'Ptolemaic' – notably Crater, Delphinus, Equuleus and Sagitta.

The first additions to the list seem to have been made by Tycho Brahe, towards the end of the sixteenth century, who proposed Coma Berenices and Antinous; the former has survived, the latter has not. Then, between 1595 and 1597, the Dutch explorers Pieter Dirkszoon Keyser and Frederick de Houtman created twelve new far-southern constellations following a trading expedition to the East Indies. These new patterns were first depicted on celestial globes by the Flemish astronomer and cartographer Petrus Plancius and fellow Dutchman Jodocus Hondius, and subsequently included on a special chart of the area around the south celestial pole by Johann Bayer in his sky atlas *Uranometria* in 1603 (see Figure 4). These patterns, which included Apus, Chamaeleon, Dorado, Grus, Hydrus, Indus, Musca, Pavo, Phoenix, Triangulum Australe, Tucana and Volans, lay in regions inaccessible to Ptolemy and are all still used today.

It seems that the celebrated Southern Cross (Crux Australis) was introduced by Petrus Plancius, and depicted by him on celestial globes

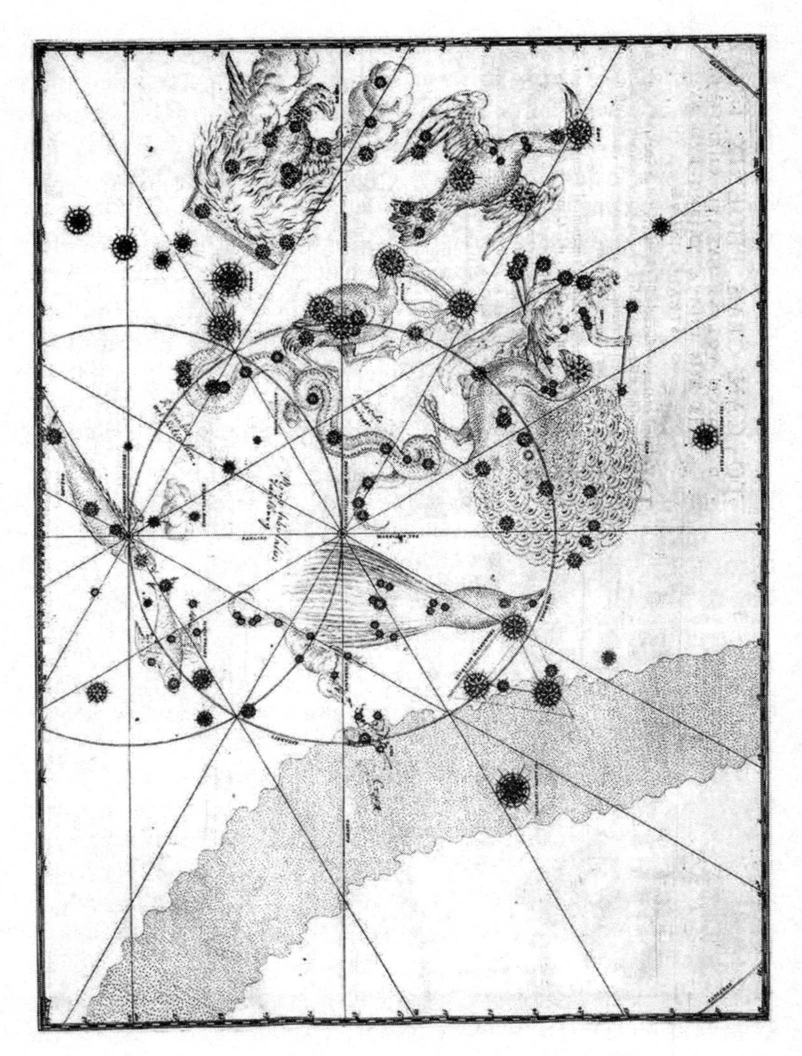

Figure 4. The special chart (no. 49) of the region around the south celestial pole from Bayer's *Uranometria*, published in 1603, which depicted the twelve new far-southern constellations created by the Dutch explorers Pieter Dirkszoon Keyser and Frederick de Houtman. (Chart supplied from the archive of the Sir Patrick Moore Heritage Trust, courtesy of the Executors.)

and a large wall map between 1589 and 1592. It was subsequently included in his sky atlas by Bayer in 1603 and its stars were certainly catalogued separately from neighbouring Centaurus (which practically surrounds it) by Frederick de Houtman in 1603. Later adopters of the constellation included Jakob Bartsch in 1624 and Augustin Royer in 1679. Indeed, the otherwise obscure French architect and astronomer Augustin Royer is often wrongly credited as being the first to portray the Southern Cross as a separate constellation.

Petrus Plancius is also credited with first introducing the constellation of Columba on a large wall map in 1592, and Camelopardalis and Monoceros on a celestial globe produced in 1612 or in 1613. A number of other constellations introduced by Plancius at this time, such as Cancer Minor (the Small Crab) and the rather cumbersome Euphrates Fluvius et Tigris Fluvius (the Rivers Euphrates and Tigris) were never officially accepted.

In 1679, Edmond Halley suggested a group to be known as Robur Caroli (Charles's Oak), but it met with no favour. Neither did Gottfried Kirch's incredibly cumbersome Gladii Electorales Saxonici (the Crossed Swords of the Electors of Saxony) of 1684, nor his Pomium Imperiale (Leopold's Orb) or Sceptrum Brandenburgicum (the Sceptre of Brandenburg) put forward in 1688.

Johannes Hevelius, in 1690, suggested quite a number of new constellations including somewhat obscure patterns such as Cerberus (the guardian dog of Hades), Musca Borealis (the Northern Fly), and Ramus Pomifer (the Apple-bearing branch). Of these suggested additions, only seven have survived: Canes Venatici, Lacerta, Leo Minor, Lynx, Scutum, Sextans and Vulpecula.

The English naturalist John Hill put forward many suggestions for new constellations in 1754. Most were, not surprisingly, perhaps, named after creatures found in the Natural World such as Aranea (the Long-Legged Spider), Dentalium (the Tooth Shell), Hippocampus (the Sea Horse) and Hirudo (the Leech). All were rejected.

Then, in 1763, Nicolas Louis de Lacaille added fourteen southern groups, most of them small and most near the south celestial pole; all are still to be found, and include Circinus, Horologium, Microscopium, Norma and Telescopium. Some of the names, however, have been shortened; Lacaille's Antlia Pneumatica (the Air Pump) is now simply Antlia; Fornax Chemica (the Chemical Furnace) is Fornax;

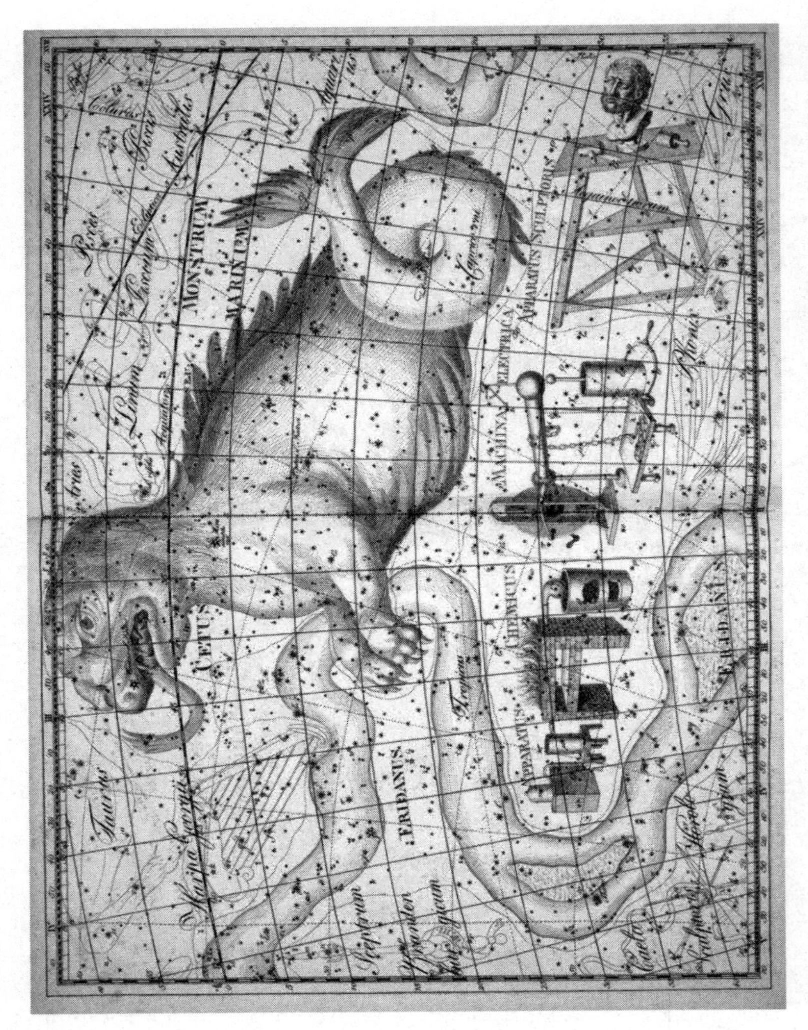

Figure 5. Johann Bode's celestial atlas *Uranographia*, published in 1801, contained many accurately scaled star maps showing the positions of the stars overlaid with artistically presented constellations. Here we see the region occupied by Eridanus, the River (left), Cetus, the Sea Monster (top), with the now discarded pattern of Machina Electrica, the Electricity Generator, below, flanked on one side by Apparatus Chemicus, the Chemical Furnace (now altered to Fornax), and on the other by Apparatus Sculptoris, the Sculptor's Studio (now shortened to Sculptor). (Image courtesy of Wikimedia Commons.)

and Reticulum Rhomboidalis (the Rhomboidal Net) has become Reticulum.

These were the last accepted additions. Pierre Charles Lemonnier proposed two more groups, Tarandus (the Reindeer) in 1743 and Turdus Solitarius (the Solitary Thrush) in 1776, both of which have vanished. Martin Poczobut, in 1777, added Taurus Poniatowski (Poniatowski's Bull), and Joseph-Jerome de Lalande suggested patterns such as Custos Messium (the Harvest Keeper) in 1779, and Quadrans Muralis (the Mural Quadrant) and Globus Aerostaticus (the Hot Air Balloon) at the end of the eighteenth century. But a master at creating unnecessary and unwanted constellations was Johann Bode who, in his great 1801 star atlas *Uranographia*, portrayed over a hundred constellations, many of them new and with names such as Frederici Honores (Frederick's Honours), Lochium Funis (the Nautical Log and Line) and Machina Electrica (the Electricity Machine). Some examples are shown in Figure 5. All of these new patterns were really too small to merit separate names and they were deservedly abandoned.

It cannot be said that the present division of the stars into eighty-eight constellations, agreed by the International Astronomical Union in 1930, is ideal, but it seems unlikely that any further modifications will be made now.

February

New Moon: 8 February *Full Moon:* 22 February

MERCURY, although it reaches greatest western elongation (26°) on 7 February, is not at all well placed for observation by those in northern temperate latitudes. For observers in southern latitudes this will be the most favourable morning apparition of the year. Figure 6 shows, for observers in latitude 35°S, the changes in azimuth (true bearing from the north through east, south and west) and altitude of Mercury on successive evenings when the Sun is 6° below the horizon. This condition is known as the beginning of morning civil twilight, and in this

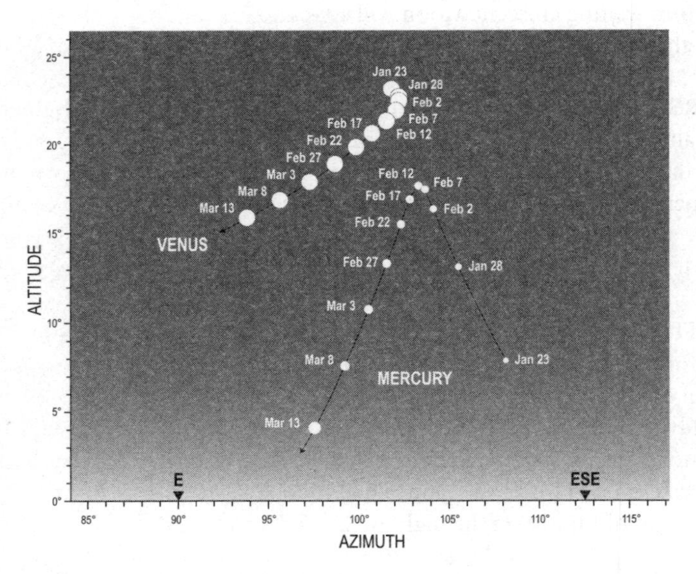

Figure 6. Morning apparition of Mercury from latitude 35°S. The planet reaches greatest western elongation on 7 February 2016. It will be at its brightest in early March, after elongation. The chart also shows the positions of the brilliant Venus, in relation to Mercury, during the month. Venus can be a useful guide to locating the much fainter planet. The angular diameters of Mercury and Venus are not drawn to scale.

latitude and at this time of year occurs about thirty minutes before sunrise. The changes in the brightness of the planet are indicated by the relative sizes of the circles marking Mercury's position at five-day intervals. It will be noticed that Mercury is at its brightest after it reaches greatest western elongation, its magnitude increasing from 0.0 to –0.3 between 7 February and the end of the month. For observers in the tropics and the Southern Hemisphere, the brilliant Venus will lie three to five degrees above Mercury in the eastern twilight sky during the second and third weeks of February and will be a helpful guide to locating the much fainter planet.

VENUS continues to be visible as a splendid object in the early morning sky, magnitude –3.9, though for observers in the latitudes of the British Isles it will only be seen low above the south-eastern horizon for a short period of time, just before dawn. The phase of the planet increases from 85 per cent to 91 per cent during the month. An incredibly thin waning crescent Moon will appear close to Venus in the dawn twilight of 6 February.

MARS continues to be visible in the mornings, and actually brightens by half a magnitude during the month, from +0.8 to +0.3. By the end of February, the planet will be rising soon after midnight from northern temperate latitudes, but over two hours earlier from the tropics and more southerly locations. The planet is moving direct in Libra. The last quarter Moon will lie about 2.7° north of Mars early on 1 February.

JUPITER, magnitude –2.4, is a brilliant object in the night sky, and visible for the greater part of the hours of darkness since it comes to opposition early in March. The planet is reasonably well placed for observers worldwide due to its declination a few degrees north of the celestial equator. It is moving retrograde some way south of the stars marking the eastern end of Leo, the Lion. The Moon, just past full, will pass only 1.7° south of Jupiter on the night of 23/24 February.

SATURN is an early morning object, moving direct against the stars of Ophiuchus during the month. By the end of February, it is rising at about 03h 30m for observers in northern temperate latitudes; about three hours earlier for those in the Southern Hemisphere. Its magnitude is steady at +0.5.

The Celestial River. On February evenings, the long, straggly line of stars marking Eridanus, the River, can be traced down to the southern horizon; it actually extends far into the Southern Hemisphere. On the whole Eridanus occupies a large but rather barren region of the southern sky with another large, dim constellation (Cetus) to its western side and Orion, the Hunter, to the other; Taurus, the Bull, lies to the north (Figure 7).

One of Ptolemy's original forty-eight constellations, Eridanus has been associated with a variety of rivers both real and mythological, but it is often identified with the River Po in Italy. According to mythology, the reckless youth Phaethon was given permission to drive the Sun-chariot for one day, but he was unable to control the winged horses as they raced across the sky. So, in order to prevent the world from being set on fire, Jupiter struck him with a thunderbolt and toppled him to his death in a river.

There is nothing at all distinctive about that part of Eridanus visible from Europe and North America. The brightest star Achernar (magnitude 0.5) has a declination of 57° south, and doesn't rise from anywhere north of latitude 33°N, so it is invisible from the whole of Europe and most of the United States, except for Florida and southern Texas (where it only just rises) and Hawaii. From Cairo, Achernar merely skirts the southern horizon when at its highest. However, it is circumpolar from Buenos Aires, Cape Town and Sydney, and from all of New Zealand. Achernar, whose name comes from the Arabic words meaning 'the end of the river', was within twelve degrees of the south celestial pole between 4000 BC and 2000 BC.

The line of generally faint stars marking Eridanus first extends westwards from Beta Eridani, also known as Kursa (magnitude 2.8), which is so close to the brilliant Rigel in the foot of Orion that it seems to belong more properly to the Orion pattern. Slightly fainter is Gamma or Zaurak (magnitude 3.0), further west and rather lower down; it is not particularly easy to identify because there are no obvious pointers to it. Next come Delta or Rana (magnitude 3.5) and the slightly fainter Epsilon or Sadira (magnitude 3.7). Sadira is one of our Sun's nearer neighbours, lying at a distance of 10.5 light years. Associated with Sadira there are certainly two asteroid-type rocky belts, one at around 450 million kilometres from the star and the other at 3,000 million kilometres. There is evidence for a gas giant planet (known as Epsilon Eridani b), intermediate in size between Uranus and Saturn, moving at

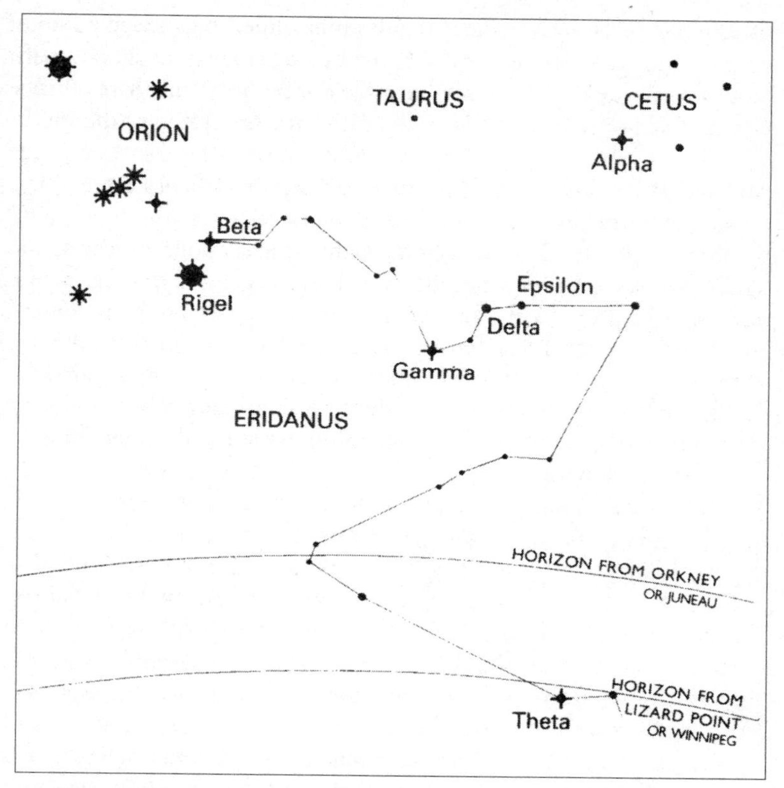

Figure 7. From the latitudes of the British Isles, the long, meandering line of stars which marks Eridanus, the River, begins near the brilliant Rigel in Orion and can be traced down to the southern horizon. The star Theta Eridani is just invisible from the south of England and from Winnipeg in Canada, but does rise from New York. However, the brilliant Achernar is much too far south to be seen from anywhere in Europe or North America. (Diagram supplied from the archive of the Sir Patrick Moore Heritage Trust, courtesy of the Executors.)

a distance of 500 million kilometres in a period of about 7 years, and it is possible that there is a second planet associated with the outermost of the two asteroid belts. In addition, Epsilon Eridani has an extensive outer disk of debris left over from the system's formation.

The ancient constellation of Eridanus ended at the star now designated Theta Eridani or Acamar (magnitude 2.9), which is just invisible from Southern England but rises from New York and anywhere south of latitude 49°N. The pattern was later extended southwards to the much brighter Achernar, which lay out of sight from Mediterranean latitudes. Theta Eridani is a fine double star; both components are white, of magnitudes 3.4 and 4.5, with a separation of 8.3 arc seconds.

An Extra Day. This year the month of February will have an extra day; twenty-nine instead of the usual twenty-eight. This is a consequence of the fact that the Earth does not go round the Sun in 365 days; it takes 365.24219 days (the length of the so-called tropical year), or, in everyday language, just under 365¼ days. This means that the civil year of 365 days is actually about one quarter of a day too short. If nothing were done, festivals such as Christmas would rotate gradually around the calendar – and we would be celebrating Christmas in June, when it is summer in the Northern Hemisphere and winter in the Southern.

The great Roman ruler Julius Caesar took advice from the Greek astronomer and mathematician Sosigenes of Alexandria and revised the 365-day Egyptian calendar. This meant giving the year 46 BC no less than 445 days to get everything back in step and, not surprisingly, it was nicknamed *annus confusionis* ('the year of confusion'). Thereafter, an extra day was tacked onto the shortest month, February, every four years and this made up for the fact that the calendar year was about a quarter of a day too short. The 366-day years were called leap years. Any year divisible by four was a leap year.

A small error still remained; this was met by making the 'century' years (1700, 1800, 1900, 2000, etc.) leap years only if they could be divided by 400. Thus 1900 was not a leap year, but 2000 was. This new calendar was named the 'Gregorian calendar' after Pope Gregory XIII, who first introduced it in 1582. The Julian calendar gained against the mean tropical year at the rate of one day in 128 years. For the Gregorian the figure is one day in 3,226 years.

By the time this Gregorian calendar was adopted in England, the error had risen to eleven days, so it was decreed that in 1752 eleven days would be 'omitted' from the calendar – so the day after 2 September 1752 was 14 September 1752. A favourite trick question is: 'What happened in England on 10 September 1752?' The answer? Nothing, because there was no such day!

March

New Moon: 9 March *Full Moon:* 23 March

Equinox: 20 March

Summer Time in the United Kingdom commences on 27 March

MERCURY may be seen by those observers in equatorial and southern latitudes as a morning object during the first week of the month, located a little over six degrees below the brilliant Venus in the eastern twilight sky about twenty-five minutes before sunrise. Figure 6, given with the notes for February, shows, for observers in latitude 35°S, the changes in azimuth and altitude of Mercury at this time. Thereafter it is too close to the Sun for observation as it moves towards superior conjunction on 23 March.

VENUS continues to be visible as a brilliant object in the early morning sky, magnitude –3.9. However, observers in northern temperate latitudes will find the planet coming to the end of its period of visibility: in particular, those as far north as the British Isles will have lost it in the glare of the rising Sun after the first week of the month.

MARS begins the month moving direct in Libra, but crosses the border into neighbouring Scorpius on 13 March. The planet is well placed for observation near the meridian before dawn. The last quarter Moon will appear quite close to Mars early on 1 March and the waning gibbous Moon will lie more or less between, and slightly to the north, of Mars and Saturn on 29 March. Mars starts the month only just brighter than Saturn, which lies about ten degrees to the east, but Mars brightens from magnitude +0.3 to –0.5 during March as its distance from the Earth decreases.

JUPITER is at opposition on 8 March and is visible all night long. The planet is moving retrograde (i.e., towards the west) in Leo, some way south of the triangle of stars marking the hindquarters of the Lion, and is a brilliant object at magnitude –2.5. Figure 8 overleaf shows the path of Jupiter against the background stars during 2016. The planet's northerly declination of +5° 58′ means that it has moved noticeably south compared with last year or 2014, but still quite favourable for observers in northern temperate latitudes, although it is well placed from locations worldwide. At opposition Jupiter's apparent diameter is 44.4 arc seconds and it is 664 million kilometres (412 million miles) from the Earth. The evenings afford plenty of opportunities for observing the movements of the four Galilean satellites, Io, Europa, Ganymede and Callisto. The almost full Moon will make a striking configuration with Jupiter on the night of 21/22 March.

SATURN is moving direct in Ophiuchus. By the end of the month it still only rises at about midnight for observers in northern temperate latitudes, but almost three hours earlier for those in the Southern Hemisphere. The planet reaches its first stationary point on 25 March and thereafter its motion is retrograde. Its brightness increases very slightly from magnitude +0.6 to +0.5 during March. The last quarter Moon will lie 3.6° north of Saturn early on 2 March.

First Spacecraft Views of Halley's Comet. The last return of Halley's comet in 1985–86 was the first return of the Space Age, and an 'armada' of spacecraft from four space agencies were utilized in studies of the comet: two Russian (Vega 1 and Vega 2), two Japanese (Sakigake and Suisei), two American (ICE and Pioneer 7) and one European (Giotto). All of the comet Halley probes were successful. The European mission, named Giotto in honour of the Italian painter Giotto di Bondone who, in 1304, depicted Halley's comet as the 'Star of Bethlehem' in one of his frescoes in the Scrovegni chapel in Padua, was targeted to pass through the comet's inner coma and image the nucleus from close range. However, prior information sent back by Russian Vega missions was invaluable; images of Halley from their cameras enabled the Giotto team to determine very accurately the position of the comet and to determine the trajectory that they wanted Giotto to follow.

Exactly thirty years ago this very month, on the night of 13–14 March 1986, the Giotto spacecraft passed within 600 kilometres of the comet's nucleus. It carried a camera, the HMC (Halley Multicolour Camera) and this functioned until fourteen seconds before closest

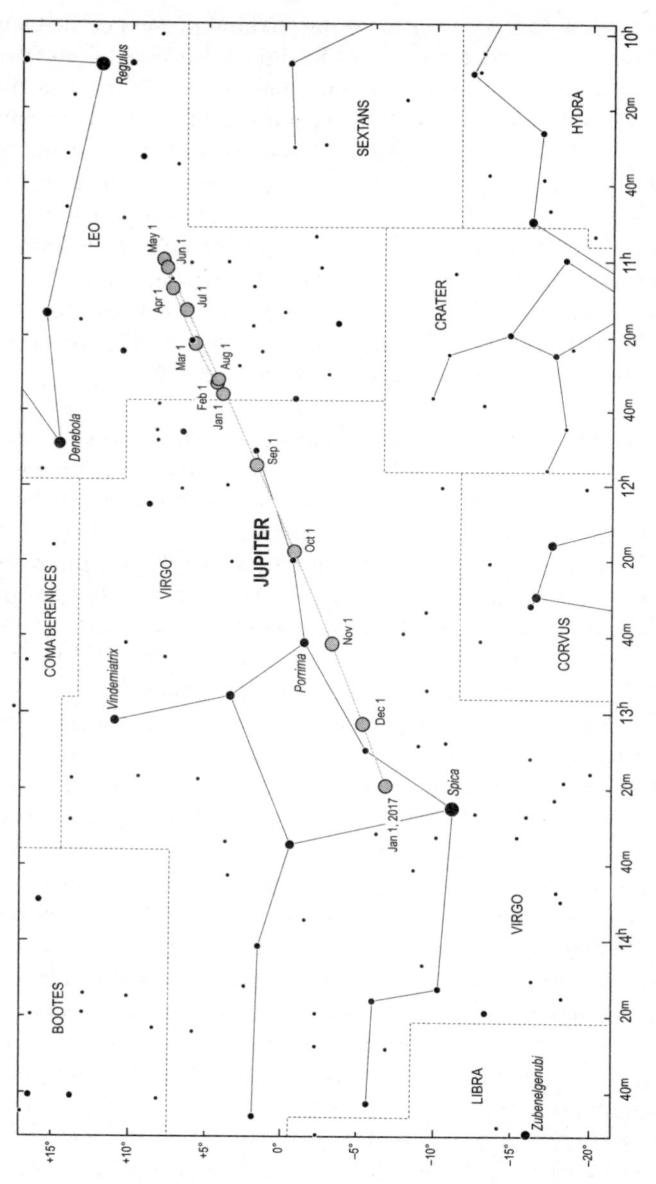

Figure 8. The path of Jupiter as it moves against the background stars of Leo and Virgo, during 2016. The planet is at opposition in Leo on 8 March.

approach to the nucleus, when it was made to gyrate by the impact of a dust particle – probably about the size of a grain of rice – and communications were temporarily interrupted. In fact, the camera never worked again, and the closest image was obtained at a distance of 1675 kilometres from the nucleus. The nucleus itself measured 16 km x 8 km x 8 km, and was shaped rather like a peanut with a total volume of over 500 km^3. If the mean density is only about 0.3 g cm^{-3}, then the mass is around 150,000 million tonnes and it would take about 40,000 million comets of this mass to equal the mass of the Earth.

The nucleus had a very dark surface layer, with a reflectivity of only 2 to 4 per cent. Prior to the encounter, scientists had expected the comet's nucleus to be composed of bright reflective ices, so they programmed Giotto's camera to centre the images on the brightest object it could see. This turned out to be bright jets of gas and dust spurting from the sunward side of the nucleus, but, fortunately, the unexpectedly dark nucleus could still be seen off to one side of the jets (Figure 9 overleaf). Water ice appeared to be the main constituent of the nucleus (about 80 per cent) with frozen carbon monoxide making up another 15 per cent followed by frozen carbon dioxide, methane and ammonia. Giotto's dust mass spectrometer showed that there were two main types of dust particles – one almost entirely dominated by the light CHON elements (organic grains made of carbon, hydrogen, oxygen and nitrogen), and the other rich in elements such as magnesium, calcium and silicon.

The shape of the terminator showed that the central region was smoother than the ends; a bright patch 1.5 kilometres in diameter was assumed to be a hill, and there were features which appeared to be craters, around 1 kilometre across. Dust jets were active, although from only a small area of the nucleus on the sunward side. The sunward side was found to have a temperature of 57°C, far higher than expected, and from this it was inferred that the ice nucleus is coated with a layer of warmer, dark dust. The icy nucleus was eroded at around one centimetre per day near perihelion, and at each return the comet must lose around 350 million to 500 million tons of material. The rotation period was found to be 53 hours with respect to the long axis of the nucleus, with a 7.3-day rotation period around the axis; the nucleus was, in fact, precessing rather in the manner of a toppling gyroscope.

Figure 9. A composite image of the nucleus of comet 1P/Halley. This image is composed of sixty-eight images of varying resolution acquired by the European Space Agency's Giotto spacecraft during its fly-by in March 1986. The data at the brightest point on the nucleus is at the highest resolution (50 metres). The Sun comes from 30° above the horizontal to the left and is 17° behind the image plane. The night side of the nucleus can be seen silhouetted against a background of bright dust. Jets can be seen originating from two areas on the nucleus. The bright region seen within the night side of the nucleus is thought to be a hill or mountain about five hundred metres high. Other surface details may be discerned in the illuminated region. (Image courtesy of ESA/MPAe.)

Johannes Fabricius. This month is the four hundredth anniversary of the death of the astronomer Johannes Fabricius, on 19 March 1616. Johannes was the eldest son of David Fabricius who was a Lutheran pastor and astronomer in Osteel, north-west Germany. On graduating from Leiden University in the Netherlands in 1611, Johannes returned home with one or more telescopes. Early one morning in March 1611, Johannes pointed one of the instruments at the rising Sun and saw several dark spots upon it. Calling his father, they investigated the spots together, observing the solar disk by projecting its image using a camera obscura. They noted that the spots appeared to move from day to day. Spots would appear on the eastern edge of the solar disk, steadily move to the western edge, disappear, and then sometimes re-appear at the eastern edge again after a period of time which was the same as that taken for them to cross the visible hemisphere. This suggested that the Sun rotated on its axis, an idea that had been put forward earlier by Giordano Bruno and Johannes Kepler, but not substantiated with hard evidence.

David and Johannes Fabricius described their sunspot observations in a 22-page booklet with the engaging title of *De Maculis in Sole Observatis, et Apparente earum cum Sole Conversione Narratio* (*Narration on Spots Observed on the Sun and their Apparent Rotation with the Sun*), published in June 1611. It was printed in Wittenberg, ready for the Frankfurt book fair that autumn, but it remained unknown to other observers for some time and was subsequently overshadowed by the independent discoveries and publications about sunspots by Christoph Scheiner and Galileo Galilei.

It is also unfortunate that, although it was the first publication on the subject of sunspots, only a small part of the booklet actually described the observations made by Johannes and his father, and they neither gave times or dates for the observations nor included a drawing of the spots.

Not much else is known about Johannes Fabricius except that he died at the very young age of twenty-nine. The following year his father David was killed by being hit over the head with a spade following an argument about the theft of a goose!

April

New Moon: 7 April *Full Moon:* 22 April

MERCURY, after the first few days of the month, is visible as an evening object for the following three weeks for those observers in equatorial and more northerly latitudes. For observers in northern temperate latitudes this will be the most favourable evening apparition of the year. Figure 10 shows, for observers in latitude 52°N, the changes in azimuth (true bearing from the north through east, south and west) and altitude of Mercury on successive evenings when the Sun is 6° below the horizon. This condition is known as the end-of-evening civil twilight and in this latitude and at this time of year occurs about thirty-five minutes after sunset. The changes in the brightness of the planet are indicated by the relative sizes of the circles marking Mercury's position at five-day intervals. It will be noticed that Mercury is at its brightest before it reaches greatest eastern elongation (20°) on 18 April. Its magnitude on 5 April is −1.2, while by 24 April, it is +1.3.

VENUS, magnitude −3.9, is already too close to the Sun for observation from the latitudes of the British Isles. Even further south it is coming to the end of its period of visibility, though those in southern latitudes will still be able to see it right through the month, low down above the eastern horizon shortly before sunrise.

MARS is now a conspicuous object in the early morning skies, rising before midnight, but rather more favourably placed for observers in the tropics and the Southern Hemisphere on account of its declination about 21° south of the celestial equator. Its magnitude increases from −0.5 to −1.4 during the month. Mars begins the month moving direct in Scorpius, but crosses into neighbouring Ophiuchus on 3 April. It reaches its first stationary point on 17 April and thereafter moves retrograde in Ophiuchus. The Moon, just past full, will pass about 5° north of Mars early on 25 April.

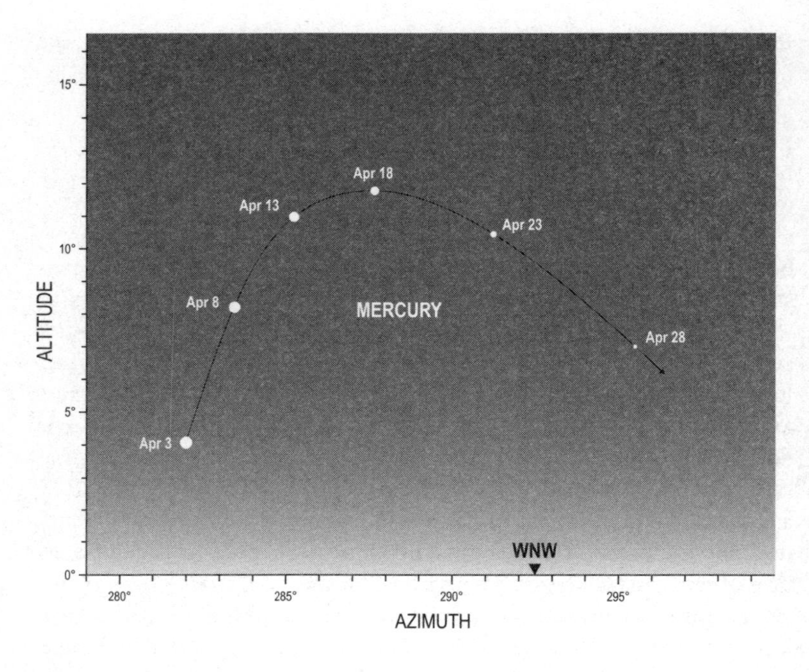

Figure 10. Evening apparition of Mercury from latitude 52°N. The planet reaches greatest eastern elongation on 18 April 2016. It is at its brightest in early April, before elongation.

JUPITER continues to be visible as a lovely object in the south-eastern sky as soon as darkness falls. As it was at opposition early in March, it is still visible until the early morning hours. Reference to Figure 8, given with the notes for March, shows that Jupiter is moving very slowly retrograde in Leo. The planet fades slightly from magnitude –2.5 to –2.3 during the month. The waxing gibbous Moon will make a lovely pairing with Jupiter on the night of 17/18 April.

SATURN is also moving retrograde in Ophiuchus, slightly less than ten degrees east of the considerably brighter Mars: Saturn yellowish; Mars reddish. Saturn continues to brighten slowly from magnitude +0.4 to +0.2 during the month. By the end of April, Saturn is rising before midnight for observers in northern temperate latitudes, but nearly three hours earlier for those further south.

Retrograding of Planets. During the second half of April, Mars seems to move in an east-to-west or retrograde direction against the background stars. Consequently, looking at the apparent movement of the planet during the year, it appears to perform a 'zig-zag'. Jupiter has been retrograding since 8 January, but after 9 May it again resumes its usual direct or west-to-east motion.

The fact that the planets do not move uniformly against the stars has been known since ancient times, and it was, in fact, one of the great stumbling blocks in the way of the old Ptolemaic theory, according to which the Earth lies in the centre of the universe and the orbits of all celestial bodies must be true circles – because the circle is the perfect form and nothing short of perfection can be tolerated in the heavens. Ptolemy of Alexandria, who brought this system to its greatest level of development (AD 120–180), was forced to introduce a cumbersome system of large and small circles (Figure 11) in which a planet moved in a small circle or epicycle, while the centre of this circle (the deferent) itself moved around the Earth in a perfect circle.* Yet the true explanation of retrograding is quite simple.

A superior planet, such as Mars or Jupiter, has an orbital velocity less than that of Earth, and is moving in a larger orbit. Therefore, for some time before and after opposition, the Earth will be overtaking the superior planet 'on the inside track', as it were, and the effect of this is to produce apparent retrograding. It does not, needless to say, indicate any real change in the motion of the planet. Before and after retrograding, the planet will reach a stationary point. This month, Mars reaches this point on 17 April, prior to retrograding before opposition in May; Jupiter did so on 8 January (before opposition) and will reach its second stationary point next month, on 9 May (after opposition). The inferior planets, Mercury and Venus, behave in a superficially different manner, but they can also move in a retrograde direction.

The Old Pole Star. Most people can identify the present North Pole Star, Polaris (magnitude 2.0), in Ursa Minor, the Little Bear, but how

* When Ptolemy, who was an excellent mathematician, realized that this comparatively simple arrangement could not explain the actual observed movements of the planets in the sky, he was compelled to introduce extra epicycles, thus making the entire system clumsy and unwieldy. However, the Ptolemaic theory was almost universally accepted by scientists up to the time of Copernicus in the sixteenth century.

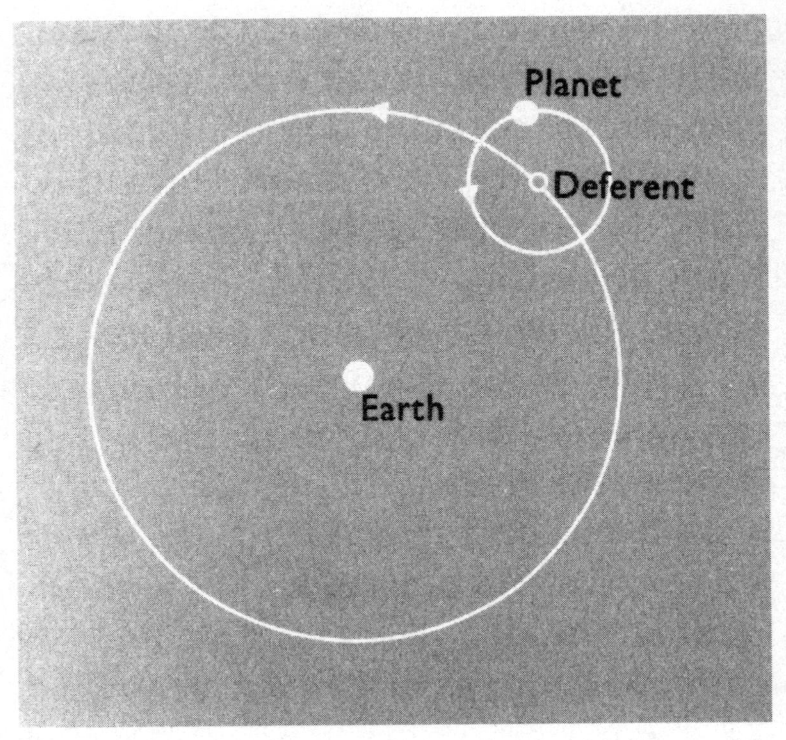

Figure 11. In the Ptolemaic system, the Earth lies at the centre of the Universe and the orbits of all celestial bodies are perfect circles. According to this somewhat cumbersome system for describing the motion of the planets, each planet moved around a small circle called the epicycle, while the centre of this circle (called the deferent) itself moved around the Earth in a perfect circle. (Diagram supplied from the archive of the Sir Patrick Moore Heritage Trust, courtesy of the Executors.)

many people can find the old Pole Star? At the time when the Egyptian Pyramids were being built, the north celestial pole lay near Thuban, in the constellation of Draco, the Dragon. In fact, the Earth's north (and south) celestial pole takes about 25,800 years to describe a complete circle on the celestial sphere. The reason is that the axis of rotation of the spinning Earth is precessing slowly (like the wobbling motion of a gyroscope or child's spinning top), due to gravitational perturbations by the Sun, Moon and planets.

Though Thuban has been given the Greek letter Alpha, it is only the eighth brightest star in the Dragon; its magnitude is 3.7 whereas the leader of the constellation, Eltamin or Gamma Draconis, located in the Dragon's head, is of magnitude 2.2. However, Thuban is a fairly considerable star; it is 260 times as luminous as the Sun, and 305 light years away. It is pure white, and of spectral type A0, so that its surface is much hotter than that of the Sun.

Identifying Thuban is easy enough, as shown in Figure 12. Find Mizar in the Great Bear, and then Kocab or Beta Ursae Minoris in the Little Bear; Thuban lies more or less midway between them, and there are no other bright stars close to it.

The closest approach of Polaris to the north celestial pole will be in March 2100. Its declination will then be +89° 32′ 51″. Polaris is actually a Cepheid variable star – the closest to the Sun – but having a very small amplitude. It is a multiple-star system, comprising the main star, which is a supergiant, two smaller companions, and two distant components.

There is no bright South Pole Star now; the pole lies near the

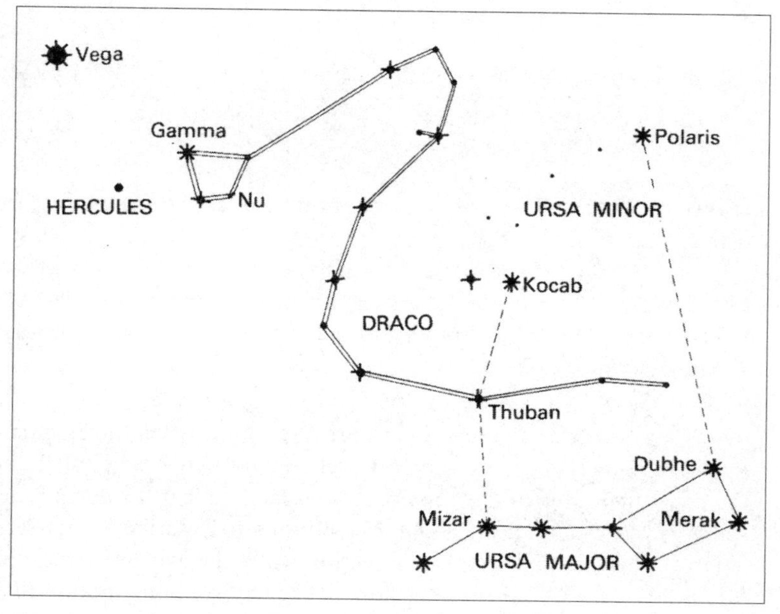

Figure 12. Finding the old Pole Star, Thuban (Alpha Draconis), using the stars Mizar in Ursa Major and Kocab in Ursa Minor; Thuban lies roughly midway between the two. (Diagram supplied from the archive of the Sir Patrick Moore Heritage Trust, courtesy of the Executors.)

obscure Sigma Octantis, magnitude 5.5. In Pyramid times there was a brighter South Pole Star, Alpha Hydri (magnitude 3.0), but the star was several degrees from the actual pole. For southern observers things will be decidedly better by AD 5000, when the pole will be fairly close to Miaplacidus or Beta Carinae, magnitude 1.7.

The Jovian-Plutonian Gravitational Effect. The former greatly missed Editor of this Yearbook, Sir Patrick Moore (Figure 13), had a very keen and rather wicked sense of humour and was well known as a practical joker. Forty years ago this very month, on 1 April 1976 (April Fool's Day), he perpetrated perhaps his greatest prank of all. That morning, he introduced the unsuspecting listeners of BBC Radio 2 to an unusual and previously unknown alignment of the planets – the Jovian-Plutonian Gravitational Effect. Moore explained that at exactly 9.47 a.m. that day Pluto would pass behind giant Jupiter, and the powerful combination of the two bodies' gravitational pull would cause a temporary reduction in Earth's gravity. He urged his radio listeners to

Figure 13. The late Sir Patrick Moore, pictured in the early 1970s with two of his favourite cars, the ancient Ford Prefect, which he called 'the Ark' (left), and an MG Magnette. (Picture supplied from the archive of the Sir Patrick Moore Heritage Trust, courtesy of the Executors.)

jump in the air at exactly that time in a bid to experience 'weightlessness'.

Such was Moore's reputation as a scientist and astronomer – he had presented his BBC TV programme *The Sky at Night* every four weeks since 1957 – that by 9.48 a.m. that morning the BBC switchboard was jammed with thousands of callers ringing to say that they had felt the decrease in gravity. One woman claimed that she and eleven friends had been sitting around her dining-room table when they had suddenly been 'wafted from their chairs and orbited gently round the room', and a woman from the Netherlands rang in to say that she and her husband had floated around the room together. There was also a man who claimed that he had risen from the ground so quickly that he had hit his head hard on the ceiling and thought he should receive compensation.

All of this was quite incredible because Moore made up the entire story as an April Fool's hoax! Jupiter and tiny Pluto (which is far smaller even than Earth's Moon) are so far away that any gravitational effects on Earth would be utterly negligible. The so-called Jovian-Plutonian Gravitational Effect never existed. The reason so many people were fooled is simply because Moore was utterly believable and his reputation was second to none.

May

New Moon: 6 May *Full Moon:* 21 May

MERCURY attains its greatest western elongation early in June. During May, the lengthening duration of twilight renders observation impossible for observers in the latitudes of the British Isles, but nearer the Equator, and in the more populous regions of the Southern Hemisphere, Mercury can be seen as a morning object low above the eastern horizon at the time of the beginning of morning civil twilight, for the last few days of the month. A transit of Mercury across the face of the Sun occurs on 9 May, when the planet passes through inferior conjunction, and this event will be visible from the British Isles – weather permitting, of course! The track crosses the northern part of the solar disk, the mid-time being about 14h 57m UT. The transit is visible from most of Asia and the Pacific Ocean, North and South America, Greenland, Europe and Africa. From the United Kingdom ingress starts at about 11h 12m UT and egress finishes at about 18h 41m UT. For more detailed information see the note by Nick James elsewhere in this Yearbook.

VENUS is no longer visible from northern temperate latitudes but can still be seen for the first three weeks of the month from the tropics and southern latitudes, low above the eastern horizon before sunrise. Its magnitude is –3.9.

MARS reaches opposition in Scorpius on 22 May, magnitude –2.0, and thus is available for observation throughout the night. The eccentricity of the orbit of the planet is such that its closest approach to the Earth (seventy-five million kilometres) does not occur until 30 May, eight days after opposition (this is almost the maximum difference). The planet's retrograde motion, which carried it from Ophiuchus into Scorpius at the beginning of May, carries it on into Libra at the end of the month. The full Moon passes about 5.5° north of Mars on the night of 21/22 May, passing north of Saturn the following evening.

JUPITER reaches it second stationary point on 9 May and thereafter resumes its direct (i.e. eastwards) motion once more. The planet will be easily recognized as darkness falls, a brilliant object south of the main stars of Leo. Now two months past opposition, the planet continues to fade slightly, from magnitude –2.3 to –2.0 during the month. With the planet now situated just north of the celestial equator, it remains visible from all latitudes until after midnight at the end of May, although slightly longer from locations in northern Europe and North America. The waxing gibbous Moon will be placed south and west of Jupiter on the evening of 14 May.

SATURN will be at opposition in early June, so it rises not long after sunset and is observable until dawn, the period of visibility being longer for those in equatorial and Southern Hemisphere latitudes. The planet brightens very slightly from magnitude +0.2 to 0.0 during May. It continues to move retrograde among the stars of Ophiuchus, but much brighter than any of the stars in that part of the sky.

Ursa Major at the Zenith. There must be few people who cannot recognize Ursa Major, the Great Bear. With its seven leading stars making up the familiar 'saucepan-shaped' outline usually nicknamed the 'Plough' in Britain and the 'Big Dipper' in North America, it cannot be mistaken (Figure 14). On May evenings, as soon as darkness falls in northern Europe, Canada and the northern United States, the pattern is virtually overhead, making it even easier to identify. It lies not far from the north celestial pole, so that from cities such as London or New York it is circumpolar; it can always be seen somewhere whenever the sky is sufficiently clear and dark.

To be accurate, Ursa Major contains many stars as well as those of the Plough. It is a large constellation, and there are some interesting objects in it, including M97, the Owl planetary nebula (unfortunately, much too faint to be seen with small telescopes), and some interesting galaxies, including the spiral M81 and the irregular M82 which is a well-known radio source.

The seven Plough stars are Alpha (Dubhe), Beta (Merak), Gamma (Phad or Phekda), Delta (Megrez), Epsilon (Alioth), Zeta (Mizar) and Eta (Alkaid or Benetnasch). Of these, five make up a moving cluster whose members share a common motion through space. The two unconnected stars are Alkaid and Dubhe, which are moving across the

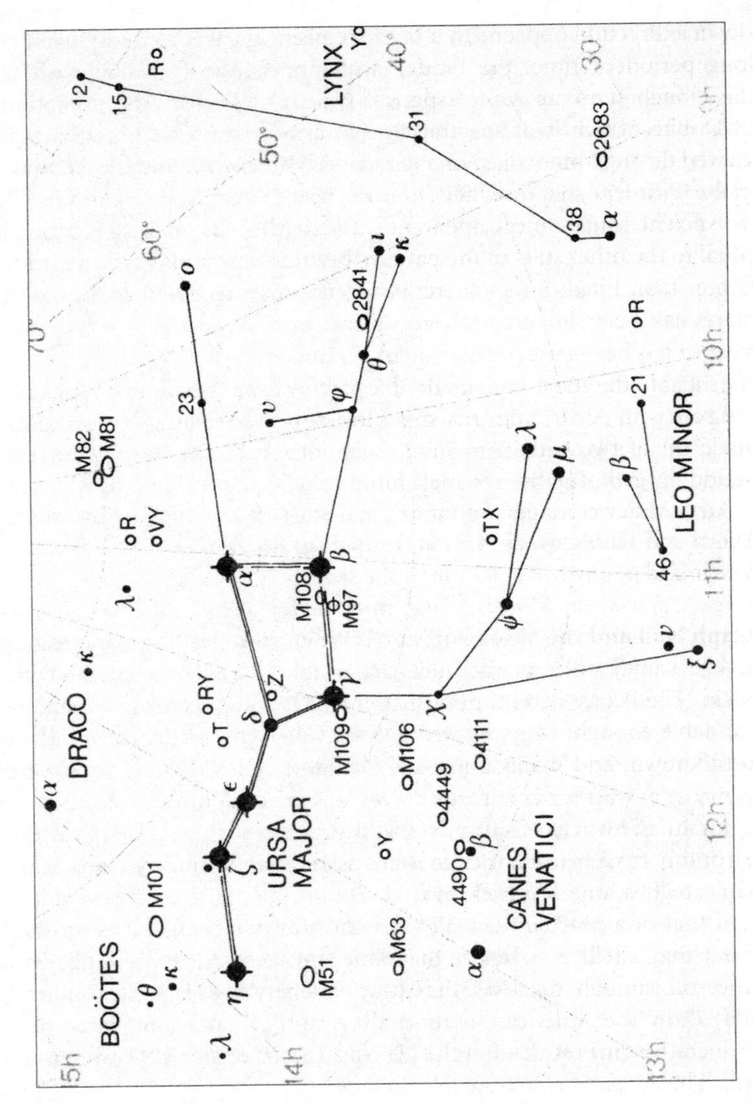

Figure 14. The pattern of Ursa Major, the Great Bear, contains many stars in addition to those forming the familiar saucepan-shaped asterism known as 'the Plough'. It is a large constellation, and there are many interesting deep-sky objects in it, particularly the galaxies M81, M82, M101, M108 and M109 and the Owl planetary nebula M97. (Diagram supplied from the archive of the Sir Patrick Moore Heritage Trust, courtesy of the Executors.)

sky in a direction opposite that of the remaining five. Over a sufficiently long period of time, the Plough will lose its familiar shape. All of the Plough stars are white (spectral type A or B) with the exception of Dubhe, which is of spectral type K, and is distinctly orange when viewed through binoculars or a telescope. Merak and Dubhe are known as the 'Pointers', because they show the way to the Pole Star, Polaris.

Ancient astronomers apparently rated Megrez, in the Plough, as equal to the other stars in the pattern, but it is now almost a magnitude fainter than Phad. Either there has been a real fading or, more likely, there has been an error in recording or interpretation. However, Megrez has been suspected of slight variability in modern times. Mizar is probably the most famous double star in the sky. It forms a naked-eye pair with Alcor, and in a small telescope Mizar itself is seen to be made up of two rather unequal components. It is a binary, but the period of revolution is extremely long.

Ursa Major contains no star of the first magnitude: the brightest are Alioth and Dubhe (mag. 1.8) and Alkaid (mag. 1.9), but it is as easily recognized as any constellation in the sky.

Asaph Hall and the Moons of Mars. When Alfred Tennyson referred to 'the snowy poles of moonless Mars' in the revised version of his poem 'The Palace of Art', published in 1842, the statement was understandable enough: when he wrote those words, no attendants of Mars were known, and it was universally believed that the polar caps were snowy or at least icy in nature.

Jonathan Swift, in Gulliver's *A Voyage to Laputa,* described how the astronomers of his remarkable flying island had discovered two Martian satellites, one of which moved around the planet in a period less than that of a Martian day. Voltaire, in *Micromegas,* had also written about two satellites – but at that time no telescope in existence was powerful enough to show the dwarf attendants which were found in 1877 by the American astronomer Asaph Hall with the great new 26-inch (66-cm) refractor at the US Naval Observatory in Washington, DC. The largest refracting telescope in the world at the time, this instrument was ideally suited to the task of searching for small satellites. Yet at first Hall had no success and he was about to give up the hunt when his wife Angelina persuaded him to carry on.

The following night, on 11–12 August 1877, Hall detected a tiny point of light which, he thought, might well be a satellite; but the mists rising from the Potomac River cut short his observations, and the next few nights were cloudy. Then, on 17 and 18 August, he was able to verify the original satellite and also to discover the second. We know them today as Phobos and Deimos – and thanks to various spacecraft, we now even have detailed maps of their rough, cratered surfaces (Figures 15a and 15b). They remain among the smallest known satellites in the Solar System, and are quite irregular in shape (being almost certainly captured asteroids), so that they are utterly unlike our own large, massive Moon.

There is no real mystery about the Swift and Voltaire descriptions, which were not, in any case, meant to be taken seriously. Venus had no

Figure 15a. Phobos (27 x 22 x 18 kilometres), the larger of Mars's two moons, imaged by the High Resolution Imaging Science Experiment (HiRISE) camera on NASA's Mars Reconnaissance Orbiter from a distance of about 6,800 kilometres. The most prominent feature is the large crater Stickney (lower right), with a diameter of 9 kilometres. A series of troughs and crater chains is obvious on other parts of Phobos. Although many appear radial to Stickney in this image, recent studies indicate that they are not related to it. Instead, they may have formed when material ejected from impacts on Mars collided later with Phobos. (Image courtesy of NASA/JPL-Caltech/University of Arizona.)

satellite; Earth had one; Jupiter, much further from the Sun, had four – and so how could Mars possibly manage with less than two? However, it is pleasant to note that modern astronomers have given the two largest craters on Deimos the names of Swift and Voltaire!

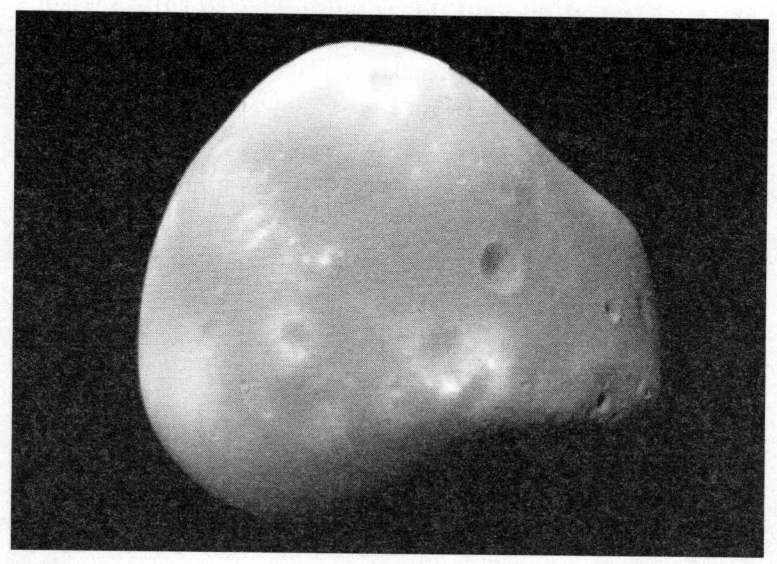

Figure 15b. Deimos (15 x 12 x 11 kilometres), the smaller of Mars's two moons, imaged by the HiRISE camera on NASA's Mars Reconnaissance Orbiter. Deimos has a smooth surface due to a blanket of fragmental rock or regolith, except for the most recent impact craters. (Image courtesy of NASA/JPL-Caltech/University of Arizona.)

June

New Moon: 5 June *Full Moon:* 20 June

Solstice: 20 June

MERCURY is at greatest western elongation (24°) on 5 June and is visible as a morning object in the eastern sky before dawn for the first three weeks of the month. During this period its magnitude brightens from +0.7 to −1.2. It is not visible to observers in the latitudes of the British Isles because of the long duration of twilight.

VENUS is too close to the Sun for observation this month because it passes through superior conjunction on 6 June. Indeed, for a short while on that date no Earth-based instrument could detect it since it is actually occulted by the Sun.

MARS, its magnitude fading from −2.0 to −1.4 during June, is still in Libra, but is setting by 1 a.m. from northern temperate latitudes by the end of the month; observers in the Southern Hemisphere will have the planet in view for a further three hours. Mars reaches its second stationary point on 30 June and then resumes its direct (eastwards) motion back towards the border with Scorpius.

JUPITER continues to move eastwards among the faint stars of Leo, some way south of the main pattern, and a brilliant object dominating the western sky of an evening. By the end of June, the planet will be setting a little before midnight for those living in the latitudes of the British Isles, and only slightly earlier for those living further south. Jupiter fades from magnitude −2.0 to −1.9 during June. The almost first quarter Moon will appear about 1.5° south of Jupiter on the evening of 11 June and the pair will make a pleasing spectacle in the twilight sky at dusk.

SATURN, magnitude 0.0, is at opposition in Ophiuchus on 3 June. From northern temperate latitudes, the planet becomes visible in the south-eastern sky as soon as darkness falls and is observable all night long. The planet will be much more favourably placed for observation from the tropics and the Southern Hemisphere since its declination is –20.5°, well south of the celestial equator. At opposition the planet is 1,349 million kilometres (838 million miles) from the Earth. Saturn is a lovely sight in even a small telescope, with the rings beautifully displayed at an angle of 26.1° as viewed from the Earth, almost the maximum angle at which they can be observed (Figure 16); the circumstances will be marginally better at opposition in 2017. The almost Full Moon will pass 3.3° north of Saturn on the night of 18/19 June.

Surveyor 1. Although the Russians were the first to successfully soft land a probe on the surface of the Moon when Luna 9 touched down in the Oceanus Procellarum (the Ocean of Storms) on 3 February 1966, the first American success was Surveyor 1, which landed in the Mare Nubium (the Sea of Clouds) fifty years ago this month, on 2 June 1966. Surveyor 1 was a complete success and demonstrated the technology necessary to achieve landing and operations on the lunar surface, an important first step in the Surveyor programme – a series of planned robotic landings on the Moon in support of the coming crewed *Apollo* landings. In the event, the Americans achieved five successes out of the seven attempted Surveyor landings; only Surveyors 2 and 4 crashed.

Although no instrumentation was carried specifically for scientific experiments, a lot of useful scientific information was obtained by Surveyor 1. It carried two television cameras – one mounted on the bottom for approach photography (which was not used) and the survey television camera. One objective was to obtain TV pictures of a spacecraft footpad after the landing, the surface material immediately surrounding it and the lunar topography, and to obtain data on radar reflectivity, bearing strength and temperatures of the lunar surface.

Surveyor 1 was launched on 30 May 1966 on an Atlas/Centaur rocket directly into a lunar impact trajectory. After a midcourse correction, the spacecraft reached the Moon about sixty-three hours after launch. At an altitude of 75.3 km above the lunar surface the main retrorocket ignited for 40 seconds and was jettisoned at an altitude of 11 km, having slowed the craft from a velocity of 2,612 m/s to just 110 m/s. Descent continued with the altimeter and Doppler radar controlling the smaller vernier engines until, just 3.4 m above the surface, these

Figure 16. Saturn's equator is tilted relative to its orbit by 27 degrees, so as Saturn moves along its orbit, first one hemisphere, then the other, is tilted towards the Sun. These images, taken with the Wide Field Planetary Camera 2 on-board the Hubble Space Telescope from 1996 to 2000, show Saturn's rings opening up from just past edge-on to nearly fully open as it moved from spring towards summer in its southern hemisphere. The first image in this sequence, at lower left, was taken soon after the spring equinox in Saturn's southern hemisphere. By the final image in the sequence, at upper right, the tilt is nearing its extreme, approaching the summer solstice in the southern hemisphere. The rings were edge-on again in 2009 and have been opening up since that time as the planet moved from spring towards summer in its northern hemisphere. (Image courtesy of NASA and The Hubble Heritage Team (STScI/AURA) Acknowledgment: R. G. French (Wellesley College), J. Cuzzi (NASA/Ames), L. Dones (SwRI), and J. Lissauer (NASA/Ames).)

engines switched off and the spacecraft dropped onto the lunar surface at a speed of only 3 m/s. It was 2 June 1966 at 06h 17m UT. The craft came down on a flat area inside a crater north of Flamsteed Crater in south-west Oceanus Procellarum.

After a brief period of engineering tests, the TV cameras were switched on and pictures were transmitted of the spacecraft footpad and surrounding lunar terrain and surface materials (Figures 17a and

17b). Some 10,338 photos were returned prior to nightfall on 14 June. The spacecraft also acquired data on the radar reflectivity and bearing strength of the lunar surface and measurements of lunar surface temperatures. Surveyor 1 survived its first lunar night and on 7 July, during its second lunar day, more photos were returned. On 13 July, after a total of 11,240 pictures had been transmitted, Surveyor 1's mission was terminated due to a sudden drop in battery voltage. Surveyor 1 was a great success and a wonderful start to the Surveyor programme, which continued until late February 1968.

Figure 17a. This Surveyor 1 self-portrait was one of 144 TV pictures taken by the soft lander spacecraft during its first day of operation on the lunar surface. The disk-shaped object at upper left is one of Surveyor's three feet. Attached members are parts of the landing leg. Beyond the foot is an area where the foot disturbed the lunar surface, apparently making an indentation with a pushed-up ridge of granular material around it. (Image courtesy of NASA/Lunar & Planetary Institute/Universities Space Research Association.)

Figure 17b. The boulder-strewn surface of the Moon's Ocean of Storms, as seen with Surveyor 1's TV camera, shows the outside of a crater rim along right centre of the horizon. The crater falls away beyond the horizon and to the right of the area covered in the picture. The distance from the spacecraft to the horizon is estimated at several hundred metres. Boulders on the horizon near the upper left of the photo may be one to two metres in length. The smallest rock fragments seen are several centimetres across. Rocks, which appear to be broken solid material, apparently were scattered from the crater towards the site where Surveyor landed. The Sun is almost overhead and shines from the upper right to the lower left. This 600-scan-line TV picture was transmitted to Earth on the morning of 5 June 1966 as Surveyor 1 began its fourth day of operation on the lunar surface. (Image courtesy of NASA/Lunar & Planetary Institute/Universities Space Research Association.)

The Father of the Big Bang Theory. This month we mark the fiftieth anniversary of the death of the great pioneering cosmologist, the Belgian Abbé Georges Lemaître, on 20 June 1966. Born in Charleroi, Belgium in 1894, the young Lemaître had a strong interest in both science and theology and after serving as an artillery officer during the First World War he studied astrophysics and was subsequently ordained as a priest in 1923. Lemaître then studied solar physics at the

University of Cambridge from 1923 to 1924, where he met Arthur Eddington, who described Lemaître as a 'very brilliant student, wonderfully quick and clear-sighted, and of great mathematical ability'. Lemaître spent the next two years at the Massachusetts Institute of Technology in the US, visiting most of the major centres of astronomical research, and becoming greatly influenced by the ideas of astronomers Edwin Hubble and Harlow Shapley (who was his supervisor at MIT) about the expanding universe.

On returning to Belgium, Lemaître became Professor of Astrophysics at the Catholic University of Leuven, near Brussels, in 1927, where he formulated his own ideas about the origin and evolution of the Universe (Figure 18). This led him to publish, in 1927, an initially largely unnoticed paper, entitled '*Un Univers homogène de masse constante et de rayon croissant rendant compte de la vitesse radiale des*

Figure 18. The Belgian priest, astronomer and professor of physics, Georges Lemaître, teaching at the Catholic University of Leuven, circa 1933. (Image courtesy of Wikimedia Commons.)

nébuleuses extragalactiques' ('A homogeneous Universe of constant mass and growing radius accounting for the radial velocity of extragalactic nebulae') in the *Annales de la Société Scientifique de Bruxelles*, in which he provided a compelling solution to the equations of General Relativity for the case of an expanding universe.

By 1930, Edwin Hubble had demonstrated through his observations with the 100-inch Hooker telescope at Mt Wilson, California that distant galaxies all appeared to be receding from us at speeds proportional to their distances. At this juncture Arthur Eddington became aware of Lemaître's 1927 paper – in which he had derived and explained the relation between the distance and the recession velocity of galaxies (Hubble's Law) – and arranged for it to be translated into English. Taken together with Hubble's observations, Lemaître's work convinced most astronomers that the universe was indeed expanding, although this was an idea which contradicted the then widely accepted theory of a static universe. But many, including Eddington, still found it impossible to accept the inference that the universe had begun at a finite time in the past; they wanted to believe that the universe had always existed.

In 1931, Lemaître put forward the startling idea that the expansion of the observable universe began with the explosion of a single particle at a definite point in time. He argued that if the universe is expanding now then it must have been smaller in the past, and if we could go far enough back in time, then all the matter in the universe would have been packed together in an extremely dense state – a single particle or 'primeval atom', as Lemaître called it. This primeval atom exploded, sending its material outwards in all directions, giving rise to space and time and the expansion of the universe that continues today. Lemaître published a more detailed version of his ideas in '*L'univers en expansion*' in 1933.

Lemaître's theory was the first expression of the currently accepted 'Big Bang' theory. Although widely accepted, the cosmologist Fred Hoyle did not agree with it and he coined the term 'big bang' as a derisory term for Lemaître's theory in a 1950s radio broadcast. Nevertheless, although his basic ideas were subsequently improved – particularly by George Gamow in 1946 – the Big Bang theory has stood the test of time and Georges Lemaître can justifiably be considered the father of that theory.

July

New Moon: 4 July *Full Moon:* 19 July

MERCURY is in superior conjunction on 7 July. It becomes visible as an evening object for the second half of the month, though not for observers in northern temperate latitudes where the continuing lengthy twilight makes observation impossible. Observers further south should refer to Figure 22, given with the notes for August. From the tropics and southern temperate latitudes, Mercury will lie only half a degree north of Venus on the evening of 16 July and will be a guide to locating the much fainter planet (Venus magnitude −3.9, Mercury −1.1) on the 16th and 17th, although both planets will be only about 4° above the west-north-western horizon at the end of evening civil twilight, around 30 minutes after sunset.

VENUS, magnitude −3.9, becomes visible in mid-July, in the evenings, low above the western horizon after sunset for observers in the tropics and southern temperate latitudes. Within a week its region of visibility has spread to include those as far north as the Mediterranean. However, the long duration of summer twilight in northern Europe means that observers here will be lucky to see the planet at all before early August.

MARS is now well past opposition and is setting around midnight from northern temperate latitudes by the end of the month, but those living in the Southern Hemisphere will be able to observe the planet for an additional three hours. Mars remains in Libra, but is now moving direct (eastwards) once again. The planet fades from magnitude −1.4 to −0.8 during July as its distance from Earth increases from 86 million to 108 million kilometres. The waxing gibbous Moon will pass nearly 7° north of Mars on the evening of 14 July.

JUPITER is still a lovely object in the western sky in the evenings, moving slowly eastwards in Leo, and fading slightly from magnitude −1.9 to

–1.7 during July. From northern temperate latitudes the planet will be setting only one and a half hours after the Sun by month end; some two hours later from the Southern Hemisphere. The waxing crescent Moon will make a pleasing configuration with Jupiter in the evening twilight on 8 and 9 July.

SATURN, just past opposition, is visible in the south-south-east from northern temperate latitudes as darkness falls and is observable for most of the night. From more southerly locations the planet is situated much higher in the sky. Saturn is moving retrograde in Ophiuchus about 6° north of the reddish Antares in Scorpius. It fades slightly from magnitude +0.2 to +0.3 during the month. The almost full Moon will appear quite close to Saturn on the night of 15/16 July.

PLUTO, officially a dwarf planet, reaches opposition on 7 July, in the constellation of Sagittarius, at a distance of 4,804 million kilometres (2,985 million miles). It is visible only with a moderate-sized telescope since its magnitude is +14.

The Star Clouds of Sagittarius. For British and North American observers, July evenings provide the best times for seeing the lovely star clouds of Sagittarius, the Archer. Not only are they beautiful, but they are also significant. When we look at them, we are looking in the direction of the very centre of our Galaxy, the Milky Way.

Sagittarius, the Archer (Figure 19 overleaf), is not a particularly prominent constellation. Unlike its neighbour, Scorpius, it contains no first-magnitude stars. A system of allotting Greek letters to stars was introduced by Johann Bayer in 1603; in general, the brightest star in a constellation is given the letter Alpha, the second Beta, and so on down to Omega, the last letter in the Greek alphabet. Bayer's system is convenient, and is still in use. Unfortunately, the strict brightness sequence is not always followed, but with Sagittarius, the alphabetical order can only be described as chaotic! The brightest stars are Epsilon (magnitude 1.9), Sigma (2.0), Zeta (2.6), Delta (2.7), Lambda (2.8), Pi (2.9) and Gamma (3.0), with both Alpha and Beta considerably fainter. Incidentally, both Alpha and Beta (which is a wide naked-eye double star) are too far south to rise in Britain or the northern United States.

Sagittarius has no really distinctive shape. Deneb in Cygnus, Altair in Aquila, and Sagittarius lie more or less in a straight line, with Altair

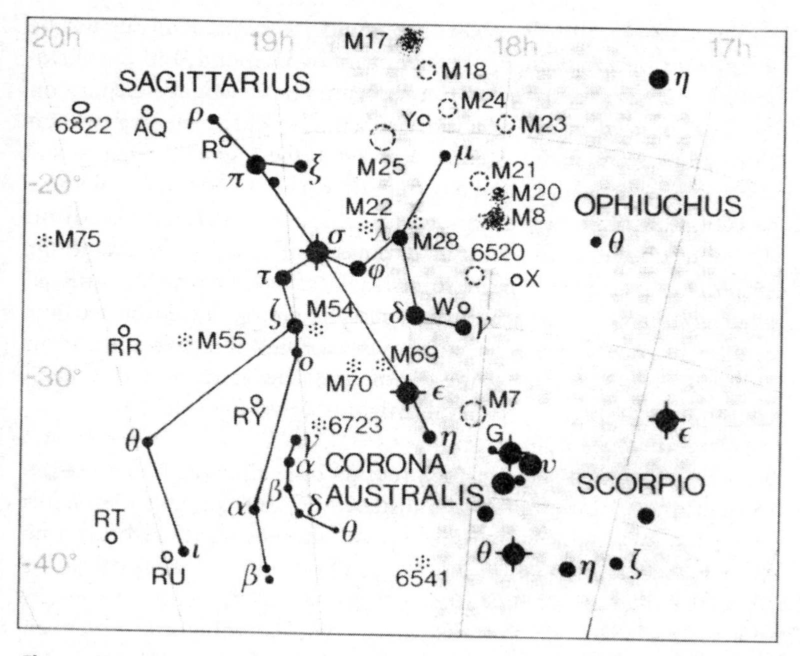

Figure 19. Sagittarius, the Archer, may be seen low down in the southern sky on July evenings from northern temperate latitudes, although its most southerly stars remain below the horizon. The brightest stars are Epsilon, Sigma, Zeta, Delta, Lambda, Pi and Gamma, with both Alpha and Beta much fainter still. The stars Phi, Delta, Epsilon and Zeta form the shape of the body of a 'teapot', while Lambda marks the top of the conical lid; Gamma^2 is the tip of the spout while Sigma and Tau make up the handle (with Phi and Zeta). (Diagram supplied from the archive of the Sir Patrick Moore Heritage Trust, courtesy of the Executors.)

in the middle, and this is probably the best way to locate the constellation. The outline is really nothing like that of an Archer; some people have compared it with an old-fashioned 'teapot', with its handle to the east and its spout to the west. The stars Phi, Delta, Epsilon and Zeta form the body of the teapot, while Lambda marks the top of the conical lid; Gamma^2 is the tip of the spout while Sigma and Tau make up the handle (with Phi and Zeta). To complete the idea of a steaming teapot, in clear and dark skies, a particularly dense area of the Milky Way can be seen rising in a north-westerly arc above the star Gamma^2, like a plume of steam rising from the spout.

There are plenty of interesting telescopic objects in Sagittarius, including the Omega or Horseshoe Nebula (Messier 17) in the northern part of the constellation, not far from Gamma Scuti; the Lagoon Nebula (Messier 8) near Mu Sagittarii with, less than two degrees away from it, the Trifid Nebula (Messier 20); and the bright globular cluster Messier 22, between Mu and Nunki. Unfortunately, Sagittarius lies well to the south of the celestial equator – it is the southernmost of the Zodiacal constellations – and so it is never seen at its best from latitudes such as those of London or New York. Of course, this acts to the advantage of observers in the Southern Hemisphere, who are able to see both Sagittarius and Scorpius in their true glory. Sagittarius is in the Zodiac, although it contains no planets at the present time; however, Pluto (now a dwarf planet) is moving slowly through the constellation.

The star clouds of Sagittarius are rich indeed, and the observer with a small rich-field telescope will find endless enjoyment in sweeping around them. This is, in fact, the richest part of the Milky Way. For centuries now we have known that our Galaxy is a flattened disk with a central bulge, and that the Sun lies not far from the main plane – at a distance of about 26,000 light years from the galactic centre, on what is known as the Sagittarius arm of the Galaxy. The whole system is rotating, and the Sun takes some 225 million years to complete one orbit, a period known as the 'cosmic year'.*

The real galactic centre is hidden, simply because there is too much dust in the way, and this dust blocks out light waves as effectively as a thick fog will blot out the landscape. However, radio waves are not blocked; and they were first detected in 1931 by Karl Guthe Jansky, an American physicist who worked for Bell Telephone Laboratories. Jansky was particularly interested in sources of static that might interfere with transatlantic radio telephone transmissions, and he built a strange-looking antenna mounted on a turntable so that it could be rotated in any direction; it was nicknamed 'Jansky's merry-go-round' (Figure 20). He picked up plenty of static, mainly from nearby and distant thunderstorms, but he also found a weak, persistent hiss which rose and fell once a day, and which repeated not every twenty-four hours, but every twenty-three hours and fifty-six minutes. This was characteristic of something associated with the distant stars. After

* Just one 'cosmic year' ago, the Earth was in the midst of its Triassic Period, a warm period with vast deserts and warm seas; dense forest even grew in the warm polar regions.

many weeks of work, Jansky managed to identify the source. The radio noise was strongest in the direction of the Sagittarius star clouds in the centre of the Milky Way.

This was the beginning of the science of radio astronomy, which has become such a vital part of astrophysical research. It is ironic that Bell Labs never allowed Jansky himself to investigate the radio waves from the galactic centre in more detail, even though he wanted to, and it was only after World War II that professional radio astronomy began.

Touchdown on Mars! Forty years ago this month, on 20 July 1976, the American lander Viking 1 became the first successful operational spacecraft to soft land on Mars. Both Viking 1 and its sister craft Viking 2 came down in the reddish parts of Mars; Viking 1 in Chryse Planitia (22.4°N, 47.5°W) and Viking 2 (on 3 September 1976) in the more northerly Utopia Planitia (48°N, 226°W). The first picture from

Figure 20. Karl Jansky built an antenna that was designed to receive radio waves at a frequency of 20.5 MHz (wavelength about 14.5 metres). It was mounted on a turntable that allowed it to rotate in any direction, earning it the nickname 'Jansky's merry-go-round'. By rotating the antenna, one could find the direction of any received radio signal. (Image courtesy of National Radio Astronomy Observatory.)

Viking 1, taken immediately after touchdown, showed a rock-strewn landscape, and the overall impression was that of a barren, rocky desert, with extensive dunes as well as pebbles and boulders (Figure 21). The colour was formed by a thin veneer of red material, probably limonite (hydrated ferric oxide) covering the dark bedrock. The sky was initially said to be salmon-pink, although later pictures modified this to yellowish-pink. Temperatures were low, ranging between –96°C after dawn to a maximum of –31°C near noon. Winds were light. The

Figure 21. This view is the centre part of the first panoramic image of Chryse Planitia taken by camera 1 on the Viking 1 Lander. The image was taken on 23 July 1976, three days after Viking 1 landed. The bright curving lines in the sky are not clouds; they are caused by internal reflections in the camera housing. Features on the horizon are about three kilometres away. The dark rock at the centre of the frame, nicknamed 'Big Joe', is about three metres across and eight metres from the lander. The centre of the frame shows fine-grained material forming a small dune field. The material forms horseshoe-shaped scour marks and wind tails around the rocks, which are about ten centimetres across. (Image P-17428 courtesy of NASA/NSSDC.)

strongest at around 10 a.m. local time, but even then were no more than 22 km/h breezes; later in the sol (the Martian day, equivalent to 24 hours and 37 minutes) they dropped to around 7 km/h, coming from the south-west rather than the east. The pattern was fairly regular from one sol (Martian day) to the next.

The Viking 2 site, in Utopia, was not unlike Chryse, but there were no large boulders, and the rocks looked 'cleaner'. There were no major craters in sight; the nearest large formation, Mie, was over 200 km to the west. Small and medium rocks were abundant, most of them vesicular (vesicles are porous holes, formed as a molten rock forms at or near the surface of a lava flow, so that internal gas bubbles escape). There were breccias and one good example of a xenolith – a 'rock within a rock' probably formed when a relatively small rock was caught in the path of a lava flow and was coated with a molten envelope which subsequently solidified. The temperatures were very similar to those at Chryse.

One main aim of the Viking missions was to search for life. Soon after arrival, Viking 1 collected samples by using a scoop, drew them inside the spacecraft, analysed them and transmitted the results to Earth. There were three main experiments:

Pyrolytic release

Pyrolysis is the braking up of organic compounds by heat. The experiment was based on the assumption that any Martian life would contain carbon, one species of which, carbon 14, is radioactive, so that when it is present it is easy to detect. The sample was heated sufficiently to break up any organic compounds which were present.

Labelled release

This also involved carbon 14, and assumed that the addition of water to a Martian sample would trigger off biological processes if any organisms were present.

Gas exchange

It was assumed that any biological activity on Mars would involve the presence of water, and the idea was to see whether providing a sample with suitable nutrients would persuade any organism to release gases, thereby altering the composition of the artificial atmosphere inside the test chamber.

However, the results of all three experiments were inconclusive, and this was also the case when they were repeated from Viking 2. The investigators had to admit that they still could not say definitely whether or not there was any trace of life on Mars.

August

New Moon: 2 August *Full Moon:* 18 August

MERCURY reaches its greatest eastern elongation (27°) on 16 August and is visible as an evening object, though not to observers as far north as the British Isles. For observers further south this will be the most favourable evening apparition of the year. Figure 22 shows, for observers in latitude 35°S, the changes in azimuth (true bearing from the north through east, south and west) and altitude of Mercury on successive evenings when the Sun is 6° below the horizon. This condition is known as the end-of-evening civil twilight and in this latitude and at this time of year occurs about thirty minutes after sunset. The changes in the brightness of the planet are indicated by the relative sizes of the circles marking Mercury's position at five-day intervals. It will be noticed that Mercury is at its brightest before it reaches greatest eastern elongation. Mercury's magnitude is –0.1 at the beginning of the month but it has faded to +1.0 a few days before the end of August. Throughout its period of visibility, Mercury will be located in the same part of the sky as the much brighter Venus and Jupiter. From elongation (16 August) until the end of the month, Mercury will lie a few degrees south of Jupiter; for the last few days of August, Venus joins in the gathering.

VENUS, magnitude –3.9, is moving slowly out from the Sun and is visible low above the western horizon in the evenings after sunset. For observers in the latitudes of the British Isles, there will be little chance of seeing the planet until mid-August. Venus passes only one degree north of the first-magnitude star Regulus (Alpha Leonis) on 5 August. That evening, Venus, Mercury, the crescent Moon and Jupiter will be strung out in a line across the western twilight sky at dusk, but the spectacle will only be visible to those in the tropics and further south. Later in the month, on 27 August, Venus will pass a scant four arc minutes (one eighth the diameter of the full Moon) north of Jupiter, an apparent separation that is five times smaller than in the spectacular

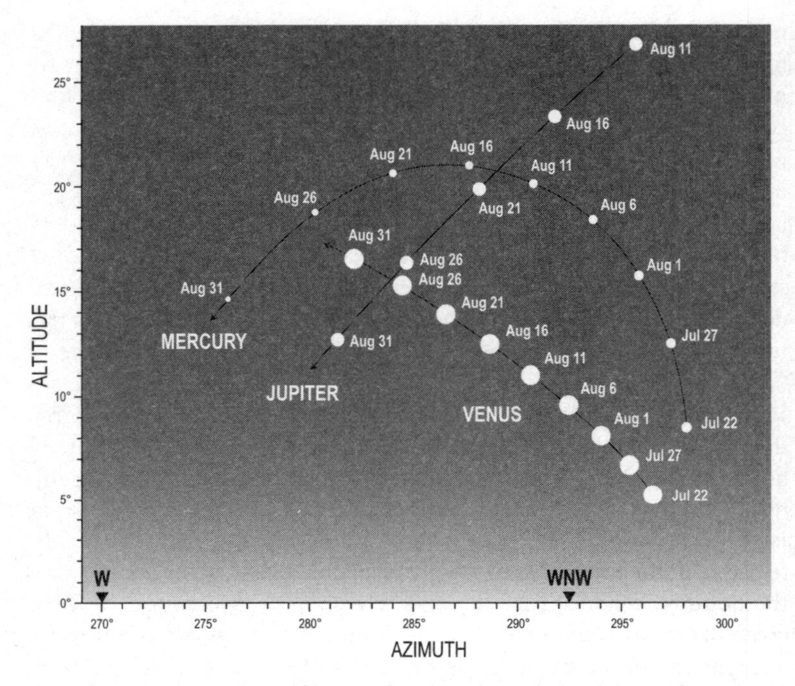

Figure 22. Evening apparition of Mercury from latitude 35°S. The planet reaches greatest eastern elongation on 16 August 2016. It is at its brightest in late July, before elongation. The chart also shows the positions of the planets Jupiter and Venus, in relation to Mercury, during its period of visibility. Jupiter is a useful guide to locating the much fainter Mercury, particularly from mid-August until the end of the month; the brilliant Venus is a useful guide for the last few days of August. The angular diameters of Mercury, Jupiter and Venus are not drawn to scale.

conjunction of 30 June 2015. Although the two planets appear extremely close together in the sky on 27 August, Venus will be more than four times nearer to Earth than Jupiter. On occasions, Venus can actually occult Jupiter, although that hasn't happened since 3 January 1818 – it seems unlikely that anyone saw it on that occasion – and it won't happen again until 22 November 2065, when the two planets will be only eight degrees from the Sun!

MARS continues to be visible to observers in the south-western sky in the evenings. By the end of August it will be setting well before midnight from the latitudes of the British Isles, but is rather better placed for those living further south. The planet continues to fade from magnitude −0.8 to −0.3 during the month. Mars, which has been in the constellation of Libra since the end of May, is moving eastwards more rapidly – on 2 August it passes into the northern part of Scorpius; into neighbouring Ophiuchus on 21 August; back into Scorpius six days later; and back into Ophiuchus again on 2 September. On the evenings of 23 and 24 August, Mars will be just two degrees north of the reddish star Antares in Scorpius, with the yellowish Saturn just over four degrees further north. Both planets will actually be in Ophiuchus at the time.

JUPITER, magnitude −1.7, continues to be visible as an evening object though only visible for a short while low above the west-north-western horizon. The planet's eastwards motion carries it from Leo into neighbouring Virgo on 9 August. Observers in the latitudes of the British Isles and suffering from the long evening twilight are unlikely to be able to see it after about the first week of August, but from the tropics and the Southern Hemisphere it remains visible all month. A very thin crescent Moon will lie just below Jupiter in the evening twilight on 5 August; from the tropics, Mercury and Venus will be seen some way below. As mentioned above, Jupiter will be incredibly close to the more brilliant Venus on 27 August.

SATURN reaches its second stationary point on 13 August and thereafter resumes its direct motion once more. The planet is in Ophiuchus and is visible as an evening object in the south-western sky from northern temperate latitudes, setting about three hours after the Sun by month end; some three hours longer from further south. Its brightness decreases very slightly from magnitude +0.4 to +0.5 during the month. The waxing gibbous Moon will lie a few degrees north and east of Saturn on the evening of 12 August.

Mars and Antares. Mars, the planet of the god of war, is always distinguished by its strongly reddish colour, particularly when it is bright, as it has been since early April. However, because of the planet's very variable distance from the Earth and its relatively small size (approximately half the diameter of the Earth), its apparent brightness is very changeable. At its best – at so-called perihelic oppositions – it can outshine any star, and any planet apart from Venus; this will next be the case in July 2018 when its magnitude will be –2.8. When at its faintest, however, Mars fades to the second magnitude, and looks very much like a star. Unwary observers have occasionally mistaken it for a nova, or 'new star'.

Mars is now past opposition, and this August the distance of the planet increases from 109 million kilometres to 134 million kilometres, so it fades noticeably from magnitude –0.8 to –0.3 during the month. Consequently, it is no longer as striking as it was back in May, although it still ranks with the brightest stars in the sky – comparable with Alpha Centauri, Arcturus and Vega, but inferior to Sirius and Canopus.

As has been noted above, Mars will appear close to the reddish star Antares (Alpha Scorpii) on the evenings of 23 and 24 August; Mars will be significantly brighter at magnitude –0.4 compared with Antares at magnitude 1.0 (slightly variable). The name Antares or 'Ant-Ares', means 'the rival of Mars' – 'Ares' being the Greek name for the god of war – because the colours of the two are superficially very alike. Looking at them this month as they are seen fairly close together, it requires a conscious mental effort to realize that Mars is one of the smaller planets in the Solar System, while Antares is a colossal supergiant star far superior to our Sun in both size and luminosity.

Although Antares is ranked only fifteenth in order of brightness among the stars in the night sky, its great distance of 550 light years shows that it is highly luminous. If our Sun were placed at the same distance as Antares, it would appear almost 10,000 times fainter, but because the surface layers of Antares are relatively cool – only about 3300 degrees Celsius – the star radiates a considerable amount of its light in the infrared part of the spectrum. If we take this into account, then Antares is probably about 60,000 times more luminous than the Sun, although there are uncertainties here. Antares is so large that it is possible for astronomers to measure the size of its disk, which gives a diameter of over 1,000 million kilometres. So, if Antares were placed at the centre of our Solar System, in place of the Sun, its vast globe would extend outwards two thirds of the way to Jupiter's orbit – extending to the middle of the Asteroid Belt.

Both Mars and Antares are well worth looking at this month, although they will be rather low in the south-western sky from the latitudes of the British Isles in the mid-evening – but much better placed for observers in the tropics and the Southern Hemisphere.

The Tears of St Lawrence. The annual Perseid meteor shower is among the most reliable of the year, producing an abundance of fast, bright meteors (Figure 23). This August, Perseid maximum occurs during daylight hours from the UK (around 12h UT on 12 August) – favouring observers in western parts of North America – but good activity should be seen on the nights of 11/12 and 12/13 August from northern Europe. The Moon will be a waxing gibbous in Ophiuchus, setting at around midnight, so observing conditions will be most favourable in the early morning hours before dawn.

Perseid meteors may be detected as early as late July and until the end of the third week of August. The shower's activity displays a marked 'kick' around August 8/9 and steadily increasing observed rates can be expected in dark skies, from this date onward until the peak. Observers watching after midnight on August 11/12 should experience increasing activity towards dawn, as the shower radiant (the region in the sky from which Perseid meteors appear to emanate), at RA 03h

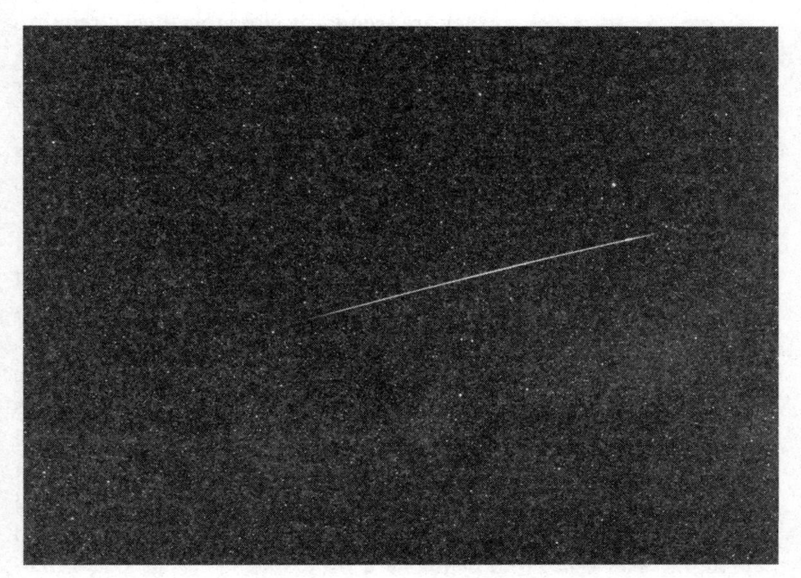

Figure 23. A bright Perseid meteor streaks from near Alpha Cephei, travelling north of the Milky Way and into Lyra, ending in a small terminal burst just south of the brilliant Vega, on the evening of 12 August during the 2013 Perseid meteor shower. (Image courtesy of John Mason.)

11m, Dec. +58°, near the Double Cluster, on the Perseus-Cassiopeia border, climbs higher in the eastern sky. Activity will be starting to decline by the time darkness falls on 12/13 August. With cloudless skies, and in a dark viewing site, observers can expect to see between fifty and seventy meteors each hour near the peak. Even in light-polluted towns or cities observed rates may still be around ten an hour in the early morning hours when the radiant is high.

The history of the Perseid stream goes back nearly two thousand years, the first recorded shower being noted by Chinese observers on 17 July AD 36. The next documented return occurred in late July 714 and there are many Chinese, Japanese and Korean records since that time. The shower was first noted in Europe from the year 811 and records are abundant from 830. We are, therefore, dealing with a meteoroid stream of considerable antiquity. The existence of the August Perseids as a major annual shower was probably recognized in about 1835 by Adolphe Quetelet in Brussels, Edward Claudius Herrick in the United States and several others, and it has been systematically observed ever since.

For many years the Perseids were known traditionally as the 'Tears of St Lawrence', in memory of the Spanish martyr who was broiled on a grid iron because of his religious beliefs on 10 August 258, close to the date of maximum of the meteor shower. In 1839, Georg Adolph Erman suggested that the meteoroidal dust which causes the shower was arranged in a ring through which the Earth passed every August. Later, between 1864 and 1866, the great Italian astronomer Giovanni Schiaparelli computed the orbit of the Perseid stream and showed that it was almost identical with that of the periodic comet 109P/Swift-Tuttle. This was the first time that any mathematical proof had been found to link meteors with comets and from that time onward the study of meteoroid streams became of greater importance.

Comet Swift-Tuttle itself was discovered on 16 July 1862 by Lewis Swift (from Marathon, New York) and three days afterwards, independently, by Horace Parnell Tuttle (from Harvard University). The comet was a prominent naked-eye object for three weeks in August to September with a tail on 27 August as much as 25° long. The initial orbit was shown to be an ellipse by Theodor von Oppolzer, the comet being assigned a return period of 123 years. More recently Brian G. Marsden of the Smithsonian Astrophysical Observatory recalculated the orbit and assigned a new period of 120 years. This being the case, one might have expected a return of comet Swift-Tuttle around 1981–82, but the comet never showed up. Because of this, there was speculation that the comet had disintegrated.*

However, to some surprise, the comet was rediscovered in September 1992 by the Japanese astronomer Tsuruhiko Kiuchi and became visible with binoculars. From revised calculations based on the observations made in 1992 and those of 1862, it has been shown that the comet is identical with Comet Koegler, observed from Beijing in 1737 by the Jesuit missionary Ignatius Koegler between 7 and 16 July of that year.

In a paper in the *Monthly Notices of the Royal Astronomical Society*, published in 1994, Kevin Yau, Donald Yeomans and Paul Weissman investigated the past and future motion of comet 109P/Swift-Tuttle. Two of its previous returns prior to the telescopic period, in AD 188 and 69 BC, have now been identified in Chinese records. No other observations have been found between AD 188 and 1737. These unobserved returns are easily explained, as the comet did not approach the Earth closely enough to reach naked-eye visibility. The calculations showed that the comet's return period has varied between 127.4 and 136.5 years and it is not due back again until 2126.

Logically, since the Perseid stream is an old one, we should expect the material to be fairly evenly spread around the entire orbit. This is because the older a meteoroid stream becomes the more broadening occurs due to collisions between the dust particles within the stream. This occurs mainly near perihelion where the particle density in the stream is highest. Hence the oldest streams are the widest with the result that the shower duration is typically a week or more and the radiant is generally rather diffuse. Rates are typically fairly constant from year to year. The August Perseids are an example of such a shower. The great width of the stream results in Perseid meteors being visible for a three-week period overall, with the period of maximum activity occurring between 10 and 13 August every year.

*The current Editor of this Yearbook and the late, great Sir Patrick Moore had a long-running wager on this matter. Sir Patrick believed that Comet Swift-Tuttle had been missed in 1981–82 or had disintegrated, but the current Editor disagreed. Sir Patrick promised the Editor a bottle of Irish whiskey if the comet was eventually recovered. When it duly returned in 1992, Sir Patrick, the great gentleman that he was, produced the promised bottle of whiskey and we enjoyed it together!

September

New Moon: 1 September *Full Moon:* 16 September

Equinox: 22 September

MERCURY is at inferior conjunction on 13 September and therefore not suitably placed for observation at first, but after the first three weeks of the month it becomes visible as a morning object for observers in northern and equatorial latitudes. For observers in northern temperate latitudes this will be the most favourable morning apparition of the year. Figure 24 shows, for observers in latitude 52°N, the changes in azimuth (true bearing from the north through east, south and west) and altitude of Mercury on successive mornings when the Sun is 6° below the horizon. This condition is known as the beginning-of-morning civil twilight and in this latitude and at this time of year occurs about thirty-five minutes before sunrise. The changes in the brightness of the planet are indicated by the relative sizes of the circles marking Mercury's position at five-day intervals. It will be noticed that Mercury is at its brightest after it reaches greatest western elongation (18°) on 28 September, the period of visibility extending well into October. Mercury brightens rapidly from magnitude +1.1 on 22 September to –0.6 by the end of the month. An incredibly thin waning crescent Moon will lie close to Mercury on the morning of 29 September.

VENUS continues to be visible as an evening object, low above the western horizon after sunset. Its magnitude is –3.9. Observers in the latitudes of the British Isles may only be able to glimpse the planet very low above the south-western horizon, despite its increasing elongation from the Sun, because the planet moves rapidly southwards in declination. From equatorial and more southerly latitudes, Venus will be setting almost two hours after the Sun by month end. The waxing crescent Moon will lie just above Venus on the evening of 3 September. The phase of Venus decreases from 92 per cent to 86 per cent during the month.

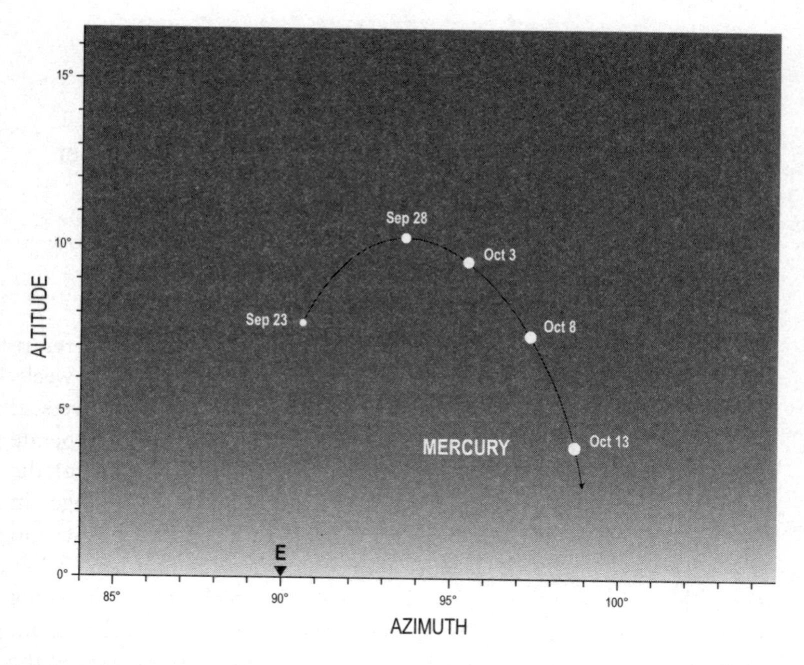

Figure 24. Morning apparition of Mercury from latitude 52°N. The planet reaches greatest western elongation on 28 September 2016. It will be at its brightest in mid-October, after elongation.

MARS remains visible in the south-western sky in the evenings, but continues to fade slowly, from magnitude –0.3 to –0.1 during September. The planet begins the month in Scorpius, but its rapid eastwards motion carries it into Ophiuchus on 2 September and Sagittarius on the 22nd. The path of Mars from 1 September until the end of the year is shown in Figure 25. The far southerly declination of the planet means that from northern temperate latitudes, the planet sets in the mid-evening by the end of September, but for observers in the tropics and the Southern Hemisphere the planet remains visible until midnight or later. Through a telescope, the disk of the planet now appears decidedly gibbous, with a phase of 85 per cent this month.

JUPITER, magnitude –1.7, is only likely to be glimpsed by observers in southern and equatorial latitudes for the first ten days of the month,

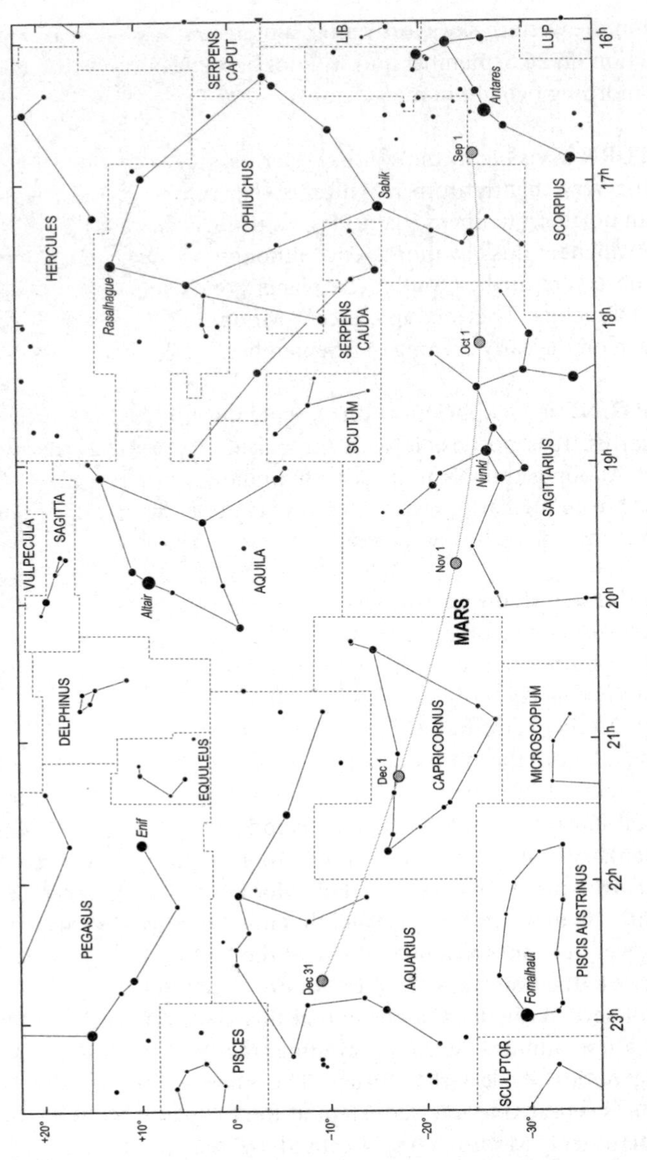

Figure 25. The path of Mars as it moves from Ophiuchus through Sagittarius, Capricornus and into Aquarius between 1 September and the end of the year. The motion of the planet is direct (i.e., eastwards) throughout this period.

low in the western sky shortly after sunset. Jupiter passes through conjunction on 26 September and will not be visible until it emerges from the morning twilight in mid-October.

SATURN is visible in the south-western sky as soon as darkness falls, fading very slightly from magnitude +0.5 to +0.6 during September. From northern temperate latitudes, Saturn will be getting rather low in the twilight at dusk by month end, although will be much better placed for observers further south. The planet continues to move eastwards in Ophiuchus. The waxing crescent Moon will be about 4° north of Saturn on the early evening of 8 September.

NEPTUNE is at opposition on 2 September, in the constellation of Aquarius. It is not visible with the naked eye since its magnitude is +7.8. At opposition Neptune is 4,330 million kilometres (2,691 million miles) from the Earth. Figure 26 shows the path of Neptune against the background stars during the year.

The Harvest Moon. In northern temperate latitudes, the Full Moon of September is generally referred to as the Harvest Moon, although the name is more correctly given to the Full Moon which occurs nearest to the autumnal equinox, and this Full Moon may, on occasions, come in early October. This month, Full Moon occurs on 16 September, some six days before the autumnal equinox. Although the Full Moon always looks rather larger than normal when it is rising – a consequence of an optical illusion known as the Full Moon illusion (see Monthly Notes for January) – it so happens that the Moon will be at perigee (closest to the Earth) just two days after Full Moon, on 18 September. Consequently, it may appear very slightly larger than normal at this time. However, it is not the apparent size of the Moon which is of particular interest at the time of the Harvest Moon.

For several nights in succession at this time of year the Moon rises only a few minutes later each evening, the delay from night to night being noticeably less than usual. The phenomenon of the Harvest Moon is repeated to a lesser extent at the following Full Moon (called the 'Hunter's Moon') and is caused by the phenomenon that on autumn evenings the ecliptic is at its lowest in the sky and is inclined at a very shallow angle to the horizon.

The average daily motion of the Moon is about thirteen degrees, and

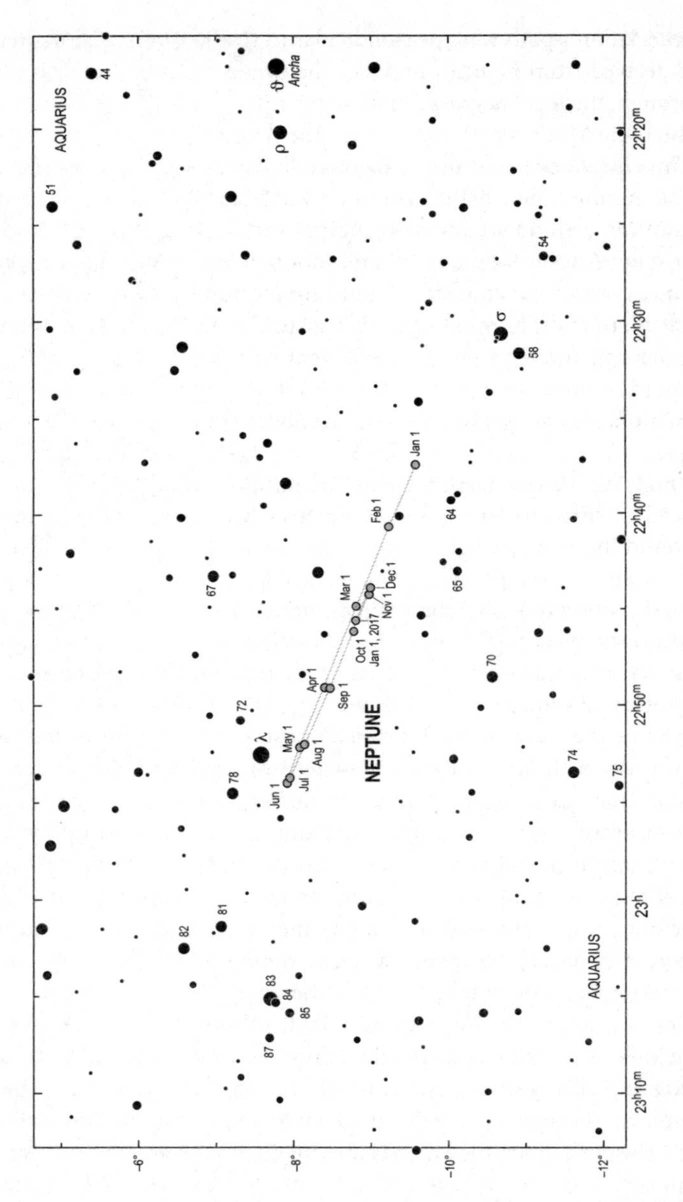

Figure 26. The path of Neptune against the stars of Aquarius during 2016. The planet is at opposition on 2 September.

if the Moon's path were perpendicular to the horizon, moonrise would be delayed from night to night by the time taken by the Earth to turn through thirteen degrees, that is, by fifty-two minutes. If the angle which the Moon's path makes with the horizon is small, then the delay is much reduced, and under favourable conditions may be only about nine minutes. Since the Moon travels right round the sky once a month, the effect occurs every month, but our attention is drawn to it most effectively at the time of Full Moon, which rises at the convenient time of sunset. It is thus in the autumn months that we are more aware that there is more moonlight than usual, a fact that was of benefit to farmers in the days before the advent of tractors with headlamps. It should be noted that there is no Harvest Moon in the tropics, while the phenomenon occurs in March in the Southern Hemisphere.

Which Way is the Earth Moving? In addition to sharing the motion of the Sun through space, the Earth also has its own annual motion around the Sun. This gives rise to the apparent daily eastwards motion of the Sun against the background stars, a movement that is easily verified by observing the changing position of the Sun among the star patterns lying along the ecliptic – the apparent path of the Sun around the sky, which is the intersection of the plane of the Earth's orbit with the celestial sphere.

Since the orbit of the Earth is almost a circle, the point in the sky towards which it is moving (at a speed of about thirty kilometres per second) at any moment in time – called rather grandly 'the apex of the Earth's way' – is very nearly at right angles to an imaginary line joining the Earth to the Sun; towards a point on the ecliptic about 90° to the west of the Sun (Figure 27). Owing to the orbital motion of the Earth, the apex will move eastwards along the ecliptic at a rate of about one degree per day, the same rate as the Sun moves.

If one faces the Sun at midday, the apex is setting in the west, and does not rise in the east until midnight. At sunset, if we look south, we are looking back towards the direction from which the Earth has come – the so-called antapex. At midnight, the apex is just rising in the east. At dawn, the apex is to the south, so that we can only look in the direction towards which the Earth is moving in space between midnight and dawn.

For those with an interest in meteor observation, it is useful to be able to locate the apex of the Earth's way relative to naked-eye stars

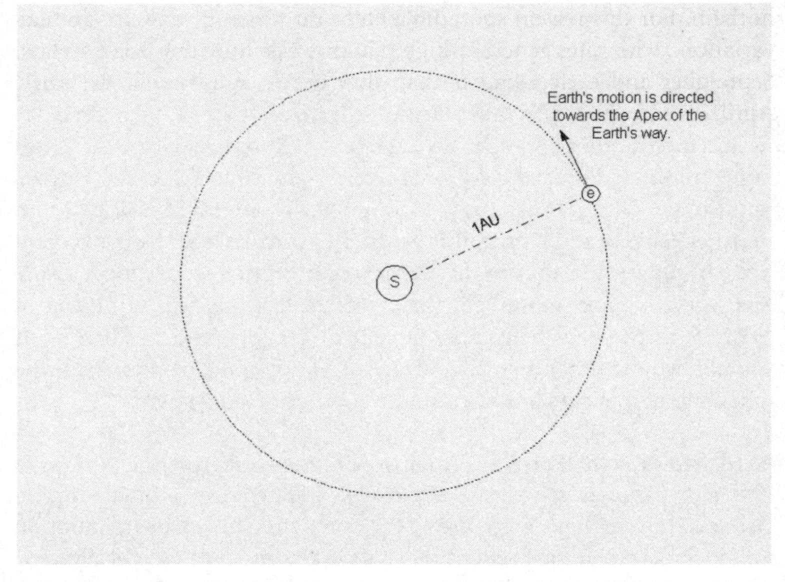

Figure 27. As the Earth orbits the Sun, the point in the sky towards which it is moving at any moment in time – called 'the Apex of the Earth's way' – is very nearly at right angles to an imaginary line joining the Earth to the Sun; towards a point on the ecliptic about 90° to the west of the Sun.

since that is the direction from which many non-shower, or sporadic, meteors will generally appear to come. There is a tendency for sporadic meteor rates to be significantly higher in the early morning hours, than earlier in the night. This so-called 'diurnal variation' in sporadic meteor rates is due to the fact that before midnight an observer is generally looking in the direction from which the Earth has come (and so sporadic meteors have to catch up the Earth from behind), whereas in the early morning hours the observer is on the side of the Earth facing its direction of travel (and the Earth is ploughing into the meteoroids more or less 'head on').

Now at the equinox on 22 September, the Sun crosses the celestial equator moving from south to north and from northern temperate latitudes at sunrise that day the ecliptic will be at its greatest elevation above the southern horizon. Consequently, the apex of the Earth's way will be due south and also at its greatest elevation above the southern horizon. For this reason sporadic meteor rates also exhibit an 'annual variation', with rates generally higher in the early morning hours in late September and early October than they are in late March and early April.

October

New Moon: 1 October and 30 October
Full Moon: 16 October

Summer Time in the United Kingdom ends on 30 October.

MERCURY continues to be visible as a morning object, low above the eastern horizon before dawn, for observers in northern and equatorial latitudes, but only for the first week of the month. Figure 24 given with the notes for September shows, for observers in latitude 52°N, the changes in azimuth and altitude of Mercury at this time. During this period its magnitude brightens slightly from −0.8 to −1.1. On 11 October, Mercury passes only a degree north of Jupiter (Mercury magnitude −1.1, Jupiter −1.7) and the slightly brighter planet may be a guide to locating Mercury, although both planets will be rather low in a bright twilight sky. It is worth noting that although the two planets appear close together in the sky at this time, Mercury is more than five times nearer to Earth than Jupiter. Mercury is too close to the Sun to be observed for the rest of the month, passing through superior conjunction on 27 October.

VENUS, magnitude −4.0, continues to be visible as a brilliant object in the western sky after sunset. The contrast in viewing the planet from different latitudes is quite marked; by the end of the month observers in the latitudes of the British Isles can only hope to see Venus for an hour after sunset, whereas those in the Southern Hemisphere can expect three hours of visibility. On 3 October, Venus and the waxing crescent Moon will make a nice pairing in the western twilight sky at dusk. On the evenings of 27 and 28 October, Venus will be located roughly midway between Antares (Alpha Scorpii) and Saturn.

MARS still has an elongation of around 80° from the Sun and remains visible in the south-western sky of an evening for observers in northern temperate latitudes, but from such locations it will only be seen at a low altitude above the horizon, since the planet is in Sagittarius and its declination is almost –26° (well south of the celestial equator) early in the month, and it sets in the mid-evening. From the tropics and the Southern Hemisphere, the planet's southerly declination means that it is much more favourably placed and it sets around midnight. Mars continues to fade, from magnitude +0.1 to +0.4 during the month.

JUPITER, magnitude –1.7, emerges from the morning twilight in mid-October for observers in northern temperate latitudes and the tropics, becoming observable above the eastern horizon shortly before dawn. However, it does not become visible to observers in the Southern Hemisphere until the end of the month. The planet is in Virgo.

SATURN, magnitude +0.6, is visible in the west-south-western sky as soon as darkness falls. From northern temperate latitudes, Saturn will be inconveniently low in the twilight at dusk by month end, although somewhat easier for observers further south. The planet continues to move slowly eastwards in Ophiuchus during the month. The brilliant Venus passes just three degrees south of Saturn on 29 October.

URANUS is at opposition on 15 October in the constellation of Pisces. Uranus is barely visible to the naked eye as its magnitude is +5.7, but it is easily located in binoculars. Figure 28 shows the path of Uranus against the background stars during the year. At opposition Uranus is 2,835 million kilometres (1,762 million miles) from the Earth.

The Great Square of Pegasus. In northern temperate latitudes, the Square of Pegasus dominates the southern part of the evening sky during the autumn months of October and November. On star maps, Pegasus, the Flying Horse, looks quite conspicuous, since its four main stars make up an obvious square (Figure 29), and you might think that it would stand out easily in the night sky. Certainly, once you have found Pegasus, it will be easily located again, but, to begin with, you must look for it carefully. This is partly because its stars are not remarkably brilliant – all four are between the second and third magnitudes – and partly because the Square is very large. The 'W' of Cassiopeia, which lies almost overhead during autumn evenings, makes a useful pointer; an imaginary line passing through its stars Gamma and Alpha will lead you straight to the Great Square.

The upper left-hand star of the Square used to be known as Delta

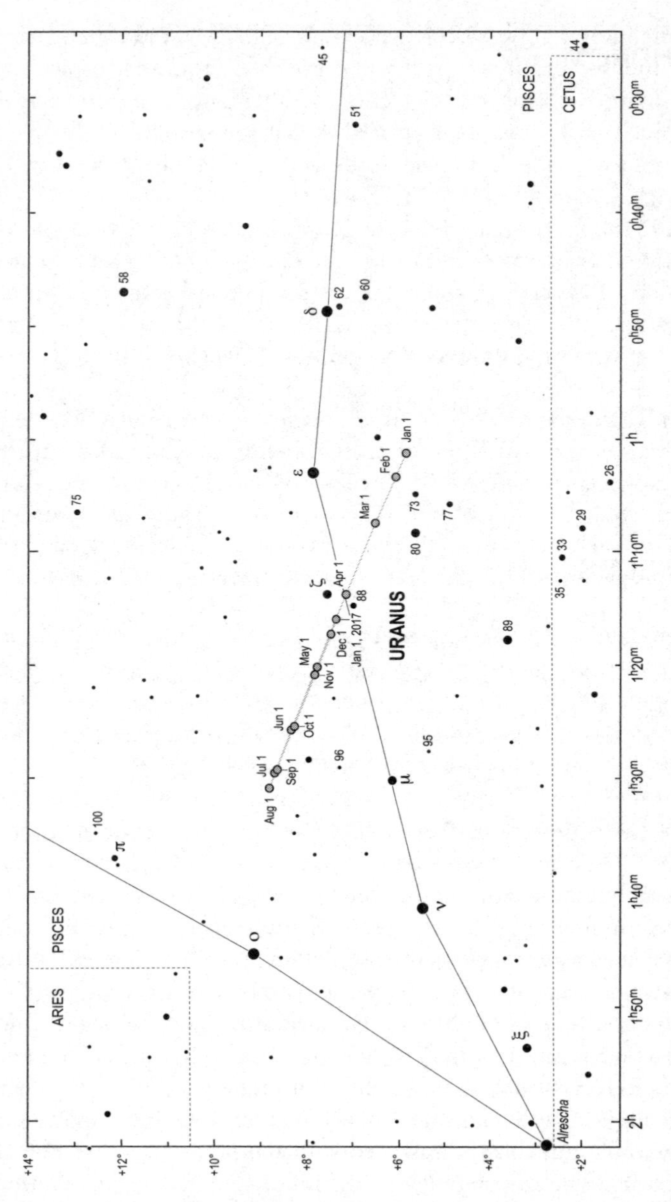

Figure 28. The path of Uranus against the background stars of Pisces during 2016. The planet is at opposition on 15 October.

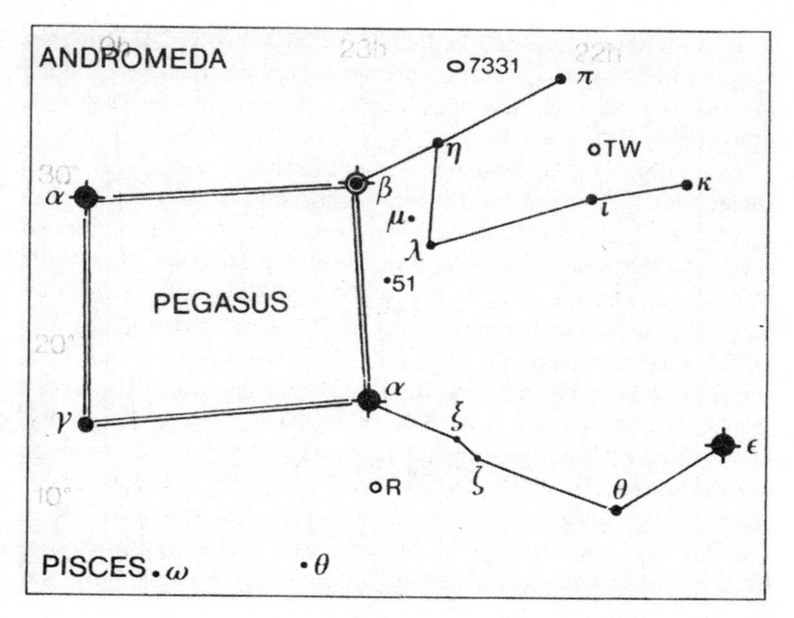

Figure 29. The four main stars comprising the Great Square of Pegasus (left) are well placed during October evenings; they are not especially bright and the Square is larger than many people expect it to be, but once located it is easy enough to find. The only other important star is Epsilon Pegasi (right), which lies well to the west of the Square. The much fainter star 51 Pegasi is just outside the Square, located slightly to the west of a line joining Beta and Alpha, and roughly halfway between them. (Diagram supplied from the archive of the Sir Patrick Moore Heritage Trust, courtesy of the Executors.)

Pegasi, but it has now been officially – and, frankly, illogically – transferred to the neighbouring constellation of Andromeda, as Alpha Andromedae; its proper name, still often used, is Alpheratz. The upper right-hand star of the Square, Beta Pegasi or Scheat, is a red giant. Like many of its kind, it is variable; the magnitude range is from 2.3 to 2.8, and there is a very rough period of 35 to 40 days. Alpha Pegasi or Markab (2.5) at the bottom right of the Square, and Gamma Pegasi or Algenib (2.9) at the bottom left, act as good comparison stars, and the light variations may be followed with the naked eye. The only other important star is Epsilon Pegasi or Enif (2.4), which lies well to the west of the Square, in a relatively isolated position roughly between Alpha Pegasi and Altair in Aquila, the Eagle.

It is interesting to count the number of faint stars visible inside the Square without optical aid. There are not a great many of them, though the area covered is considerable; anyone who counts a dozen will be doing rather well.

One rather famous faint star just outside the Square is 51 Pegasi (magnitude 5.5), which is, in theory, visible with the unaided eye but may be a bit of a challenge. It is easy enough in binoculars, located slightly to the west of a line joining Scheat and Markab, and roughly halfway between them. 51 Pegasi is about 50 light years away from Earth, and on 6 October 1995, Swiss astronomers Michel Mayor and Didier Queloz announced that they had discovered an exoplanet (dubbed 51 Pegasi b) orbiting the star. They had detected very slight, regular variations in the radial velocity of the star due to the gravitational effects of the planet using the very sensitive ELODIE spectrograph at the Observatoire de Haute-Provence in France. The exoplanet was subsequently confirmed from the Lick Observatory near San Jose in California by Geoff Marcy and Paul Butler. 51 Pegasi b was the first confirmed planetary-mass companion to be found orbiting an ordinary star like our Sun. We now know that 51 Pegasi b has a mass about half that of Jupiter and orbits incredibly close to its parent star, completing an orbit in just 4.2 days; its surface temperature must be about 1200°C.

Well below the Square, fairly close to the horizon from Britain, is Fomalhaut in Piscis Austrinus, the Southern Fish, the southernmost of the first-magnitude stars visible from Europe or the mainland United States. It is never well seen from these locations, but, fortunately, a line extended downwards from Scheat through Markab points almost directly to it. The best time to look for Fomalhaut is around 8 p.m. to 9 p.m. in mid-October; by midnight it will be too low to be seen, although from the latitude of New York it is visible for longer.

Open Star Clusters. In October the distinctive Zodiacal constellation of Taurus, the Bull, is rising in the east in the mid-evening. The pattern contains two very well-known star clusters, the Pleiades or Seven Sisters and the V-shaped Hyades near Aldebaran (Alpha Tauri). A great deal of research in astronomy and astrophysics has been directed to the study of star clusters. This work has the advantage that all the stars in a cluster are at about the same distance from us, so that they can be studied under similar conditions; for example, their apparent brightness is at once a measure of their true intrinsic brightness or luminosity, since there are no distance effects. In addition, all of the stars in the cluster will be of very similar age and initial chemical composition, so the effects of other more subtle variables on the properties of stars in the cluster (e.g. how stars of different masses evolve as they age) are more easily studied than they are for isolated stars.

The beautiful Pleiades cluster (Figure 30) is a typical example of a loose or open cluster, of which over a thousand are known in our Galaxy alone. Such clusters are typically found only in galaxies in which active star formation is taking place. Open clusters have no regular shape, contain from a few hundred to a few thousand stars that were formed from the same giant molecular gas and dust cloud, and they are nearly always found in or near the plane of the Milky Way. In such a cluster, the stars are loosely gravitationally bound to each other and move through space as a unit, sharing the same motion, so that they travel in more or less parallel paths.

Open clusters are usually less than a few hundred million years old and in the course of their long journey around the Galaxy the cluster will be disrupted, as stars are lost following close encounters with other cluster members and by chance encounters with passing clusters and gas clouds as they orbit. As a result a cluster will be broken up, and the rate at which this occurs will depend on how tightly the stars were packed in the cluster in the first place. It has been calculated that most clusters of this kind would be disrupted completely in the course of one or two trips around the Galaxy. Thus we know that the open clusters as we see them today could not have been in existence when the majority of the stars of the Galaxy were formed many thousands of millions of years ago.

The youngest open clusters are usually still contained within the cloud of dust and gas from which they formed. High-energy ultraviolet radiation from young, massive stars within the cluster ionizes the surrounding gas to produce a glowing emission nebula known as an HII region. The HII refers to ionized hydrogen (HI is used for neutral hydrogen) since the gas clouds are primarily composed of hydrogen. Over time, stellar winds blowing from the stars in the cluster will disperse the cloud of gas and dust. Typically, about one tenth of the mass of a gas cloud will coalesce into stars before stellar winds drive the remainder away.

The Pleiades cluster consists of about a thousand, mainly hot, blue-white stars, packed into a region of space about forty light years across, lying about four hundred light years from us. The nearby cluster of the

Figure 30. The beautiful Pleiades cluster in Taurus, the Bull, which is nicely on view during evenings from October to December, is a typical example of a loose or open cluster. The cluster, which contains over 1,000 confirmed members, is dominated by a number of highly luminous, hot, blue-white stars that formed about 100 million years ago. The delicate wisps of nebulosity (known as reflection nebulae) seen around some of the brighter stars are part of an interstellar dust cloud through which the cluster members happen to be passing at the present time. They are no longer thought to be caused by dust left over from the formation of the star cluster itself. (Image courtesy of NASA, ESA, AURA/Caltech, Palomar Observatory, with thanks to D. Soderblom and E. Nelan (STScI), F. Benedict and B. Arthur (U. Texas), and B. Jones (Lick Obs.).)

Hyades (Figure 31) covers a larger area, packed less tightly, and only about 150 light years distant. About a hundred stars are known to belong to the Hyades cluster, but Aldebaran is not one of them – this bright reddish star is in the foreground, less than half as far away. All of the members of the Hyades cluster share a common motion through space. As seen from the Earth, the paths of the stars converge towards a point in Orion slightly east of Betelgeuse. On average the Hyades stars have a proper motion of 0.12 arc seconds per year and a radial velocity of +40 kilometres per second. The main core of the cluster lies thirty-three degrees from the convergent point.

Figure 31. The V-shaped star cluster known as the Hyades, also in Taurus, the Bull, is a very different example of an open cluster in comparison with the Pleiades. There is no dust or gas and no brilliant blue-white stars, but there are some red giant stars. The Hyades cluster is more than six times older than the Pleiades. The reddish, first-magnitude star Aldebaran or Alpha Tauri (below centre of frame), which appears to be part of the Hyades cluster, is, in fact, in the foreground, less than half as far away as the cluster members. (Original image courtesy of Todd Vance and licensed under the Creative Commons Attribution-ShareAlike 2.5 Generic license.)

The Hyades cluster is very different from the Pleiades. There is no dust or gas here, and no brilliant blue-white stars, but there are some red giant stars. The Hyades is an old cluster, with an age of about 625 million years, compared with the Pleiades which are only about 100 million years old.

November

Full Moon: 14 November *New Moon:* 29 November

MERCURY is slowly emerging from the evening twilight and for the second half of the month is visible to observers in the tropics and the Southern Hemisphere as an evening object in the west-south-western sky for a short while after sunset; it will not be visible from locations as far north as the British Isles. Its magnitude is –0.5. Mercury will lie about 3.5° south of Saturn (magnitude +0.5) at dusk on 23 November.

VENUS, magnitude –4.1, is still visible for several hours after sunset in the western sky, for observers in equatorial and more southerly latitudes, while those in northern temperate latitudes may only see it for a much shorter period of time, low above the south-western horizon. The phase of the planet decreases from 78 per cent to 69 per cent during November.

MARS is still an evening object, although for observers in northern temperate latitudes it is fairly low in the south-western sky and sets in the mid-evening; for those living in the tropics and further south it is visible until midnight. The planet begins the month in Sagittarius but moves into Capricornus on 8 November. It continues to fade slowly, from magnitude +0.4 to +0.6 during the month. The waxing crescent Moon will be about 6° north of Mars on the evening of 6 November.

JUPITER, magnitude –1.7, continues to be visible as a brilliant morning object in the east-south-eastern sky, rising over three hours before the Sun in mid-November from northern temperate latitudes. The planet is slowly becoming more favourable for those living further south as its southerly declination increases, reaching –5.5° by month end; the planet is in the constellation of Virgo. Jupiter's four Galilean satellites are readily observable with a small telescope or even a good pair of binoculars provided that they are held rigidly. The waning crescent Moon will be close to Jupiter on the morning of 25 November and the pair will be a lovely sight in the dawn twilight sky.

SATURN, magnitude +0.6, is visible to observers in the tropics and the Southern Hemisphere in the west-south-western sky as soon as darkness falls in early November, but even from these locations will be inconveniently low after mid-November as it draws in towards conjunction with the Sun next month. The planet remains in Ophiuchus.

The Leonid Meteor Storm of 1866. This month, from about 15–20 November and especially during the early morning hours of 17 November, it is likely that a few very swift meteors will be seen emanating from a radiant within the 'Sickle' marking the head of Leo the Lion – although bright moonlight will hamper observations this year. Such Leonid meteors may be seen every year, but the shower is most famous for the periods of enhanced activity that have, with a few exceptions, occurred at intervals of about thirty-three years. This enhanced activity is caused by a dense swarm of dust particles in the vicinity of the parent comet. At such times, the Leonids have produced some of the most remarkable meteor showers of the past thousand years, with peak observed rates of many thousands of meteors per hour. However, these great meteor storms do not occur unfailingly every thirty-three years due to the influence of planetary perturbations on the orbit of the Leonid stream, and the uneven dust distribution near to the parent comet.

Great Leonid meteor storms were witnessed on 11–12 November 1799 and again on 12–13 November 1833. On the basis of his careful analysis of these two great displays and 11 former Leonid showers, Professor Hubert A. Newton, a mathematician at Yale College and his gifted student Josiah Gibbs established the length of the cycle between significant Leonid showers as 33.25 years, but noted that extraordinary displays could occur over a period of two to three years, and lesser displays for five or six years. He predicted that the most likely date of the next major Leonid shower would be in November 1866.

Accordingly, 150 years ago this very month on the night of 13–14 November 1866, a spectacular Leonid shower occurred on schedule exactly as Newton had predicted. At Birr Castle in Ireland, Robert Stawell Ball (later Sir Robert), who acted as assistant astronomer as well as tutor to the children of William Parsons, third Earl of Rosse, was observing nebulae with the great Rosse reflecting telescope when his work was interrupted. Writing twenty years later, Ball described the scene:

> It was about ten o'clock at night when an exclamation from an attendant by my side made me look up from the telescope, just in time to see a fine meteor dash across the sky. It was presently followed by another, and then again by others in twos and threes . . . At this time the Earl of Rosse (then Lord Oxmantown) joined me at the telescope and . . . we decided to . . . ascend to the top of the wall of the great telescope from whence a clear view of . . . the heavens could be obtained. There, for the next two or three hours, we witnessed a spectacle which can never fade from memory. The shooting stars gradually increased in number until sometimes several were seen at once. Sometimes they swept over our heads, sometimes to the right, sometimes to the left, but they all diverged from the east . . . All the tracks of the meteors radiated from Leo . . . It would be impossible to say how many thousands of meteors were seen . . .

The Times of Thursday, 15 November includes a selection of letters to the Editor reporting observations of the great meteor display which had taken place in the early hours of 14 November. The astronomer John Russell Hind described the event as witnessed from Twickenham by four observers, including himself:

> . . . From midnight to 1 o'clock a.m., Greenwich time, 1120 meteors were noted, the number gradually increasing. From 1 a.m. to 1h 7m 5s no less than 514 were counted, and we were conscious of having missed very many, owing to the rapidity of their succession. At the latter moment there was a rather sudden increase to an extent which rendered it impossible to count the number, but after 1.20 a decline became perceptible. The maximum was judged to have taken place about 1.10, and at this time the appearance of the whole heavens was very beautiful, not to say magnificent. Beyond their immense number, however, the meteors were not particularly remarkable, either as regards brilliancy or the persistence of the trains, few of which were visible for more than three seconds . . . From 1.52 to 2, 9300 were registered . . .

From London, G. J. Symons remarked in *The Times* that the meteors were most numerous at 01.12 GMT, when they were falling at a rate of 100 per minute; in fact, 'the sky was scored in all directions with their trains'. Many observers produced detailed accounts of the shower, attempting to fix the position of the radiant point among the stars of Leo, and counting the number of meteors seen in successive intervals of time in an effort to accurately establish the time of shower maximum.

At the Radcliffe Observatory, Oxford, the Reverend R. Main and two colleagues undertook to note the location and appearance of individual meteors seen, but after 00.30 GMT the numbers increased so rapidly that it was impossible to do anything more than count them. They estimated that the peak occurred at 01.10 GMT, when 123 meteors were counted in a single minute. The Reverend F. Howlett, observing from near Canterbury, noted that meteors were most numerous in the eight minutes from 01.02 to 01.10, when they averaged 200 per minute.

More than eight thousand meteors were counted by eight members of staff at the Royal Observatory, Greenwich. The time of maximum can be fixed quite precisely as between 01.05 and 01.15 GMT. At this time, although the radiant lay little more than 20° above the eastern horizon, the Greenwich observers were recording around 120 meteors per minute (Figures 32a and 32b overleaf). The Greenwich team also plotted, on a star chart, the tracks of many of the meteors they observed in order to define the position of the shower radiant. Professor Alexander S. Herschel observed the shower from Glasgow and collected together fifteen independent determinations of the position of the apparent radiant point, including one of his own. He found that twelve of the measures were situated within an area bounded by a circle of less than three degrees in angular diameter, the other three measures having been rejected as inaccurate.

Using the radiant co-ordinates and time of maximum activity determined during the 1866 Leonid storm, together with the derived orbital period of 33.25 years, it was possible to calculate orbital elements for the Leonid stream particles. In late 1866 and early 1867, Giovanni Schiaparelli, Urbain Le Verrier and John Couch Adams all published orbits for the Leonid stream. Schiaparelli correctly concluded that stream orbits resembled those of comets, rather than those of planetary bodies. Le Verrier noted that, over a long period of time, planetary perturbations would cause the particles in the stream to spread

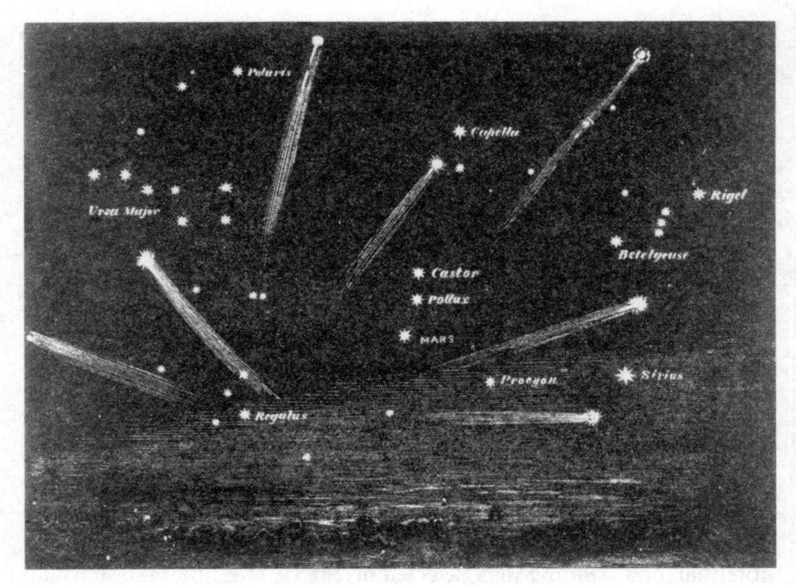

Figure 32a. Leonid meteors streaking away from the radiant point in Leo, as seen from the Royal Observatory, Greenwich on the night of 13–14 November 1866, according to a contemporary artist. At the time of maximum, the radiant would have been little more than 20° altitude in a direction slightly north of east.

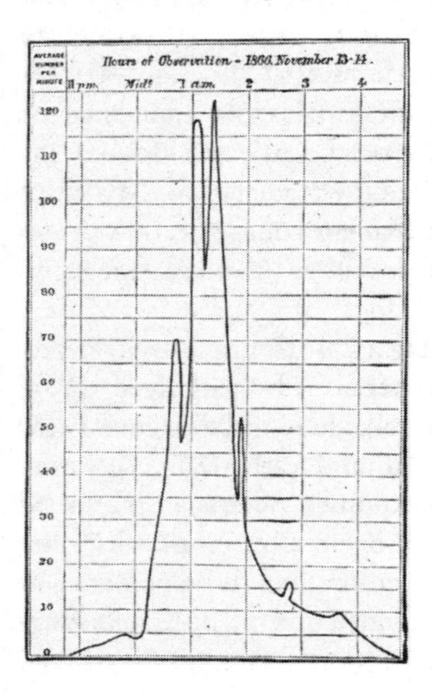

Figure 32b. Observed hourly Leonid meteor rates deduced from the observations made by a group of eight observers at the Royal Observatory, Greenwich on the night of 13–14 November 1866, between 11.00 and 04.00 GMT. The peak observed rate was close to 120 meteors per minute (equivalent to 7,200 meteors per hour) shortly after 01.00 GMT.

uniformly all around the orbit. Since the Leonids did not produce an equally strong shower every year, he reasoned that it must be a comparatively young stream, where the particles had not yet had time to spread entirely around the orbit.

In January 1867, when Theodor von Oppolzer published his orbit for comet 1866 I Tempel-Tuttle, three astronomers, Carl Peters, Schiaparelli and von Oppolzer himself, realized, almost simultaneously, that its orbit was nearly identical to that of the Leonid stream. The periodic comet now known as 55P/Tempel-Tuttle had been recognized as the parent of the Leonid meteors.

The 'Picture of the Century'. In 1966 and 1967, NASA launched five Lunar Orbiter missions for the purpose of mapping the lunar surface before the *Apollo* landings. All five Lunar Orbiter missions were successful, and 99 per cent of the Moon's surface was photographed at a resolution of sixty metres or better. The first three Orbiter missions, flown in low inclination orbits, were dedicated to imaging twenty potential lunar landing sites, selected on the basis of observations made by telescopes on Earth. The fourth and fifth Orbiter missions were devoted to broader scientific objectives and were flown in high-altitude polar orbits. Lunar Orbiter 4 photographed the entire nearside and 95 per cent of the far side, and Lunar Orbiter 5 completed the far-side coverage and acquired medium- (20 metres) and high- (2 metres) resolution images of 36 pre-selected areas.

The second of these missions, Lunar Orbiter 2, was launched on 6 November 1966 and imaged the Moon from 18–25 November, exactly fifty years ago this month. The Lunar Orbiters carried out their photography with a dual-lens camera, a film-processing unit, a readout scanner and a film-handling apparatus. The wide-angle, medium-resolution lens had a 3.1-inch (80-mm) focal length and the other, a narrow-angle high-resolution lens, had a 24-inch (610-mm) focal length; both lenses placed their frame exposures on a single roll of 70 mm film. The two cameras were bore-sighted, so that each high-resolution photo was always centred within the medium-resolution frame. The film was moved during each exposure to compensate for the spacecraft's motion, which was estimated by an electro-optical sensor. The film was then processed, scanned, and the images transmitted by radio back to Earth. The film images provided a very effective method of storing information for transmission bit by bit, at a modest rate.

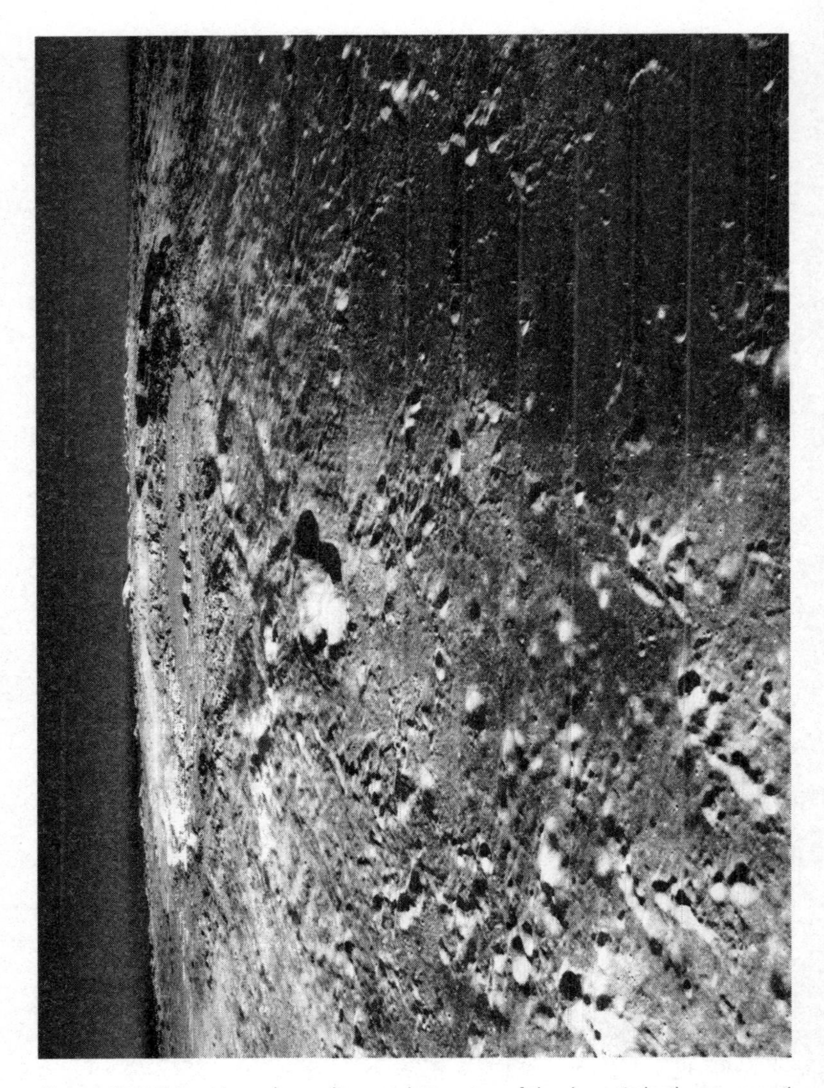

Figure 33a. This wide-angle, medium-resolution view of the dramatic landscape around the lunar crater Copernicus was called the 'picture of the century' by many people at the time of its original public release in 1966. The image was taken from an altitude of 45.7 kilometres above the lunar surface by the Lunar Orbiter 2 spacecraft on 24 November 1966, while 240 kilometres due south of Copernicus. (Image courtesy of NASA/Lunar & Planetary Institute/Universities Space Research Association.)

Figure 33b. The corresponding narrow-angle, high-resolution frame from the centre of the medium-resolution image shown in Figure 33a revealed a wealth of fine surface detail in and around the crater Copernicus, but it was the oblique angle from which the picture was taken – providing a sense of the true elevation of lunar-surface features – that aroused such great interest. (Image courtesy of NASA/Lunar & Planetary Institute/Universities Space Research Association.)

One iconic image (Frame 2162) acquired by the cameras of Lunar Orbiter 2, showing the dramatic landscape within the crater Copernicus, captured everyone's imagination at the time. Indeed, it was frequently referred to as the 'picture of the century' when it was first released in 1966. This particular image was taken by Lunar Orbiter 2 at 12.05 a.m. GMT on 24 November 1966 from an altitude of 45.7 kilometres above the lunar surface, 240 kilometres due south of Copernicus. At the time the picture was originally released, most views of the Moon's surface involved looking straight down. Little, if any, sense of the true elevation of lunar surface features was usually available. This particular image, taken from an oblique angle, changed that perception by showing the Moon to be a world with complex topography – some of it Earth-like, but much of it decidedly un-Earth-like. At the time, this detailed image of the lunar surface was described by NASA Scientist Martin Swetnick and subsequently quoted by *Time* magazine as 'one of the great pictures of the century'.

The original medium-resolution version of Frame 2162 is shown in Figure 33a, with the corresponding high-resolution version that aroused such interest also shown here in Figure 33b. Apparently, this stunning view of the crater Copernicus was acquired by Lunar Orbiter 2 as one of twelve 'housekeeping' images taken to advance the roll of film between possible astronaut landing sites being surveyed.

December

Full Moon: 14 December *New Moon:* 29 December

Solstice: 21 December

MERCURY is now so far south of the Equator, and consequently always so low above the horizon after sunset, that from the latitudes of the British Isles observers will find it incredibly difficult to detect the planet, even though it is at greatest eastern elongation (21°) on 11 December. For observers in equatorial or more southerly locations, Mercury will continue to be visible as an evening object in the south-western sky for a short while around the end of evening civil twilight, about twenty-five minutes after sunset, but only for the first two and a half weeks of the month. During this period of visibility its magnitude fades from –0.5 to +0.2. Mercury passes through inferior conjunction on 28 December.

VENUS is a magnificent object in the early evenings completely dominating the south-western sky for several hours after sunset. Its magnitude is –4.2. Venus is now moving gradually northwards in declination, passing from Capricornus into Aquarius on 31 December. By the end of the month the planet will be setting more than four hours after the Sun from northern temperate latitudes, although the altitude of the planet will be lower than for those living further south, where the period of visibility is slightly less.

MARS continues to move rapidly eastwards, passing from Capricornus into Aquarius on 15 December. The fact that it is also moving northwards in declination means that the planet is now more favourably placed for those living in northern temperate latitudes than it was in the late summer and autumn months, setting over five hours after the Sun by month end. It is no longer the conspicuous object that it was at opposition in May, its magnitude having faded to +0.9, but it is obviously brighter than any of the stars in that part of the sky. Mars is visible just to the west of due south as darkness falls, moving through the south-western sky during the mid-evening. From more southerly latitudes, the planet is high in the south-west at dusk, but sets less than four hours after the Sun. The waxing crescent Moon will lie north-east of Mars on 5 December.

JUPITER continues to be visible in the south-eastern sky as a brilliant morning object in the constellation of Virgo. Its period of visibility is gradually increasing, rising at about 01.30 hours at the end of the month from northern temperate latitudes and slightly earlier from more southerly locations. The brightness of Jupiter increases slightly from magnitude −1.8 to −1.9 during December. The waning crescent Moon will make a nice pairing with Jupiter on the mornings of 22 and 23 December.

SATURN is in conjunction with the Sun on 10 December and consequently will not be visible at all this month for observers in the latitudes of the British Isles. For those in the tropics, the planet may just be glimpsed low in the east-south-eastern sky just before dawn at the very end of the month. The planet is in Ophiuchus, magnitude +0.5.

Rigel. Orion, the Hunter (Figure 34), is coming back into the evening sky, and will be very conspicuous from now through to the end of the spring. Its two brightest stars are quite unlike each other. Betelgeuse (Alpha Orionis), marking Orion's right shoulder, is a huge red, cool supergiant of spectral type M, having a surface temperature of 3,400°C. Rigel (Beta Orionis), marking the Hunter's left foot, is also a supergiant, but blue-white in colour, of spectral type B8, with a high surface temperature of 11,200°C. Betelgeuse (magnitude +0.5, variable) is normally fainter than Rigel (magnitude +0.12), but Betelgeuse is subject to slow semi-regular variations in brightness in common with many other red supergiant stars.

Rigel is a very remote star. Its distance is difficult to measure accurately, but it is thought to be about 860 light years, so that we are now seeing it as it used to be in the time of Henry II, the first Plantagenet King of England. Yet despite its great distance, Rigel still appears as the seventh brightest star in the sky, and is only fractionally inferior to Capella in Auriga and Vega in Lyra, though Rigel never rises so high above the horizon as viewed from the latitudes of the British Isles.

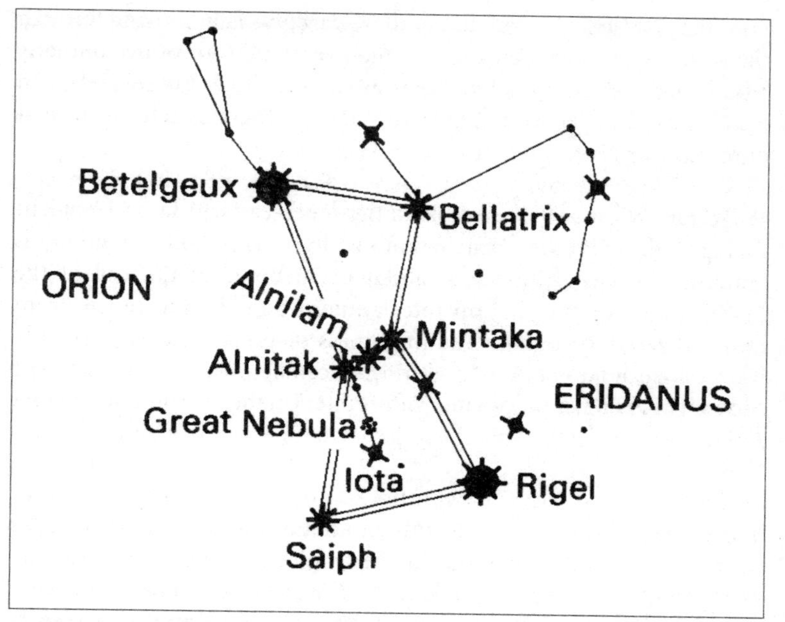

Figure 34. During December evenings, the brilliant winter constellations have again come into view. Of these, Orion, the Hunter, dominates the southern aspect of the sky. The pattern is quite unmistakeable, consisting of a rectangle of stars which includes the two brightest members – Betelgeuse (top left) and Rigel (bottom right) – with the three stars of the Belt positioned diagonally across the middle of the rectangle. (Diagram supplied from the archive of the Sir Patrick Moore Heritage Trust, courtesy of the Executors.)

When one takes account of the ultraviolet radiation it is emitting, the luminosity of Rigel must be at least 85,000 times that of the Sun. Measurements of the angular diameter of Rigel and theoretical considerations both indicate that it is about 75 times the diameter of our Sun, around 105 million kilometres across; its mass is around 14–18 times the mass of our Sun, although some estimates are greater. In any event, Rigel is probably only about ten million years old.

Like all normal stars, Rigel is producing its energy by nuclear transformations taking place inside it. As it radiates energy, it is losing mass, and is using up its available 'hydrogen fuel'. With the Sun, a relatively modest Main Sequence star of type G, the mass-loss amounts to four million tonnes per second; with Rigel, the rate of mass loss is considerably greater, and the evolutionary sequence must be run through much more quickly – so high-mass stars like Rigel die relatively young. Our Sun is not expected to alter much for around 5,000 million years in the future, but it seems that Rigel has a dead helium core – having already used up the available hydrogen in its core – and is now swelling and cooling. Eventually, it will become a cool, red supergiant star similar to the present-day Betelgeuse, by which time it will be fusing helium into carbon and even heavier elements. Since it is extremely massive, Rigel, like Betelgeuse, will also end its career in a brilliant and spectacular supernova explosion, blasting much of its material away into space.

Rigel has a fairly bright companion of magnitude 6.8, lying 9.5 arc seconds distant at a position angle of 204°. The pair are said to provide a good test for a two-inch telescope; a three-inch instrument will show it easily.

The Northern Lights – on Mars! In late December 2014, an instrument on NASA's MAVEN (Mars Atmosphere and Volatile Evolution) spacecraft, which has been in orbit around Mars since September 2014, detected evidence of widespread aurorae in the planet's northern hemisphere (Figure 35). These 'Christmas Lights', as scientists called them, encircled the planet and descended so close to the Martian equator that if the lights had occurred on Earth, they would have been visible from places as far south as Morocco, Iraq and Texas in the United States.

On Earth, auroral displays occur when electrically charged particles, originating in the solar wind, enter Earth's magnetic field and spiral downwards around the North and South Magnetic Poles, exciting atoms of oxygen and nitrogen high in the atmosphere as they collide with them at altitudes of between a hundred and four hundred kilometres above the ground. As these excited atmospheric atoms decay back to their ground state, they emit coloured light; greens and reds are due to excited atoms of oxygen; purples, deep reds and turquoise-blue to atoms of nitrogen. It is this light that we see as the aurora.

In the Northern Hemisphere the aurora is a glowing oval encircling the North Magnetic Pole (the aurora borealis or Northern Lights) and usually passing over places such as Alaska, southern Greenland, northern Iceland and the far north of Norway, Sweden and Finland. There is another auroral oval encircling the South Magnetic Pole (the aurora australis or Southern Lights). When the Sun emits a particularly

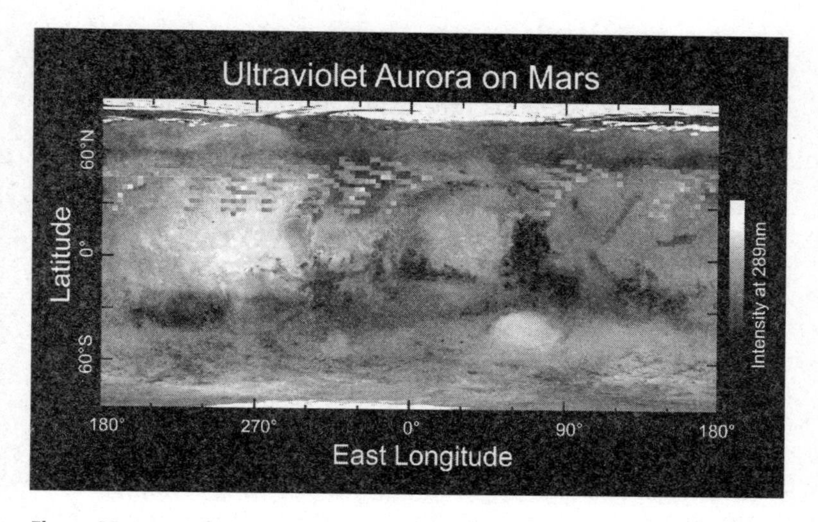

Figure 35. A map of MAVEN's Imaging Ultraviolet Spectrograph (IUVS) auroral detections in December 2014 overlaid on a map of Mars's surface. The map shows that the aurora (shown by the scattered small rectangular boxes) was widespread in the planet's northern hemisphere, and was not tied to any particular geographic location. The aurora was detected in all observations during a five-day period. (Image courtesy of University of Colorado.)

energetic burst of electrically charged particles, or when the solar wind is very disturbed, the auroral oval expands towards the Equator and it is then that the aurora may be seen from the United States, the North of England and Denmark, for example – or South Island, New Zealand, Tasmania or Southern Australia.

The MAVEN observations weren't the first time that a spacecraft has detected aurorae on Mars. In 1994, an ultraviolet camera on the European Space Agency's Mars Express observed an ultraviolet glow coming from 'magnetic umbrellas' in the planet's southern hemisphere. Unlike Earth, Mars does not have a global magnetic field that envelops the entire planet and guides any electrically charged particles entering the atmosphere. Instead, Mars has multiple umbrella-shaped magnetic fields that sprout out of the ground like mushrooms, here and there, but these are mainly in the planet's southern hemisphere. These umbrellas are remnants of an ancient global magnetic field that decayed billions of years ago.

Although one might have expected to find Martian aurorae in the canopies of this patchwork of magnetic umbrellas, it appears from the MAVEN observations that because Mars has no organized global magnetic field to guide the particles north and south, they can go anywhere they want. Magnetic fields in the solar wind drape right across Mars and deep into the atmosphere. Electrically charged particles just follow those field lines downwards, penetrating deeply into the Martian atmosphere and triggering auroral displays less than a hundred kilometres above the surface – lower than aurorae on Earth.

Because MAVEN detected the Martian aurorae with its ultraviolet camera, it is hard to say what the human eye would see. The atmosphere of Mars is more than one hundred times thinner than Earth's, and it is primarily carbon dioxide (95.3 per cent), but there are small amounts of nitrogen (2.7 per cent) and oxygen (0.13 per cent) and traces of other gases. It is the nitrogen and oxygen that may be key to the colour of any Martian aurorae. Excited nitrogen atoms in the Martian atmosphere might produce a diffuse violet glow, while oxygen might colour the sky a faint green. So one day when astronauts go to Mars, they might find that, occasionally, the Red Planet has colourful night skies.

Transit of Mercury

NICK JAMES

There will be a transit of Mercury on 9 May 2016, visible from the Americas, Europe and Africa.

When the circumstances are right both of the inner planets can move between the Earth and Sun and so can be seen to transit the Sun's disk. The orbits of Mercury and Venus are inclined by 7° and 3° respectively to the plane of the Earth's orbit (the ecliptic). These planets cross the ecliptic twice for each revolution around the Sun. The point at which a planet moves through the ecliptic plane moving north is called the ascending node. The equivalent point moving south is the descending node. A transit occurs when the planet is near a node and the Earth

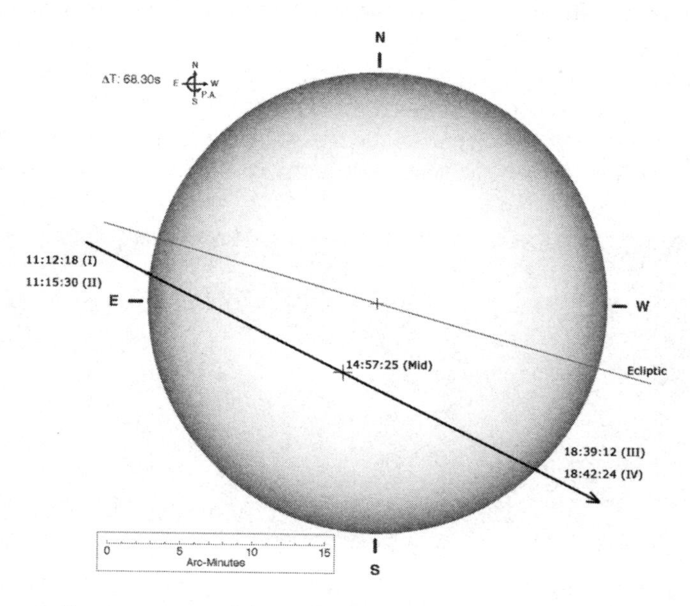

Figure 1. The circumstances of the 9 May 2016 transit for a geocentric observer. Local times will be slightly different. (Diagram courtesy of Xavier Jubier.)

is in the right place so that all three objects, the Earth, planet and Sun, line up.

The relative periods of the orbits of the Earth and Venus mean that transits of Venus are exceptionally rare. There were transits in 2004 and 2012 but the next one will not be until 2117. Transits of Mercury are much more common, occurring on average thirteen times per century. The last one was in November 2006 and the next one, after this year, will be in November 2019.

Mercury's descending and ascending node crossings take place around 8 May and 10 November each year, so transits can only take place around these dates. The eccentricity of the orbits of the Earth and Mercury mean that, compared to November, we are slightly further from the Sun in May and slightly closer to Mercury. This means that Mercury appears at its largest during May (descending node) transits and the Sun is at its smallest. Even so, the planet is only around twelve arc seconds in diameter and so observation will require optical aid.

For a geocentric observer the transit lasts from 11h 12m to 18h 42m UTC. These times vary by a few minutes depending on your location. For the UK the times are 12h 12m to 19h 41m BST and for New York 07h 13m to 14h 41m EDT. Exact circumstances for your particular location can be determined using Xavier Jubier's transit calculator at http://xjubier.free.fr/en/site_pages/transits.html.

The most interesting part of the transit occurs between the first visibility of Mercury on the solar disk (contact I) and the time that it is

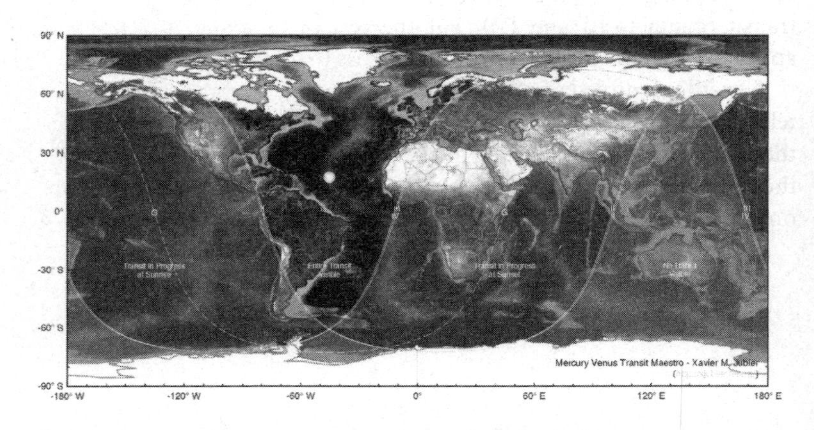

Figure 2. A map showing the visibility of the transit. (Map courtesy of Xavier Jubier.)

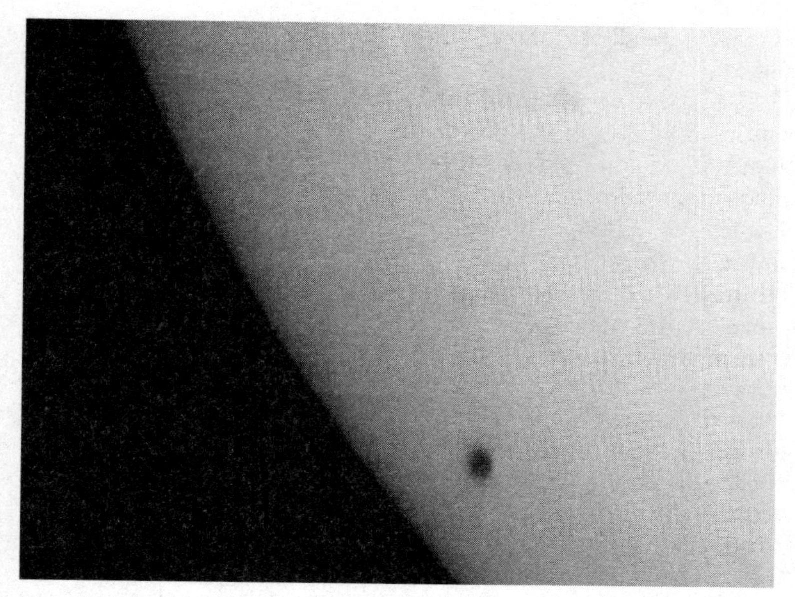

Figure 3. An image of Mercury near the Sun's limb taken on 7 May 2003. The apparent size of Mercury is the same at this transit. (Image courtesy of Nick James.)

fully on the disk (contact II). Turbulent seeing in our atmosphere tends to make the transiting planet cling on to the Sun's limb exhibiting an effect called the 'black drop'. This effect is repeated at the end of the transit (contacts III and IV). For the rest of the time Mercury will appear as a small black disk crawling across the face of the Sun.

Never look directly at the Sun without using a suitable filter. When telescopes or binoculars are used, approved filters should be fitted over the main lens or mirror of the instrument. It is not safe to use filters at the eyepiece end. Alternatively, you can project an image of the Sun onto a screen in order to observe the event.

Eclipses in 2016

MARTIN MOBBERLEY

During 2016 there will be four eclipses: a total solar eclipse, an annular solar eclipse and two penumbral lunar eclipses. All four events occur in two months, namely March and September.

1. *A total eclipse of the Sun* on 9 March will be visible along a track stretching from the Indian Ocean to the Pacific Ocean. The eclipse begins, at sunrise, in the Indian Ocean and rapidly heads east until it crosses the Kepulauan Mentawai region islands of Pagai Utara and Pagai Selatan just prior to arriving at the east coast of Sumatra. As the umbra crosses central Sumatra, totality exceeds two minutes on the centreline. The track then crosses the southern part of Bangka island,

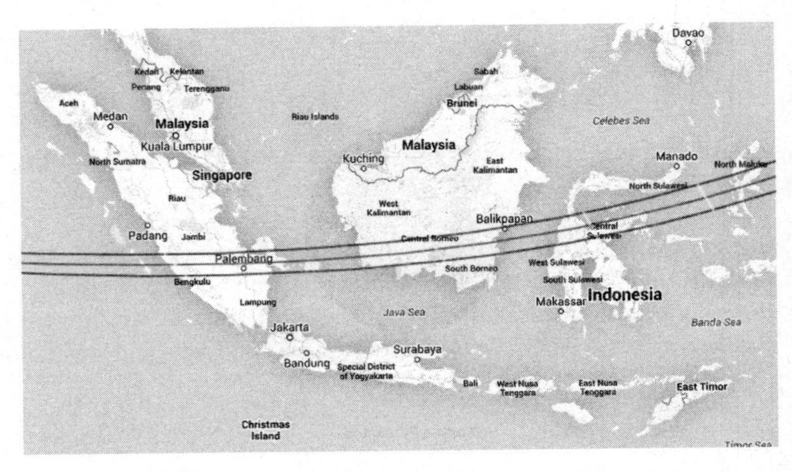

Figure 1. The track of the 9 March Total Solar Eclipse across the Sumatra, Borneo and Indonesia region from approximately 00h 19m UT on 9 March (left-hand edge) to 00h 56m UT on the same day (right-hand edge). On the centreline the duration of totality increases from around 1 minute 50 seconds to 3 minutes 22 seconds across the map from left to right. Eclipse predictions by Fred Espenak, NASA's GSFC.

the whole of Belitung island (although the north is favoured), and then crosses the Java Sea to the west coast of Borneo. After Borneo the umbral shadow flies over the Java Sea again and crosses the northern half of Sulawesi, the Molucca Sea and the central part of the smaller island of Halmahera. In central Halmahera the eclipse duration reaches three minutes and nineteen seconds, only fifty seconds short of the maximum for this eclipse. After this point the umbral track curves north, into cooler waters as it speeds across the Pacific Ocean. The maximum totality duration of four minutes and nine seconds occurs at 10° 6.5° N and 148′ 50.2′ E, with the umbral track being 155 kilometres in width. This far into the Pacific the cloud cover is usually rather better than in the tropical Indonesian region. The sunset end of this eclipse occurs roughly midway on a line between Hawaii and San Francisco, approximately 1500 kilometres north-north-east of Hawaii, so a cruise ship sailing from Hawaii to the nearest part of the track to the north is a possible option for those who plan to observe the event. This has been a strange Solar Cycle but by 2016, the activity of the Sun should be in

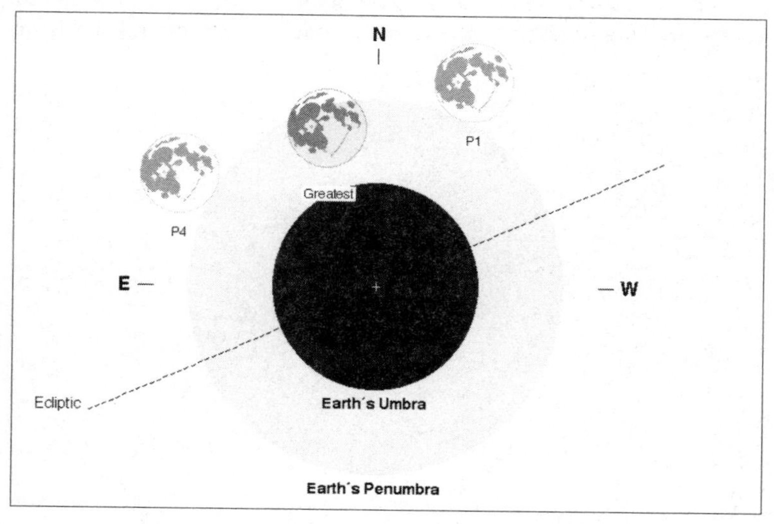

Figure 2. The track of the Moon through the Earth's penumbral shadow (outer ring) on 23 March. The Moon first touches the penumbra at P1 (09h 39m UT) and leaves the penumbra at P4 (13h 55m UT). At the point of greatest immersion in the penumbra (11h 47m UT) the lunar south pole should appear visibly dark, even to the casual observer. Eclipse predictions by Fred Espenak, NASA's GSFC.

decline, so one can expect quite a few long equatorial streamers but with a minimum of prominence activity, unless we are very lucky. Mercury and Venus will be easy targets to the west of the Sun, shining at magnitudes –0.6 and –3.9 respectively. This eclipse is the 52nd of 73 eclipses in Saros cycle 130, with the Moon being some 4.5 per cent bigger than the Sun at the point of greatest eclipse.

2. *A penumbral eclipse of the Moon* on 23 March will see the Moon pass through the northern penumbral shadow of the Earth with three quarters of the lunar disk being immersed at mid-eclipse. This should result in the southern hemisphere of the Moon darkening noticeably to the experienced observer. Sadly, no part of this eclipse is observable from the UK as it occurs close to midday UT, and so observers in the mid-Pacific will be best placed. Specifically, those observers at the longitudes of New Zealand, eastern Australia and Japan are best placed, with those on the western US seaboard also being reasonably placed.

3. *An annular eclipse of the Sun* on 1 September tracks across Africa and northern Madagascar, crossing Gabon, Congo, Democratic Republic of

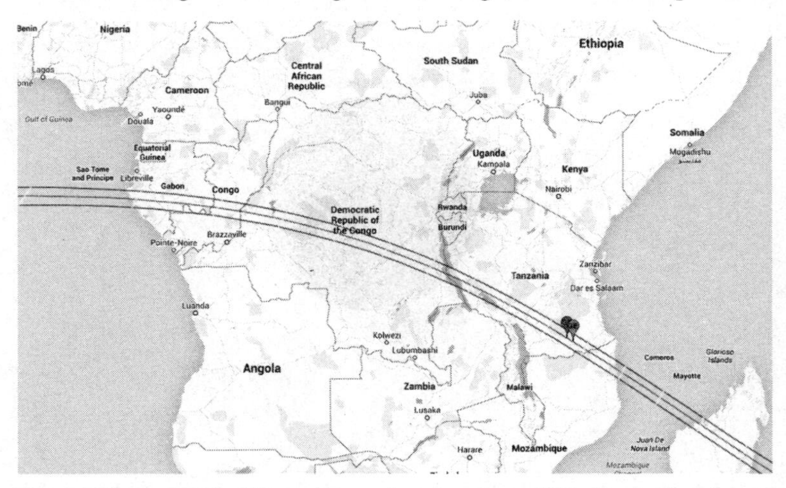

Figure 3. The track of the 1 September annular eclipse across Africa. The duration of annularity will be 2 minutes 53 seconds as the eclipse enters Gabon at 07h 39m UT where the altitude will be 33 degrees. At maximum eclipse (3 minutes 5 seconds), near the Tanzania/Mozambique border at 09h 07m UT, the altitude of the Sun will be 70 degrees just before it leaves the African coastline. Eclipse predictions by Fred Espenak, NASA's GSFC.

Congo, Tanzania, Malawi and Madagascar. It ends in the south Indian Ocean not far from the coast of SW Australia. This is the 39th of 71 eclipses in Saros cycle 135 and the greatest annular duration will be 3 minutes 5 seconds with an annular track almost 100 kilometres in width and the Moon smaller than the Sun by 2.6 per cent at this point.

4. *A penumbral eclipse of the Moon* on 16 September will see the Moon pass through the southern penumbral shadow of the Earth with 90 per cent of the lunar disk being immersed at mid-eclipse. This should result in the northern hemisphere of the Moon darkening noticeably to the experienced observer. From the UK, the eclipse peaks just after moonrise and from London, the altitude of the Moon when the eclipse peaks, at 18h 54m UT, will be only 5 degrees. At first, this might seem like a huge disadvantage, but, unless you have trees and buildings blocking the view, it could actually be to an observer's slight advantage. The Full Moon is very bright, dazzling, in fact, especially against a black sky (in this case the Sun will be almost 8 degrees below the London horizon).

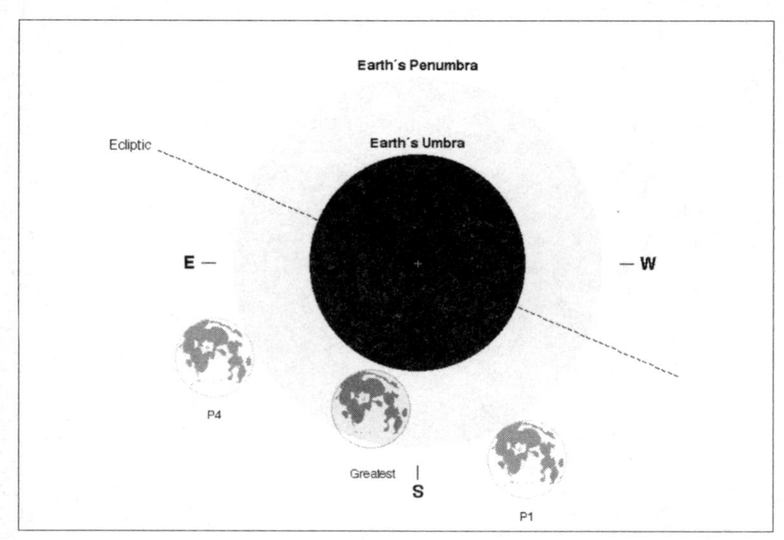

Figure 4. The track of the Moon through the Earth's penumbral shadow (outer ring) on 16 September. The Moon first touches the penumbra at P1 (16h 55m UT) and leaves the penumbra at P4 (20h 54m UT). At the point of greatest immersion in the penumbra (18h 54m UT) the lunar north pole should appear visibly dark, especially to UK observers, who see it rising and dimmed, very low in the east. Eclipse predictions by Fred Espenak, NASA's GSFC.

In penumbral lunar eclipses the darkening of the Moon can be quite subtle to the eye, but when it is low in altitude the lunar surface is dimmed considerably, making it easier to detect the penumbral shadow. The Moon leaves the penumbra at 20h 54m UT, when it has a 22-degree altitude from London, by which time even the most experienced observer will not have detected any dimming for the previous half-hour. As far as the rest of the World is concerned, the optimum place to be, for a Moon transiting in mid-eclipse, would be the Indian Ocean, or anywhere between Western Australia and South Africa. No eclipse is visible from anywhere in the US.

Occultations in 2016

NICK JAMES

The Moon makes one circuit around the Earth in just over twenty-seven days and as it moves across the sky it can temporarily hide, or occult, objects that are further away, such as planets or stars. The Moon's orbit is inclined to the ecliptic by around 5.1° and its path with respect to the background stars is defined by the longitude at which it crosses the ecliptic passing from south to north. This is known as the longitude of the ascending node. After passing the node the Moon moves eastwards relative to the stars, reaching 5.1° north of the ecliptic after a week. Two weeks after crossing the ascending node it crosses the ecliptic moving south and then it reaches 5.1° south of the ecliptic after three weeks. Finally, it arrives back at the ascending node a week later and the cycle begins again.

The apparent diameter of the Moon depends on its distance from the Earth but at its closest it appears almost 0.6° across. In addition, the apparent position of the Moon on the sky at any given time shifts depending on where you are on the surface of the Earth. This effect, called parallax, can move the apparent position of the Moon by just over one degree. The combined effect of parallax and the apparent diameter of the Moon mean that if an object passes within 1.3° of the apparent centre of the Moon, as seen from the centre of the Earth, it will be occulted from somewhere on the surface of our planet. For the occultation to be visible the Moon would have to be some distance from the Sun in the sky and, depending on the object being occulted, it would have to be twilight or dark.

For various reasons, mainly the Earth's equatorial bulge, the nodes of the Moon's orbit move westwards at a rate of around 19° per year, taking 18.6 years to do a full circuit. This means that, whilst the Moon follows approximately the same path from month to month, this path gradually shifts with time. Over the full 18.6-year period, all of the stars that lie within 6.4° of the ecliptic will be occulted.

Only 4 first-magnitude stars lie within 6.4° of the ecliptic. These are Aldebaran (5.4°), Regulus (0.5°), Spica (2.1°) and Antares (4.6°). As the nodes precess through the 18.6-year cycle there will be a monthly series of occultations of each star followed by a period when the star is not occulted. In 2016, there are thirteen occultations of Aldebaran, one every lunar month, and one occultation of Regulus.

In 2016, there will be nine occultations of bright planets. Three of Mercury, two of Venus and four of Jupiter. Only two of these events (both of Jupiter) take place at a solar elongation of greater than thirty degrees. The table lists occultations of the bright stars and planets which are potentially visible from somewhere on the Earth and where the solar elongation exceeds thirty degrees. More detailed predictions for your location can often be found in magazines or the *Handbook of the British Astronomical Association*.

Object	**Date and time of Minimum Distance (UT)**		**Minimum Distance** °	**Elongation** °	**Best Visibility**
Aldebaran	20 Jan 2016	02:37	0.5	131	North America
Aldebaran	16 Feb 2016	07:54	0.3	103	Mid Pacific
Aldebaran	14 Mar 2016	14:08	0.3	76	India/China
Aldebaran	10 Apr 2016	22:18	0.3	49	North America (daylight)
Aldebaran	2 July 2016	04:18	0.4	–31	China (daylight)
Jupiter	9 July 2016	09:34	0.8	61	Southern Indian Ocean
Aldebaran	29 July 2016	11:15	0.3	–57	North Atlantic (daylight)
Jupiter	6 Aug 2016	03:20	0.2	39	SW Pacific (daylight)
Aldebaran	25 Aug 2016	16:46	0.2	–83	Mid Pacific
Aldebaran	21 Sept 2016	22:32	0.2	–109	India
Aldebaran	19 Oct 2016	06:42	0.3	–136	Eastern North America
Aldebaran	15 Nov 2016	17:01	0.4	–164	China
Aldebaran	13 Dec 2016	04:32	0.5	168	North America
Regulus	18 Dec 2016	18:13	1.0	–117	Antarctica

Comets in 2016

MARTIN MOBBERLEY

More than sixty-five short-period comets should be detected approaching perihelion by the professional patrols in 2016, although a few of that number, namely D/Denning (1894 F1), D/Schorr (1918 W1) and D/Haneda-Campos (1978 R1) have been declared defunct and so have a D/ prefix. Essentially, a D/ prefix means that these comets have either run out of volatile material or disintegrated. However, with so many deep surveys now scouring the night sky they may yet be rediscovered. D/Harrington-Wilson (1952 B1) is another example of a defunct comet that may be recovered in 2016 although its perihelion date, if it still exists, is in February 2017. All of the short-period comets expected to return to perihelion in 2016 orbit the Sun with periods of between 3.8 and 19.6 years and many are too faint for amateur visual observation, even with a large telescope. A faint 2016 comet with a period much longer than the typical short-period range is Spacewatch (2011 KP36), with a period of 238 years.

Bright or spectacular comets usually have much longer orbital periods and, apart from a few notable exceptions like 1P/Halley, 109P/Swift-Tuttle and 153P/Ikeya-Zhang, the best performers usually have orbital periods of many thousands of years and are often discovered less than a year before they come within amateur range. For this reason it is important to regularly check the best comet websites for news of bright comets that may be discovered well after this Yearbook is finalized. Some recommended sites are:

British Astronomical Association Comet Section: www.ast.cam.ac.uk/~jds/

Seiichi Yoshida's bright comet page: www.aerith.net/comet/weekly/current.html

CBAT/MPC comets site: www.minorplanetcenter.net/iau/Ephemerides/Comets/

Yahoo Comet Images group: http://tech.groups.yahoo.com/group/Comet-Images/

Yahoo Comet Mailing list: http://tech.groups.yahoo.com/group/comets-ml/

The CBAT/MPC web page above also gives accurate ephemerides of comet positions in right ascension and declination.

As many as twenty periodic comets might reach a magnitude of fourteen or brighter during 2016 and so they should all be observable with large amateur telescopes equipped with amateur CCD imaging systems, in a reasonably dark sky, from the Northern or Southern Hemispheres. Sadly, only eight or nine comets are, at the time of writing, predicted to reach, or, in a few cases, exceed, magnitude 11, and so visual targets for large amateur telescopes will be few and far between unless some brighter prospects are discovered in late 2015 or 2016. In recent years, the amateur astronomer Terry Lovejoy from Queensland, Australia has discovered some splendid comets less than a year from their peak, which the professional patrols have missed, so a binocular magnitude 'Comet Lovejoy' in 2016 cannot be ruled out. The current cometary highlights for the coming year are likely to be Catalina (2013 US10), reaching magnitude 9 in January and, for Southern Hemisphere observers at the extreme end of the year, 45P/Honda-Mrkos-Pajdusakova, brightening just as 2016 turns to 2017, with that comet likely to be a binocular visible object early in 2017. There are a few other comets which may entertain the devoted comet watcher, though, such as P/Ikeya-Murakami (2010 V1) and the close approach of the small comet 252P/LINEAR in March, as well as 226P/Pigott-LINEAR-Kowalski in September.

It should perhaps be explained that the distances of comets from the Sun and the Earth are often quoted in Astronomical Units (AU) where 1 AU is the average Earth-Sun distance of 149.6 million kilometres or 93 million miles.

The brightest cometary prospects for 2016 are listed below in the order they reach best visibility, which often coincides with reaching perihelion. The first comets in the list reached perihelion at the end of the previous year (2015) and the last in the list reach perihelion next year (2017). The comet 29P/Schwassmann-Wachmann is nowhere near its 2019 perihelion during the year but is best placed in July in the southern sky constellation of Sagittarius. This enigmatic comet will often be too faint for visual detection, even in large amateur telescopes, but it can rise to eleventh magnitude when in outburst. Despite being magnitude 14 or brighter some returning comets are poorly placed, the very worst example being 157P/Tritton, which manages to hide behind the Sun for the entire year! I include it in the list below purely for completeness.

Comet	Period (years)	Perihelion	Peak Magnitude
22P/Kopff	6.40	2015 Oct 26	12 in Jan
Catalina (2013 US10)	Long	2015 Nov 15	7 or 8 in Jan
10P/Tempel	5.40	2015 Nov 15	12 in Jan
116P/Wild	6.50	2016 Jan 12	12 in Jan
P/Arend (50P)	8.30	2016 Feb 7	14 in Feb
P/Ikeya-Murakami (2010 V1)	5.40	2016 Mar 10	9 in Mar
252P/LINEAR	5.34	2016 Mar 17	10 in Mar
104P/Kowal	5.89	2016 Mar 26	14 in Apr
100P/Hartley	6.35	2016 Apr 2	14 in Apr
53P/Van Biesbroeck	12.60	2016 Apr 29	14 in May
PanSTARRS (2013 X1)	Long	2016 Apr 20	11 in June
Spacewatch (2011 KP36)	238	2016 May 26	14 in June
29P/Schwassmann-Wachmann 1	14.70	2019 April 9	11 in outburst
157P/Tritton	6.29	2016 June 10	13 v. poor elong.
81P/Wild	6.41	2016 July 20	12 poor elong.
9P/Tempel	5.58	2016 Aug 2	11 in Aug
43P/Wolf-Harrington	6.13	2016 Aug 19	12 in Sept
144P/Kushida	7.57	2016 Aug 31	11 in Sept
226P/Pigott-LINEAR-Kowalski	7.32	2016 Sept 5	12 in Sept
45P/Honda-Mrkos-Pajdusakova	5.26	2016 Dec 31	7 in Dec
93P/Lovas	9.19	2017 Mar 1	14 in Dec
Vales (2010 H2)	7.53	2017 Sept 17	14 in Dec

WHAT TO EXPECT

The year 2016 could start quite favourably for comet observers if Catalina (2013 US10) has lived up to expectations. It will have reached its perihelion on 15 November 2015 at 0.823 AU from the Sun, so will be past its best from a solar-heating perspective, but it will also approach the Earth to within 0.725 AU on 17/18 January, so could still

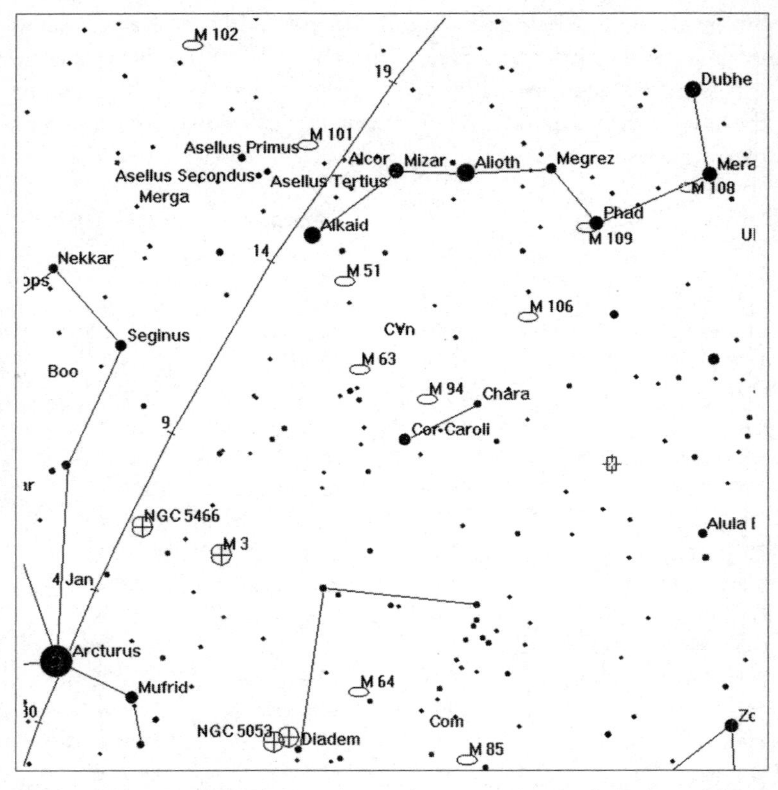

Figure 1. Comet Catalina (2013 US10) starts the year near Arcturus but curves north and west towards the tail of Ursa Major. On 16/17 January it passes two degrees west of M101.

be magnitude 7 or 8 at this time. For Northern Hemisphere observers the comet will also be very well placed, living above +80 degrees declination from 29 January to 3 February. On 1 January the comet passes very close to brilliant Arcturus in the morning sky, heading due north along the Boötes/Canes Venatici border at over four arc minutes per hour. In the pre-dawn sky of 6 January it passes one degree to the east of the magnitude 9 globular cluster NGC 5466 and just before midnight on 16/17 January it passes two degrees west of the large galaxy M101 in Ursa Major. At 109 million kilometres from the Earth the comet is only 6 light minutes away, compared to 19 million light years for the galaxy. Put another way, the comet is 1.7 trillion times closer!

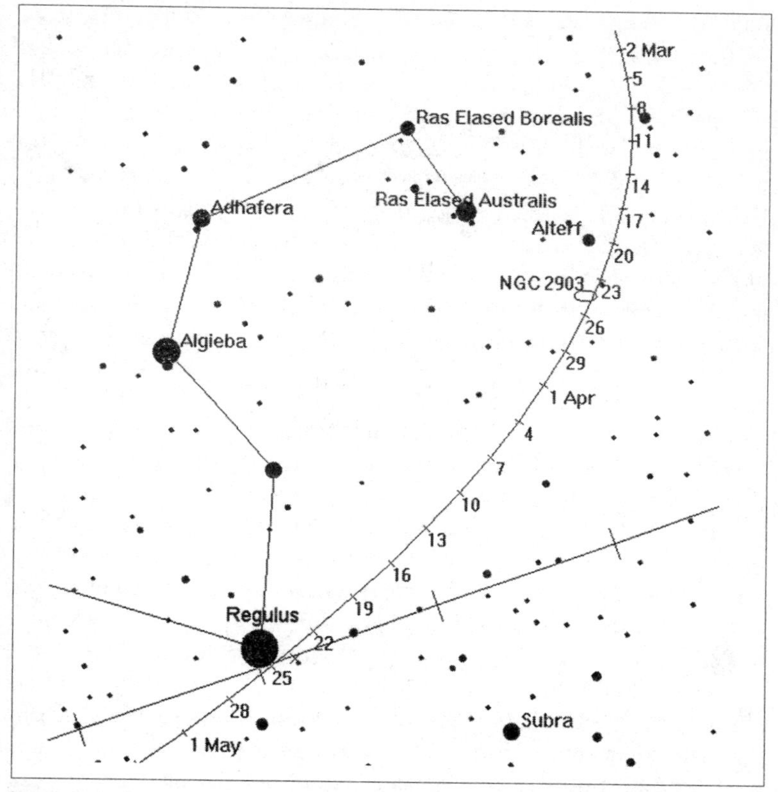

Figure 2. Comet P/2010 V1 (Ikeya-Murakami) tracks south and east near Leo's 'Sickle' from mid-March to early May. On 24/25 March, it passes the popular galaxy NGC 2903 and on 25 April, it passes under Regulus. The straight line is the ecliptic plane.

During the following weeks, 2013 US10 tracks north through Draco and Camelopardalis. On 22 February, by which time the comet may have faded to magnitude 10, it passes the open cluster NGC 1502 in Camelopardalis and a month later, at around magnitude 13, it passes the open clusters NGC 1528 and 1545 on 22 March and 28 March, respectively.

The early months of 2016 should also see another ninth-magnitude comet well placed in Northern Hemisphere skies, namely P/2010 V1 (Ikeya-Murakami). This comet was discovered visually by those two Japanese amateur astronomers on 3 November 2010 using 25 and 46 cm reflectors. This was Karou Ikeya's seventh visual comet discovery. In the 1960s, competing with his countryman Tsutomu Seki, Ikeya discovered and co-discovered a string of five bright comets and also co-discovered the ultra-long period Ikeya-Zhang, four decades later, in 2002. So, with 2010 V1, Ikeya's visual discoveries span almost half a century. This is the first observed return of P/2010 V1 (Ikeya-Murakami), which orbits the Sun every 5.4 years. Comet Ikeya-Murakami starts the year in the pre-dawn sky on the northern boundary of Leo, where Leo meets Leo Minor on the western side of both constellations. It slowly moves south, almost hitting Leo's western border with Cancer, before curving back south and east. On 10 March, the comet reaches perihelion at a distance of 1.57 AU from the Sun when it is 0.65 AU behind the Earth, so any tail will be rather foreshortened from our viewing angle. Even so, this does mean that the comet will be well placed and high up from UK skies at midnight. On the evenings of 24 March and 25 March, the comet is scheduled to have a very photogenic encounter with the splendid galaxy NGC 2903 in Leo, passing within a quarter of a degree of that wonderful deep-sky target. With both objects being around ninth magnitude it should be an interesting encounter. For the next month the comet heads south and east and on 25 April, now faded to magnitude 10 or so, it passes half a degree below bright Regulus.

Comet 252P/LINEAR is another interesting and potentially visual (in a modest telescope) cometary target for telescope users in the Southern Hemisphere during March. However, it will only be bright for a very short time, as this is a small comet simply passing very close to the Earth. How close? Well, its orbit at the time of writing places it 0.036 AU from us from the night of 20/21 March to 21/22 March. Sadly, for UK observers, its sky location around –70° in declination, means this Earth fly-by will be totally unobservable from Northern Hemisphere locations. But if you live in the far south, such as Australia or New Zealand, watching this tenth-magnitude comet whizz through the constellations of Apus, Triangulum Australe and Ara at around twenty-four arc seconds per minute, could be fascinating. This will be the fifth-closest cometary encounter of all time, as shown in the table below and, at its closest, the comet will pass within 5.4 million kilometres of Earth (3.4 million miles), which is roughly fourteen times the average distance of the Moon from the Earth. The comet moves north after its close approach but, by the time it is well placed from UK latitudes, it will probably have faded below magnitude 13.

Comet	Distance (AU)	Date Closest
D/1770 L1 (Lexell)	0.0151	1770 July 1.7
55P/1366 U1 (Tempel-Tuttle)	0.0229	1366 Oct 26.4
C/1983 H1 (IRAS-Araki-Alcock)	0.0312	1983 May 11.5
1P/837 F1 (Halley)	0.0334	837 Apr 10.5
252P/LINEAR	0.0360	2016 Mar 21.7
3D/1805 V1 (Biela)	0.0366	1805 Dec 9.9
C/1743 C1	0.0390	1743 Feb 8.9
7P/Pons-Winnecke	0.0394	1927 June 26.8
C/1702 H1	0.0437	1702 Apr 20.2
45P/Honda-Mrkos-Pajdusakova	0.0601	2011 Aug 15.34

The nine undisputed closest cometary approaches in history, along with this year's approach of 252P/LINEAR. There is some evidence that C/1491 B1 came as close as 0.0094 AU on 20 February 1491, and also that C/1014 C1 came as close as 0.0376 AU on 25 February 1014, which, if true, would place those comets in first and eighth place in this table.

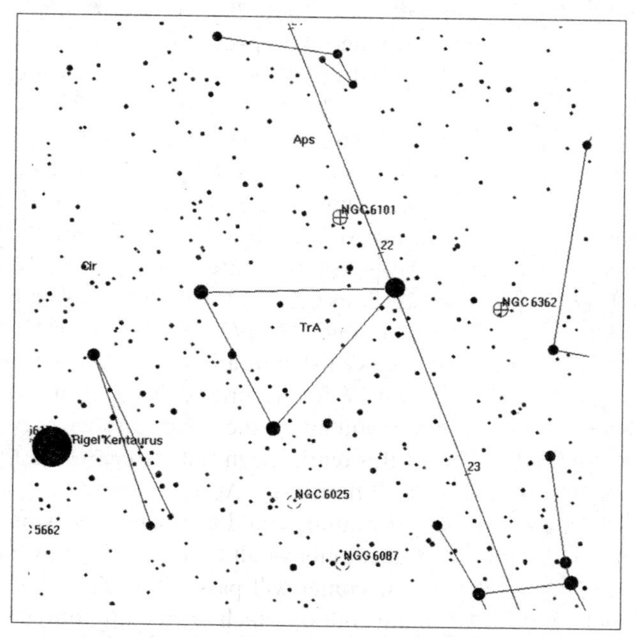

Figure 3. For Southern Hemisphere observers comet 252P/LINEAR passes rapidly through the constellations of Apus, Triangulum Australe and Ara from 21/22 March to 22/23 March. South is up in this chart.

Comet 2013 X1 (PanSTARRS) is a fairly active comet which will be within amateur CCD range for much of 2016. It is a long-period comet, discovered three years ago, which finally reaches perihelion on 20 April and should peak at magnitude 10 or 11 in May. Unfortunately, despite being fairly active, it does not approach the Sun very closely, which is counter-intuitive as its angular elongation from the Sun is quite small for much of the year. In fact, the perihelion distance of 2013 X1 is a mere 1.3 AU and it is a dismal 2.0 AU from the Earth at that time. The closest approach to the Earth is on 21/22 June at 0.64 AU, some two months after perihelion, by which time the comet is definitely a Southern Hemisphere object. It should maintain a magnitude of 11 or so throughout June as it moves from Aquarius, through Piscis Austrinus, Microscopium, Sagittarius and Telescopium. Early in the month of June, around the fourth, Southern Hemisphere viewers will be able to

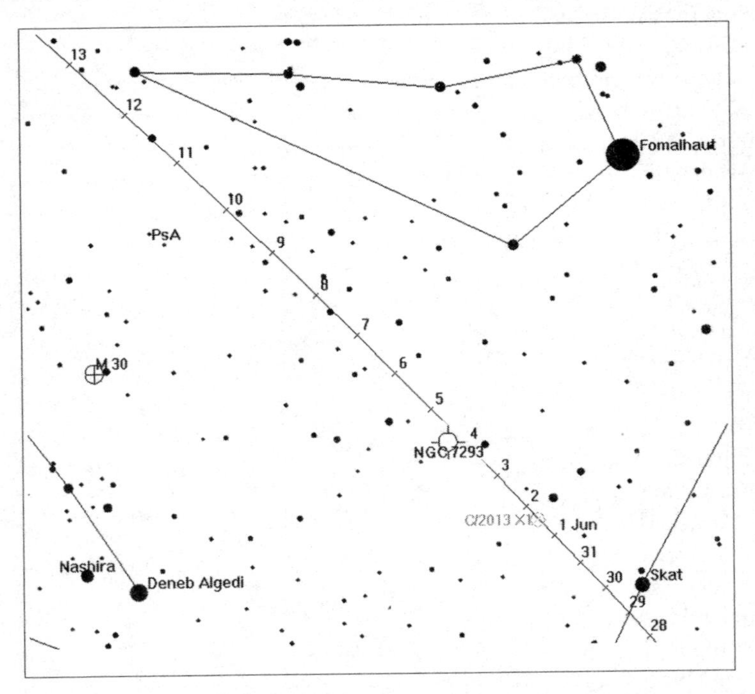

Figure 4. From late May to mid-June comet 2013 X1 (PanSTARRS) can be found near to the bright star Fomalhaut in Piscis Austrinus. On the fourth, it is close to the Helix Nebula NGC 7293. South is up in this chart.

image 2013 X1 passing from north-east to south-west of the famous Helix Nebula, NGC 7293, in Aquarius, approaching within a quarter of a degree of the outer nebulosity.

Another comet that might attract interest from CCD imagers, even though it will not be bright, is Comet Spacewatch (2011 KP36). This object was originally given an asteroidal designation but its remarkable 238-year orbit is too long for it to be considered a standard short-period comet and too short for it to be regarded as a long-period comet. It may reach magnitude 14 from August to October, but at perihelion (26 May) it is a huge 4.9 AU from the Sun and a similar distance from the Earth. Clearly this is a big, but distant comet almost as far from the Sun as Jupiter. For the whole year the comet stays close to the Pisces/Cetus border region and in late September it is close to opposition, so it transits around midnight and is closest to Earth (4.0 AU) just below the ecliptic.

Comet 29P/Schwassmann-Wachmann is an object that entertains amateur astronomers every year with its regular outbursts in brightness. The most impressive displays see the comet rise to eleventh magnitude. While this might not seem particularly remarkable, one has to consider that this is a comet that never comes closer to the Sun than 5.8 AU. If the comet were placed at the standard absolute-magnitude distance of 1 AU from Sun and Earth, with the standard 5 and 10 log-brightness rules being applied, its greatest outbursts might end up being zero magnitude! This is an extraordinary object, unique in our Solar System and only tamed by its huge distance from the Sun. Unfortunately for UK observers the comet has been deep in the southern sky for a while now and 2016 is no different. It will be best placed in July, crawling westwards through the increasingly dense star fields of eastern Sagittarius, sometimes invisible visually, even in large amateur telescopes, and sometimes very obvious as a faint nebulous smudge.

Comet 81P/Wild returns to the inner Solar System every 6.4 years and arrives at perihelion on 20 July this year. Unfortunately, this return is a very unfavourable one with the comet being on the far side of the Sun, with respect to the Earth, as it brightens to magnitude 12.

Comet 9P/Tempel is one of the few comets to have been visited and imaged by space probes. A projectile from the Deep Impact probe struck its surface on 4 July 2005 and the comet was revisited by the Stardust spacecraft on 15 February 2011. This was therefore the first time that a comet had been visited twice. The comet reaches perihelion

(1.54 AU) on 2 August but should maintain a brightness of magnitude 11 from late June to mid-August as it is moving away from Earth as it approaches perihelion. Unfortunately, UK observers will lose the comet in the summer twilight of early June (the problem becoming severe for those in the far north by late May) as the Leo/Virgo border region where it is located sinks into the evening twilight murk. Therefore, May could be the best month to find this comet at eleventh magnitude in eastern Leo, not far from Denebola. Southern Hemisphere comet observers fare better after perihelion when 9P/Tempel should be an eleventh-magnitude object passing from Virgo to Libra by the seventeenth of the month. There is an interesting passage of the comet 9P Tempel through the galaxy fields of NGC 3801, 3802, 3803, 3806 and 3790 on the evening of 6 May. Minor planet 6 Hebe is just a quarter of a degree below the comet on this date. On the next evening, 7 May, 6 Hebe is just five arc minutes south-east of the comet's coma, and

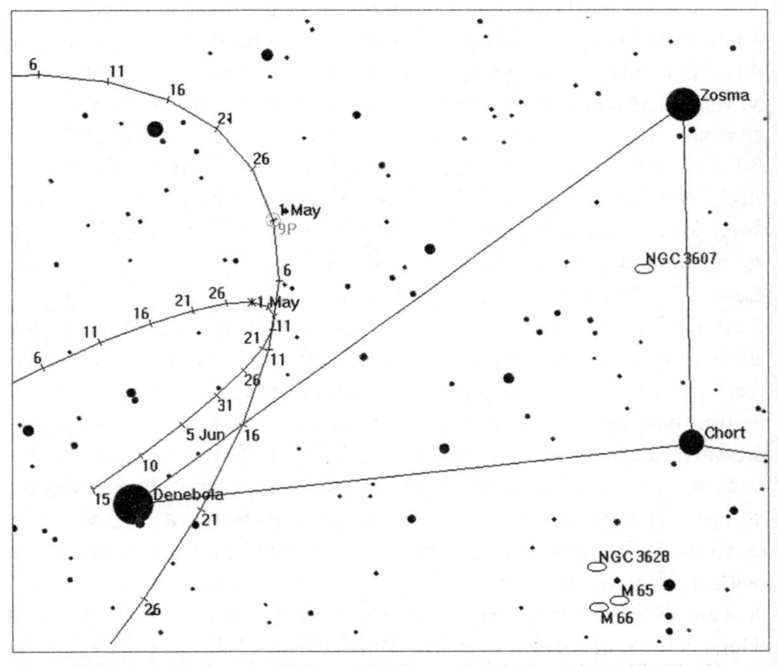

Figure 5. The tracks of comet 9P/Tempel (wide curve) and minor planet (6) Hebe (narrow ellipse) intersect around 6 May and 7 May near to the bright star Denebola in eastern Leo. See chart in the Minor Planets section for a close-up view.

possibly immersed in its tail! Both comet and asteroid should be around eleventh and tenth magnitude respectively on these dates. Despite both objects being in the inner Solar System, they are, in reality, nowhere near each other. At 152 million kilometres from the Earth, 9P/Tempel is roughly halfway between us and 6 Hebe, which is some 338 million kilometres distant.

Comet 43P/Wolf-Harrington is another regular visitor to the inner Solar System, with a period of 6.13 years. The comet reaches perihelion (1.36 AU) on 19 August, but unfortunately its small, angular elongation from the Sun makes it a tricky object, even after mid-September, in the pre-dawn Northern Hemisphere twilight, when it will struggle to reach eleventh magnitude in south-eastern Cancer. By the time its elongation becomes healthy it will be heading south and fading rapidly in Sextans.

144P/Kushida was discovered by the Japanese amateur astronomer Yoshio Kushida at the Yatsugatake South Base Observatory in Japan in January 1994. He is also the co-discoverer of another periodic comet, 147P/Kushida-Muramatsu. 144P/Kushida has an orbital period of 7.57 years and so this return is its third since discovery. The comet reaches perihelion (1.43 AU) on 30 August and its track through the pre-dawn Northern Hemisphere twilight is disappointingly similar to that of 43P/Wolf-Harrington, with the two comets barely escaping the twilight, some 5 degrees apart in the sky! In fact, 144P is likely to be purely a CCD target as it moves, at around magnitude 13, through southern Leo, in late September and early October.

226P/Pigott-LINEAR-Kowalski is a comet with an unusual name and an unusual history, but whether or not it performs this autumn is hard to predict. It was originally discovered back in 1783, on 19 November, two years after the discovery of Uranus by Herschel, by Edward Pigott of York, England. So, this is that very rare phenomenon, a comet originally discovered by an Englishman and still observable today. Comet Pigott was in the constellation of Cetus at the time of its 1783 discovery and Edward Pigott described it as looking like a nebula, with a diameter of about two to three arc minutes. Numerous well-known astronomers of that era observed it too, including William Herschel, Charles Messier, Pierre Méchain and John Goodricke. After 21 December 1783, the comet was lost, despite being thought to have a short-period orbit, for a staggering 220 years, until the LINEAR system detected a new comet in January 2003, which seemed to tie in with the

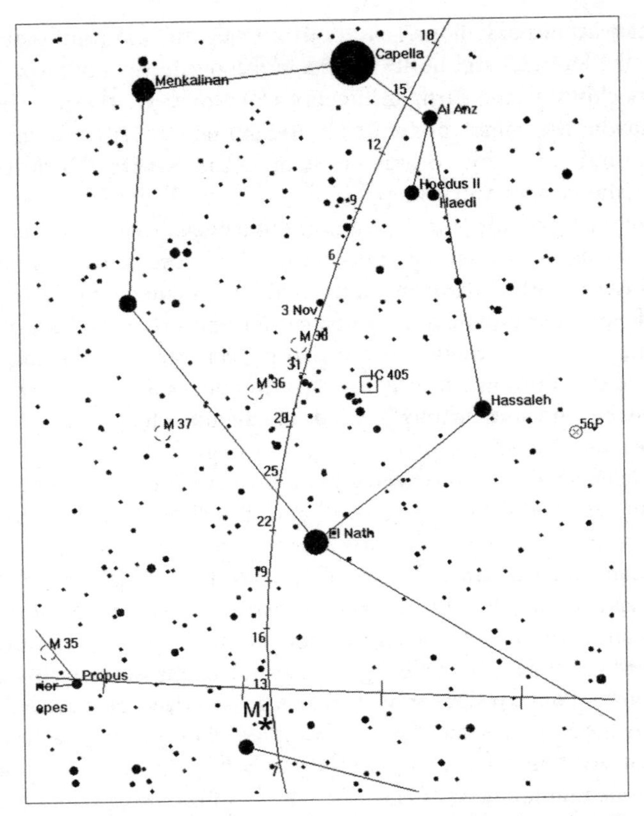

Figure 6. Comet 226P/Pigott-LINEAR-Kowalski heads north through northern Taurus and then Auriga, from early October to mid-November. On 9 October it will be close to the supernova remnant M1 (base of image) and in mid-November it passes close to the brilliant star Capella. The straight line near the base is the ecliptic plane.

original object. The story did not end there though, because it passed very close to Jupiter in 2006 (0.056 AU), which altered the orbit considerably. Rich Kowalski of the Catalina Sky Survey was the first to find the comet on its 2009 return, hence the complex triple-barrelled name. This return looks like being quite a favourable one, although opinions seem to vary about how good. The most pessimistic predictions have 226P reaching magnitude 15 this autumn, but the most optimistic see it reaching magnitude 10. Given the comet's disappearance for 220 years it is clearly unpredictable, so anything might happen. In 2016, the comet reaches perihelion (1.78 AU) on 5 September, but the distance from the Sun will still be 1.9 AU in mid-November, at which time it will be closest to the Earth (0.99 AU on 14 November). On the date of perihelion, the comet, as seen from the latitude of London, will be in Orion and some four degrees west of Bellatrix, roughly 20 degrees above the eastern horizon by 2 a.m. on 5 September. Things keep improving from that point on and the comet will be well placed at midnight from UK skies and at a decent northerly declination from mid-October onwards. When closest to Earth, on 14 November, the comet will have a declination of +44 degrees and be situated south-west of Capella, roughly midway between that brilliant star and the enigmatic variable star Epsilon Aurigae. Perhaps the most interesting encounter will be the comet's passage past the Crab Nebula, Messier 1, on 9 October, when it will be just ten arc minutes west of the supernova remnant. A lot depends on how bright this long-lost comet decides to be at this return.

The final comet worthy of an individual mention, 45P/Honda-Mrkos-Pajdusakova, may be one of the best comets at the start of next year, although it will only be visible from Southern Hemisphere skies until then. It was first discovered by Minoru Honda, from Japan, on 3 December 1948 and then by the Czech observers Mrkos and Pajdusakova on 7 December and 6 December, respectively (Pajdusakova was not sure she had seen a new comet until her colleague Mrkos also spotted it the next day). Comet 45P is an interesting object; it has a small perihelion distance of 0.53 AU, achieved on the last day of the year, but it can also approach the Earth very closely. I previously mentioned comet 252P/LINEAR and its close Earth approach this year of 0.036 AU, the fifth-closest Earth approach in history, but in that same table you will see that 45P's 2011 approach of 0.0601 AU comes in at tenth place. 45P returns every 5.26 years, a shift in time of 3 months each return, with respect to the Earth's orbital position, so now and again very close Earth encounters are experienced where the orbit of Earth and comet almost cross. This time around the closest point will be next year, on 11 February 2017, when Earth and comet are just 0.08 AU apart, some 42 days after perihelion. Therefore, for the first six weeks of next year, the comet should maintain a magnitude of 7 or so as the fade caused by moving away from the Sun is counteracted by the comet moving closer to the Earth. However, that excitement is for next

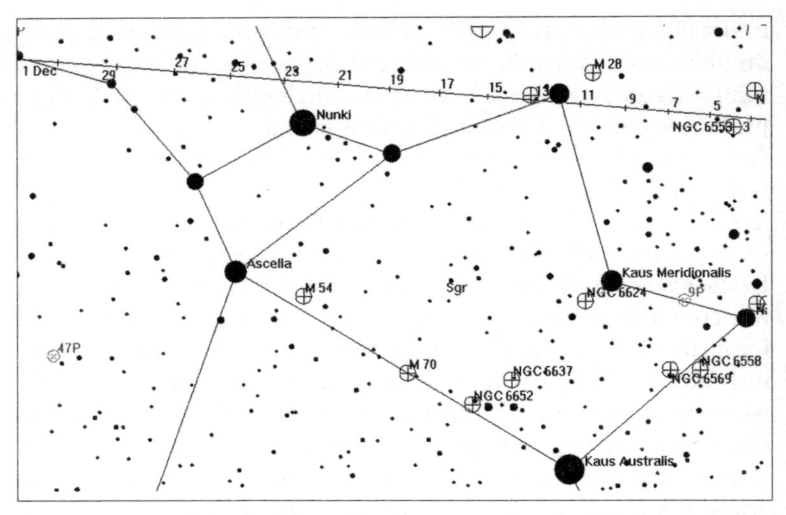

Figure 7. Comet 45P/Honda-Mrkos-Pajdusakova passes above the 'lid' of the Sagittarius 'teapot' during mid-November. During the month it will probably rise from magnitude 16 to magnitude 12 and then brighten explosively in December as it sinks into twilight.

year, but what happens this year? Well, CCD imagers in the Southern Hemisphere should be able to image 45P above the Sagittarius 'teapot' lid in November as it brightens rapidly from around magnitude 16 to 12 during the month.

In the first few weeks of December the same observers, if they have a good horizon, should be able to witness the pre-perihelion approach, waiting until the sky is just dark enough to see 45P moving from eastern Sagittarius to Capricornus on 14 December and heading into central Capricornus by the end of the year. The angular elongation from the Sun shrinks from 40 to 32 degrees during the month and it will be too close to the horizon in bright twilight to be observed in the last week of December. However, those Southern Hemisphere observers who persevere may catch an 8th-magnitude comet 45P around 20 December, some 14 degrees above the evening twilight horizon with the Sun 14 degrees down.

Of course, it goes without saying that the phase of the Moon and its proximity to the field where any comet lies has a huge bearing on the object's visibility. Even when a comet is high up at midnight, a Full Moon ten or twenty degrees away will trash any prospects of observing it visually, so potential comet spotters need to be very aware of the position and brightness of the Moon when planning any comet observing. CCDs are much more tolerant of a bright sky, though, and image processing can work wonders in compensating.

THE DEFUNCT COMETS

By their very nature comets are highly unpredictable performers. Their nuclei come in a variety of shapes and sizes, are covered with patches of icy material on the surface or under the surface, and they spin at various rates too. The smallest ones are highly prone to disintegration when they come close to the Sun and others can have their orbit altered drastically, now and again, by the giant planet Jupiter. Every apparition is different from the Earth-based observer's viewpoint. The comet may arrive at perihelion with the Earth poorly placed, such as when the comet is on the opposite side of the Sun, or, conversely, the comet may pass within tens of millions of kilometres of our home planet. In addition, short-period comets lose material every time they reach perihelion and, after thousands of orbits, a comet may simply have no more material to discharge. Some comets slowly fragment on each passage with more components being detected on every return. It is therefore hardly surprising that some comets are simply lost and declared 'Defunct', implying they have simply disappeared or broken up and will never be seen again. However, with deep CCD surveys of the night sky becoming ever more powerful, some of these formerly missing comets are occasionally recovered and modern computing power often enables an object with a slightly different orbit (often altered by the influence of giant Jupiter) to be revealed as an old comet, not seen for many decades. In 2016, three of these D/ category comets are predicted to return to perihelion if they still exist. In all probability they will not all be recovered, but a recovery of one or two of them cannot be ruled out. The 2016 D/ comets are D/Denning (1894 F1), D/Schorr (1918 W1) and D/Haneda-Campos (1978 R1). D/Harrington-Wilson (1952 B1) may be recovered as well, although it is more likely to reach perihelion next year. If the orbits of these comets are unaltered from their last appearance they would reach perihelion on 6 March, 6 April, 11 November and 27 February 2017, respectively. Their orbital periods when last seen were 8.3, 8.5, 6.4 and 5.5 years. In addition, if they are as active as when they were last observed, they could either be too faint for amateur observation, or as bright as eighth magnitude in the case of D/Harrington-Wilson at the end of this year. A brief history of these long-lost comets may be of interest.

W. F. Denning of Bristol, England, was sweeping for comets when he found 1894 F1 on 26 March 1894. The comet was then more than six weeks past its closest approach to both the Earth and the Sun. Various orbits of 6.8- to 7.9-year period were computed and some thought it was actually another lost comet, namely 5D/Brorsen, returning. However, it has not been seen again for the past 120 years.

While attempting to photograph the minor planet 232 Russia, the astronomer R. R. E. Schorr, of Hamburg Observatory, discovered his comet photographically with a one-metre reflector on 23 November 1918. It had passed perihelion two months earlier and had also passed closest to Earth a week before Schorr's discovery. In 1982, Brian Marsden and others re-measured all of the 1918 plates and came up with a period of 6.67 years, but it has still evaded detection for almost 100 years.

The other two comets were discovered in more recent times and are arguably the most likely to be seen again in 2016.

Toshio Haneda, of Fukushima, Japan, discovered 1978 R1 with an 8.5-cm refractor on 1 September 1978, as a magnitude-10 object. A few hours later, José da Silva Campos, in Durban, South Africa, independently discovered the same comet which he described as diffuse, with a condensation, and of ninth magnitude. Brian Marsden calculated a 5.97-year orbit for this comet in January 1979, with a perihelion distance of 1.1 AU, but the comet did not reappear at the 1984/85 or 1991 predicted times and positions, even allowing for an increased perihelion distance (1.2 AU) and period (6.3 years) computed by Nakano in Japan. It has now eluded detection for thirty-seven years. The comet and Earth orbits cross with a minimum separation of 0.14 AU based on the original 1978 data.

Harrington and Wilson discovered 1952 B1 using the 122-cm Schmidt telescope during the course of the Palomar Observatory Sky Survey on 30 January 1952. They estimated the magnitude as fifteen, and described the comet as diffuse, with a tail less than one degree long. The discoverers confirmed the comet on a plate exposed on 4 February. One year after discovery, L. Kresák pointed out that this comet could be the lost periodic comet Taylor of 1916. Some twenty-seven years after its discovery, Brian Marsden calculated a 6.36-year orbit for the comet, similar to other calculations by Nakano in 2002, which gave 6.35 years. However, the comet is still lost, although if it did come back to perihelion in February 2017, it could theoretically reach eighth magnitude this December. As for all these defunct comets we will just have to wait and see if one of the big patrols, like PanSTARRS, picks them up.

EPHEMERIDES

The tables below list the right ascensions and declinations of, arguably, the nine most promising comets in 2016, at their peak, as well as the distances, in AU, from the Earth and the Sun. The elongation, in degrees from the Sun, is also tabulated along with the estimated visual magnitude, which can only ever be a rough guess as comets are, without doubt, a law unto themselves!

C/2013 US10 (Catalina)

Date	RA (2000)			Dec.			Distance from Earth	Distance from Sun	Elongation from Sun	Mag.
	h	m	s	°	'	"	AU	AU	°	
Jan 01	14	13	58.7	+18	22	22	0.898	1.180	77.5	6.9
Jan 11	14	04	23.0	+38	57	35	0.754	1.307	96.7	6.9
Jan 21	13	26	32.1	+64	41	36	0.736	1.438	112.7	7.3
Jan 31	09	02	29.4	+81	37	10	0.866	1.572	116.0	8.1
Feb 10	04	49	25.1	+72	52	43	1.098	1.706	109.7	8.9
Feb 20	04	14	53.1	+64	06	41	1.380	1.839	100.6	9.7
Mar 01	04	08	45.4	+58	20	34	1.683	1.972	91.3	10.5

P/2010 V1 (Ikeya-Murakami)

Date	RA (2000)			Dec.			Distance from Earth	Distance from Sun	Elong-ation from Sun	Mag.
	h	m	s	°	'	"	AU	AU	°	
Jan 21	09	49	50.8	+33	01	19	0.698	1.641	154.2	9.4
Jan 31	09	46	16.9	+32	50	10	0.654	1.616	160.4	9.2
Feb 10	09	40	00.9	+32	05	27	0.628	1.597	162.5	9.0
Feb 20	09	33	12.6	+30	38	26	0.621	1.583	158.8	9.0
Mar 01	09	28	11.3	+28	30	21	0.631	1.575	151.8	9.0
Mar 11	09	26	35.2	+25	50	15	0.657	1.573	143.9	9.1
Mar 21	09	29	10.8	+22	50	08	0.698	1.576	136.2	9.2
Mar 31	09	35	47.5	+19	40	59	0.752	1.585	129.2	9.4

252P/LINEAR

Date	RA (2000)			Dec.			Distance from Earth	Distance from Sun	Elong-ation from Sun	Mag.
	h	m	s	°	'	"	AU	AU	°	
Mar 02	05	46	01.9	-32	57	15	0.121	1.014	97.2	13.0
Mar 07	05	46	51.3	-38	25	27	0.094	1.003	93.7	12.4
Mar 12	05	50	05.0	-48	13	06	0.068	0.997	91.0	11.6
Mar 17	06	06	32.7	-69	00	28	0.045	0.996	90.4	10.8
Mar 22	17	09	29.6	-68	17	05	0.036	1.001	95.7	10.3
Mar 27	17	30	43.1	-29	55	13	0.049	1.010	102.8	11.0
Apr 01	17	33	54.2	-12	02	52	0.073	1.024	107.6	11.9
Apr 06	17	34	04.8	-03	24	47	0.100	1.042	111.8	12.7
Apr 11	17	32	42.5	+01	27	22	0.128	1.064	116.0	13.3

C/2013 X1 (PanSTARRS)

Date	RA (2000)			Dec.			Distance from Earth	Distance from Sun	Elongation from Sun	Mag.
	h	m	s	°	'	"	AU	AU	°	
May 01	23	28	06.2	-01	58	59	1.745	1.323	49.0	12.3
May 11	23	21	43.9	-04	55	10	1.503	1.349	61.2	12.0
May 21	23	10	07.9	-09	18	22	1.244	1.390	75.3	11.7
May 31	22	47	55.3	-16	21	42	0.984	1.444	92.6	11.3
Jun 10	22	01	01.2	-28	04	57	0.760	1.510	115.9	10.8
Jun 20	20	14	09.0	-43	33	18	0.642	1.586	145.1	10.5
Jun 30	17	25	32.3	-48	52	26	0.707	1.670	150.7	10.9
Jul 10	15	37	38.0	-42	56	16	0.922	1.760	130.2	11.6
Jul 20	14	51	19.4	-37	06	06	1.209	1.854	112.6	12.3

C/2011 KP36 (Spacewatch)

Date	RA (2000)			Dec.			Distance from Earth	Distance from Sun	Elongation from Sun	Mag.
	h	m	s	°	'	"	AU	AU	°	
Jul 30	00	50	22.6	+02	01	43	4.395	4.906	114.6	14.6
Aug 09	00	51	53.2	+01	35	26	4.274	4.913	124.0	14.6
Aug 19	00	52	15.7	+01	00	56	4.168	4.921	133.7	14.5
Aug 29	00	51	33.6	+00	19	13	4.082	4.931	143.6	14.5
Sep 08	00	49	53.7	-00	28	12	4.018	4.942	153.7	14.5
Sep 18	00	47	28.3	-01	19	05	3.980	4.953	163.8	14.4
Sep 28	00	44	33.1	-02	10	49	3.971	4.966	172.3	14.5
Oct 08	00	41	26.6	-03	00	35	3.991	4.979	170.5	14.5
Oct 18	00	38	29.3	-03	45	34	4.041	4.994	161.1	14.5
Oct 28	00	35	59.5	-04	23	29	4.119	5.010	150.8	14.6
Nov 07	00	34	13.6	-04	52	39	4.222	5.026	140.5	14.6

29P/Schwassmann-Wachmann

(NB magnitude 15 when not in outburst)

Date	RA (2000)			Dec.			Distance from Earth	Distance from Sun	Elongation from Sun	Mag.
	h	m	s	°	'	"	AU	AU	°	
Jun 01	19	53	21.6	-25	21	05	5.172	5.933	135.1	15.3
Jun 11	19	50	26.3	-25	26	40	5.068	5.930	145.2	15.3
Jun 21	19	46	34.9	-25	33	00	4.987	5.927	155.5	15.2
Jul 01	19	42	00.4	-25	39	09	4.934	5.925	165.8	15.2
Jul 11	19	36	59.3	-25	44	13	4.908	5.922	175.0	15.2
Jul 21	19	31	51.5	-25	47	24	4.913	5.919	171.2	15.2
Jul 31	19	26	56.6	-25	48	09	4.947	5.916	161.1	15.2
Aug 10	19	22	33.8	-25	46	11	5.008	5.914	150.8	15.2
Aug 20	19	18	59.6	-25	41	30	5.096	5.911	140.5	15.3
Aug 30	19	16	25.9	-25	34	16	5.205	5.908	130.3	15.3

9P/Tempel

Date	RA (2000)			Dec.			Distance from Earth	Distance from Sun	Elongation from Sun	Mag.
	h	m	s	°	'	"	AU	AU	°	
May 01	11	40	28.4	+18	46	14	1.001	1.786	125.6	11.8
May 11	11	40	43.7	+16	50	18	1.028	1.742	117.4	11.6
May 21	11	44	55.3	+14	26	02	1.062	1.700	110.1	11.4
May 31	11	52	44.1	+11	39	02	1.099	1.663	103.7	11.2
Jun 10	12	03	47.5	+08	33	35	1.141	1.629	98.0	11.1
Jun 20	12	17	41.9	+05	13	33	1.184	1.601	93.1	11.0
Jun 30	12	34	04.7	+01	42	37	1.229	1.578	88.7	10.9
Jul 10	12	52	40.9	-01	55	47	1.277	1.560	84.9	10.9
Jul 20	13	13	17.9	-05	37	40	1.327	1.548	81.6	10.9
Jul 30	13	35	45.5	-09	18	41	1.381	1.543	78.5	10.9

226P/Pigott-LINEAR-Kowalski

Date	RA (2000)			Dec.			Distance from Earth	Distance from Sun	Elong-ation from Sun	Mag.
	h	m	s	°	'	"	AU	AU	°	
Sep 21	05	21	31.7	+13	15	55	1.346	1.783	97.6	12.7
Oct 01	05	29	05.0	+17	36	07	1.246	1.793	105.3	12.5
Oct 11	05	33	18.4	+22	36	57	1.154	1.808	114.0	12.4
Oct 21	05	33	11.2	+28	21	34	1.078	1.827	123.6	12.3
Oct 31	05	27	23.3	+34	43	14	1.022	1.851	133.5	12.2
Nov 10	05	14	31.8	+41	18	32	0.993	1.879	142.6	12.2
Nov 20	04	53	55.6	+47	27	35	0.996	1.911	148.7	12.3
Nov 30	04	26	40.7	+52	26	52	1.032	1.946	149.2	12.5
Dec 10	03	56	48.5	+55	51	02	1.098	1.984	144.5	12.7

45P/Honda-Mrkos-Pajdusakova

Date	RA (2000)			Dec.			Distance from Earth	Distance from Sun	Elong-ation from Sun	Mag.
	h	m	s	°	'	"	AU	AU	°	
Nov 16	18	38	15.9	-25	32	02	1.452	1.025	44.9	14.5
Nov 21	18	51	39.7	-25	21	24	1.401	0.953	42.8	13.8
Nov 26	19	06	00.0	-25	04	53	1.342	0.882	41.1	13.0
Dec 01	19	21	15.8	-24	41	20	1.275	0.811	39.5	12.2
Dec 06	19	37	22.6	-24	09	31	1.201	0.742	38.1	11.3
Dec 11	19	54	09.9	-23	28	16	1.118	0.678	37.0	10.4
Dec 16	20	11	17.0	-22	36	45	1.027	0.621	35.9	9.4
Dec 21	20	28	07.8	-21	34	57	0.927	0.575	34.9	8.5
Dec 26	20	43	46.1	-20	24	09	0.822	0.544	33.6	7.8
Dec 31	20	56	57.9	-19	07	00	0.713	0.533	31.8	7.3

Minor Planets in 2016

MARTIN MOBBERLEY

Some 700,000 minor planets (also known as asteroids) are known. They range in size from small planetoids hundreds of kilometres in diameter to boulders tens of metres across. More than 450,000 of these now have such good orbits that they possess a numbered designation and almost 20,000 have been named after mythological gods, famous people, scientists, astronomers and institutions. Most of these objects live between Mars and Jupiter but some 12,000 have been discovered between the Sun and Mars and more than 1,500 of these are classed as potentially hazardous asteroids (PHAs) due to their ability to pass within 7.5 million kilometres of the Earth while also having a diameter greater than roughly 200 metres. The first four asteroids to be discovered were (1) Ceres, now regarded as a dwarf planet, (2) Pallas, (3) Juno and (4) Vesta, which are all easy binocular objects when at their peak, due to having diameters of hundreds of kilometres.

In 2016, most of the first ten numbered asteroids reach opposition during the year. However, with orbital periods of, typically, four or five years, it is often the case that the Earth cannot 'overtake' an asteroid on the inside track, so to speak, in a specific twelve-month period. This is the case when opposition has occurred late in the previous year. In these instances the asteroid will not be transiting the meridian at midnight during any month in 2016 and so will not be seen at its brightest and closest during the year. So, this year the asteroid (4) Vesta does not reach opposition until January 2017 and (9) Metis does not reach opposition until February 2017. Despite this, both these minor planets are observable by the end of 2016 at a northerly declination.

As far as the remaining top ten asteroids are concerned, well, (5) Astraea is the first to reach opposition, transiting at midnight on 19/20 February. On 30/31 January it passes south-west of brilliant Regulus in Leo, travelling within a degree of that star. It then heads north and west through south-western Leo, peaking at around magnitude 8.9 in late

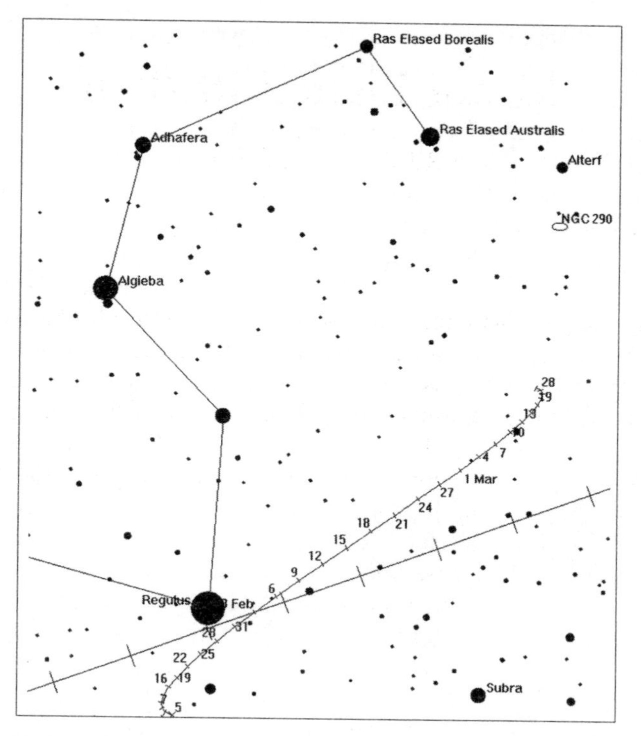

Figure 1. Minor planet (5) Astraea moves past Regulus on 30/31 January, heading north and west to the west of Leo's sickle in February and March. The straight line is the ecliptic plane.

February and travelling at around a quarter of a degree per day against the background stars.

Minor planet (10) Hygiea is the next of the first ten numbered asteroids to reach opposition, transiting the meridian around midnight, at a magnitude of 9.5, on 15/16 March, very close to the border between southern Virgo and southern Leo and about four degrees east and ten degrees south of the giant planet Jupiter. Throughout March, Hygiea moves slowly west and north through the relatively barren star fields of that Virgo/Leo border region and should be relatively easy to spot with a small telescope as it moves roughly one degree every five days.

Almost twenty degrees further to the north-east during the same months, minor planet (6) Hebe finds itself immersed in the galaxy-packed region of northern Virgo and southern Coma during March. It reaches opposition, at a magnitude of 9.8, on the night of 23/24 March, while in the south-western corner of Coma, heading for eastern Leo. On the night of 19/20 March Hebe is almost five degrees due east of the brilliant star Denebola at the eastern end of Leo. In the weeks prior to opposition, at around magnitude 10, Hebe passes close to a number of bright galaxies in Virgo and Coma, running the risk of being mistaken for a supernova. On the night of 2/3 March it skims the western edge of NGC 4313 and between 10 and 11 March it just misses the eastern tip of the galaxy NGC 4216. On the evening of 9 April Hebe passes within arc minutes of the northern edge of galaxy NGC 3934 and its partner to the south, NGC 3933. If we look ahead to six weeks past opposition, as I mentioned in the comet section (see Figure 5 in the 'Comets in 2016'

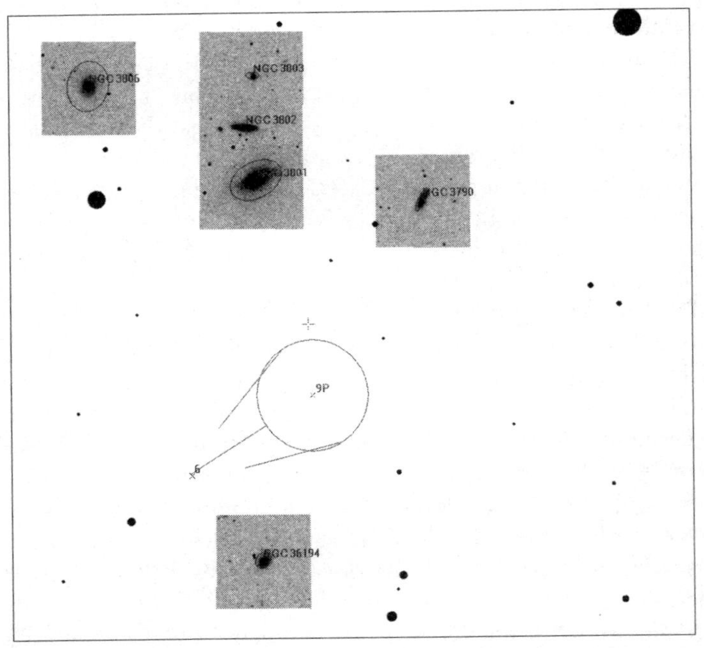

Figure 2. Minor planet (6) Hebe is just five arc minutes from the centre of comet 9P/Tempel's core on 7 May. The minor planet actually sits on the comet's tail (marked as X 6) in this simulation which shows a very narrow half-degree field. The galaxies NGC 3801, 2, 3, 6, NGC 3790 and PGC 36194 are also indicated. See also the wider field diagram (Fig. 5) in the previous Comets section.

chapter, and it is worth repeating here), there is an interesting passage of the comet 9P/Tempel through the galaxy field of NGC 3801, 3802, 3803, 3806 and 3790 on the evening of 6 May. Minor planet (6) Hebe is just a quarter of a degree below the comet on this date and, on the next evening, (6) Hebe is just five arc minutes south-east of the comet's coma. Skies permitting, it makes for an interesting photo-opportunity.

At the end of April, in eastern Virgo, near to the constellation's western border with northern Libra, the minor planet (3) Juno reaches opposition. As with all main belt asteroids reaching opposition in the spring months, (3) Juno is tracking slowly north and west, roughly parallel to the ecliptic plane, at around thirty-five arc seconds per hour. At magnitude 10.0 large tripod-mounted binoculars will pick the asteroid up, but most observers would prefer to use a small telescope. On the night of 25/26 April asteroid (3) Juno passes half a degree south of the eleventh-magnitude galaxy NGC 5691.

The minor planet (7) Iris reaches opposition at the end of May, peaking in brightness at around magnitude 9.2 on 28/29 May. However, at a declination of −23 degrees, this will primarily be an object for Southern Hemisphere observers to track down. Unlike the previous minor planets mentioned above, (7) Iris reaches opposition in a densely packed part of the night sky, in the extreme south-western corner of Ophiuchus, where it meets the head of the Scorpion. Indeed, at opposition, (7) Iris can be found just three degrees north of the brilliant red giant star Antares. On 1 June, it crosses the border into Scorpius. The stars are packed so densely here that identifying Iris might be a challenge in itself, although at magnitude 9.2 it is at least a couple of magnitudes brighter than most of the distant Milky Way background stars. On 7 June, a photo opportunity for CCD imagers occurs when (7) Iris passes just ten arc minutes to the north of the globular cluster Messier 80. On 27 June, the minor planet passes one degree north of the magnitude 2.3 star delta Scorpii, also known as Dschubba. Brilliant Mars will be just ten degrees to the west on this date.

Some two weeks after (7) Iris reaches opposition in southern skies, (8) Flora does likewise, although at a declination of −18 degrees, this is a slightly more realistic target for UK asteroid spotters who are happy to search in the murky skies just 20 degrees above the horizon from southern England. This time the constellation in question is, again, Ophiuchus and (8) Flora should peak in magnitude at around 9.4 on

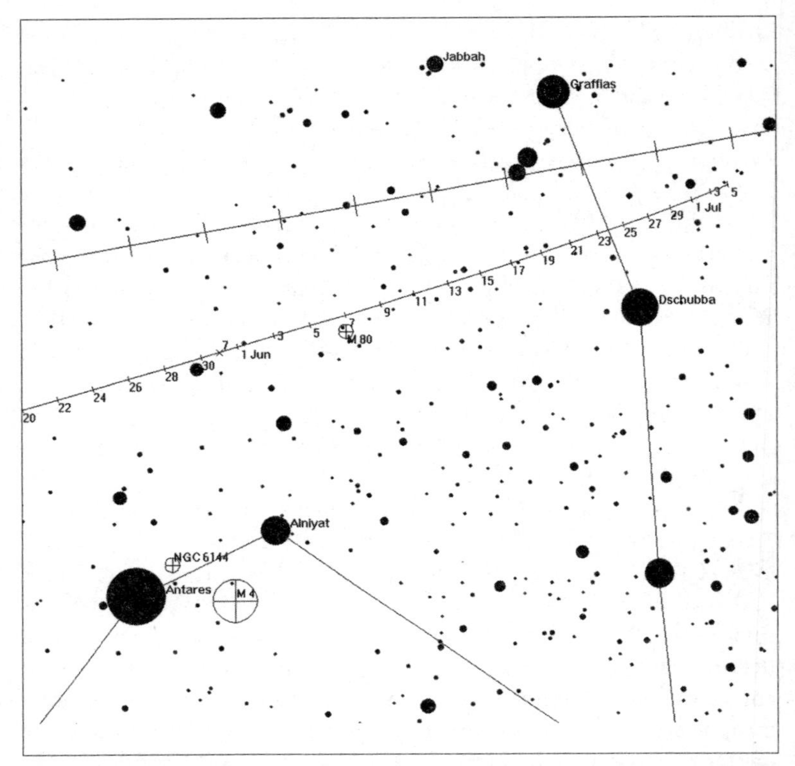

Figure 3. The track of the minor planet (7) Iris through the head of Scorpius from late May to the start of July. On 7 June, the asteroid passes Messier 80. The straight line at the top of the diagram is the ecliptic plane.

12 June. The next day it passes just half a degree to the north of the globular cluster Messier 9. By the start of August Flora will have reached the end of its westerly travel and be close to being stationary in western Ophiuchus, close to the Scorpius border. At this point the asteroid will be just three degrees east of Saturn in the southern sky.

In mid-August it will be back to the Northern Hemisphere and familiar UK skies for the opposition of minor planet (2) Pallas, which peaks, at magnitude 9.2, on 14 August. At this time of the year far Northern Hemisphere skies are arguably at their most 'user-friendly' with darkness having returned after the summer twilight months of May to July and with the evenings being relatively warm too. Pallas

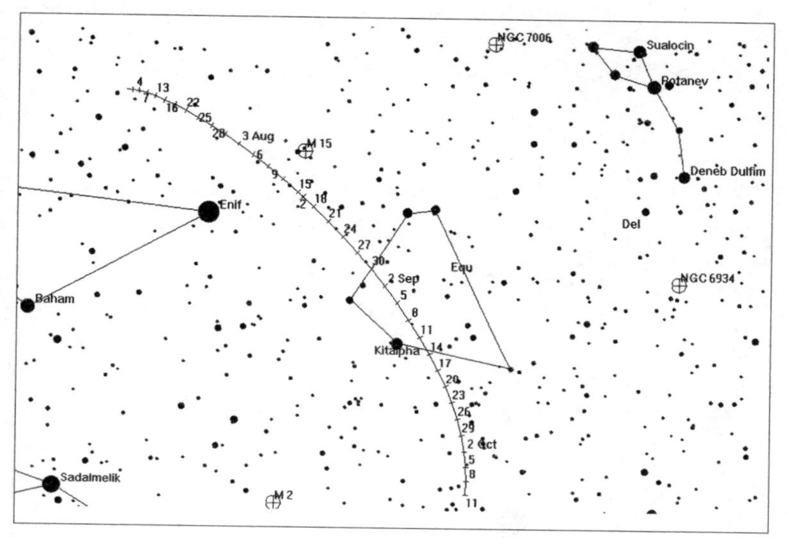

Figure 4. The minor planet (2) Pallas moves out of Pegasus and into Equuleus from July to October. On 16 August it is close to Messier 15.

moves south and west, travelling from south-western Pegasus and entering Equuleus on 21 August. On 16 August, Pallas passes slightly more than a degree and a half to the south of the splendid globular cluster Messier 15.

The last of the top ten numbered asteroids to reach opposition before 2016 ends will be the one now re-designated as a dwarf planet, due to its size, namely (1) Ceres. This former minor planet reaches opposition on the night of 21/22 October and its diameter gives it a very healthy binocular magnitude of 7.5 in mid-October. At this time Ceres will be a degree below the celestial equator, in northern Cetus, just below the Pisces border. It will also be four degrees south of the fourth-magnitude star Alpha Piscium, also known as Alrisha. Three weeks earlier Ceres passed just three degrees north of the well-known variable star Mira, also in Cetus. On the night of 6/7 October this dwarf planet will pass some twenty arc minutes north of the faint galaxies NGC 863 and 859.

As I mentioned at the start of this section, the final two asteroids which are in the top ten of the numbered minor planets, and are also visible at the end of this year, do not reach opposition until 2017. Even

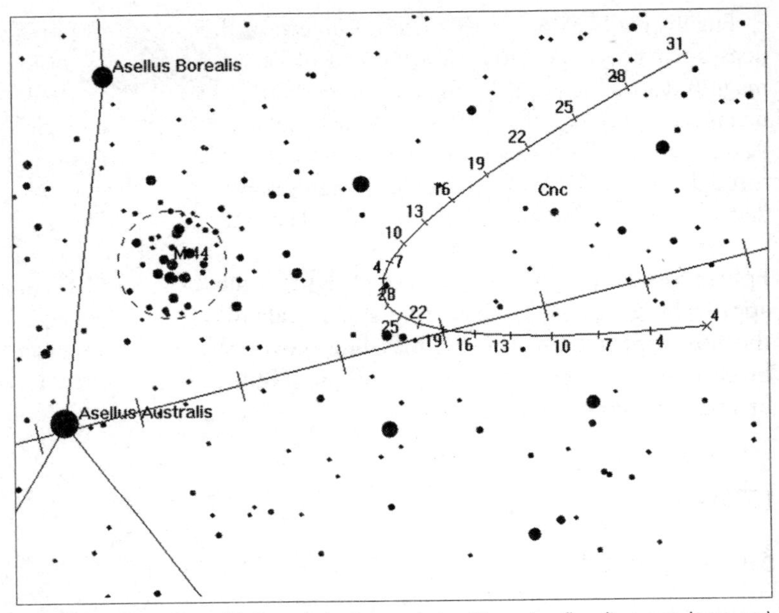

Figure 5. Minor planet (4) Vesta is in Cancer during November (heading east, lower arc) and December (heading west and north, upper arc) and is not far from the famous 'Beehive' cluster Messier 44.

so, (4) Vesta and (9) Metis are sufficiently well placed that they will be worth hunting down, and at magnitude 6.7 by year's end, Vesta is the brightest asteroid in our skies in 2016, despite, at a mean diameter of 525 km, being far smaller than Ceres, which is a far more spherical 950 km-diameter dwarf planet. The main reason Vesta is so much brighter is its albedo: it reflects 42 per cent of the light compared to only 9 per cent reflected by the much darker and larger Ceres.

In December, Vesta can be found in Cancer, travelling slowly west and north, just above the ecliptic plane and just a few degrees west of the famous star cluster known as 'The Beehive' or Messier 44. There are no other stars as bright as Vesta within half a degree of its west-north-westerly track through Cancer in December and so picking it up with big binoculars should not be too hard. Even though it does not reach opposition until mid-January in 2017, Vesta is still forty-five degrees above the horizon from southern UK latitudes by midnight in mid-December.

Finally, (9) Metis is a rather trickier object. In this case the asteroid does not reach opposition until the end of February 2017. It will be magnitude 9.1 then, but by the end of December 2016 it will only be magnitude 10.1. At that time Metis will be more than twenty-five degrees above the southern UK horizon at midnight, travelling east through Leo. On 11 December, at magnitude 10.4, it will be two degrees north of the well-known Messier galaxy M105.

Ephemerides for the best-placed bright minor planets at, or approaching, opposition in 2016, listed in calendar order, appear in the following tables. These tables only cover the period of peak brightness when the minor planets will be visible in large binoculars or small telescopes.

(5) Astraea

Date	RA (2000)			Dec.			Distance from Earth	Distance from Sun	Elong-ation from Sun	Mag.
	h	m	s	°	'	"	AU	AU	°	
Jan 31	10	05	43.8	+11	30	30	1.125	2.081	160.9	9.3
Feb 05	10	02	21.2	+12	08	50	1.109	2.082	167.0	9.1
Feb 10	09	58	33.9	+12	49	33	1.099	2.082	173.2	9.0
Feb 15	09	54	32.6	+13	31	17	1.095	2.083	179.0	8.7
Feb 20	09	50	28.7	+14	12	35	1.098	2.084	174.3	8.9
Feb 25	09	46	33.6	+14	52	09	1.107	2.086	168.2	9.1
Mar 01	09	42	57.7	+15	28	48	1.122	2.087	162.1	9.3

(10) Hygiea

Date	RA (2000)			Dec.			Distance from Earth	Distance from Sun	Elongation from Sun	Mag.
	h	m	s	°	'	"	AU	AU	°	
Mar 01	11	46	09.7	-04	17	10	1.992	2.948	161.3	9.8
Mar 06	11	42	38.0	-04	00	08	1.968	2.943	166.9	9.6
Mar 11	11	38	54.0	-03	40	33	1.951	2.938	171.9	9.5
Mar 16	11	35	04.2	-03	18	59	1.942	2.933	174.5	9.4
Mar 21	11	31	15.3	-02	56	08	1.939	2.929	171.8	9.5
Mar 26	11	27	33.8	-02	32	41	1.944	2.924	166.8	9.6
Mar 31	11	24	05.5	-02	09	19	1.955	2.919	161.3	9.7
Apr 05	11	20	56.0	-01	46	43	1.974	2.915	155.7	9.8

(6) Hebe

Date	RA (2000)			Dec.			Distance from Earth	Distance from Sun	Elongation from Sun	Mag.
	h	m	s	°	'	"	AU	AU	°	
Mar 02	12	23	15.7	+11	37	31	1.918	2.857	157.1	9.9
Mar 07	12	19	37.4	+12	28	29	1.903	2.862	161.5	9.8
Mar 12	12	15	39.1	+13	18	22	1.895	2.866	164.8	9.8
Mar 17	12	11	27.5	+14	06	00	1.895	2.870	165.9	9.8
Mar 22	12	07	09.7	+14	50	20	1.901	2.874	164.5	9.8
Mar 27	12	02	52.7	+15	30	30	1.915	2.878	161.2	9.9
Apr 01	11	58	43.5	+16	05	45	1.936	2.881	156.7	9.9

(3) Juno

Date	RA (2000)			Dec.			Distance from Earth	Distance from Sun	Elong-ation from Sun	Mag.
	h	m	s	°	'	"	AU	AU	°	
Apr 10	14	49	36.3	-02	41	14	2.325	3.266	155.9	10.2
Apr 15	14	46	05.5	-02	06	31	2.308	3.271	160.4	10.1
Apr 20	14	42	19.4	-01	32	48	2.298	3.276	164.0	10.0
Apr 25	14	38	23.3	-01	00	44	2.296	3.281	166.0	10.0
Apr 30	14	34	22.5	-00	30	54	2.300	3.286	165.7	10.0
May 05	14	30	22.3	-00	03	50	2.312	3.291	163.2	10.1
May 10	14	26	28.5	+00	19	57	2.331	3.295	159.3	10.1
May 15	14	22	46.4	+00	40	06	2.357	3.299	154.7	10.2

(7) Iris

Date	RA (2000)			Dec.			Distance from Earth	Distance from Sun	Elong-ation from Sun	Mag.
	h	m	s	°	'	"	AU	AU	°	
May 16	16	40	01.5	-24	08	52	1.895	2.879	163.4	9.7
May 21	16	35	09.6	-23	52	49	1.873	2.874	169.3	9.5
May 26	16	30	03.4	-23	35	08	1.858	2.869	175.2	9.4
May 31	16	24	50.6	-23	16	02	1.850	2.864	177.8	9.3
Jun 05	16	19	39.2	-22	55	53	1.850	2.858	172.3	9.4
Jun 10	16	14	37.7	-22	35	09	1.856	2.853	166.4	9.5
Jun 15	16	09	53.9	-22	14	20	1.869	2.847	160.5	9.7

(8) Flora

Date	RA (2000)			Dec.			Distance from Earth	Distance from Sun	Elong-ation from Sun	Mag.
	h	m	s	°	'	"	AU	AU	°	
May 26	17	38	35.3	-17	49	12	1.464	2.439	159.4	9.7
May 31	17	33	45.9	-17	51	54	1.439	2.433	165.0	9.6
Jun 05	17	28	30.1	-17	55	30	1.420	2.427	170.5	9.5
Jun 10	17	22	57.2	-17	59	59	1.408	2.420	174.5	9.4
Jun 15	17	17	17.4	-18	05	21	1.402	2.414	173.4	9.4
Jun 20	17	11	41.0	-18	11	36	1.403	2.407	168.6	9.5
Jun 25	17	06	17.8	-18	18	48	1.410	2.400	163.0	9.6
Jun 30	17	01	17.0	-18	26	59	1.423	2.394	157.3	9.7

(2) Pallas

Date	RA (2000)			Dec.			Distance from Earth	Distance from Sun	Elong-ation from Sun	Mag.
	h	m	s	°	'	"	AU	AU	°	
Aug 05	21	39	07.7	+12	13	33	2.452	3.365	149.3	9.3
Aug 10	21	35	22.8	+11	33	52	2.429	3.361	152.6	9.3
Aug 15	21	31	32.8	+10	48	47	2.411	3.357	155.1	9.2
Aug 20	21	27	42.8	+09	58	49	2.401	3.353	156.4	9.2
Aug 25	21	23	57.8	+09	04	36	2.398	3.349	156.3	9.2
Aug 30	21	20	22.9	+08	06	53	2.403	3.344	154.9	9.2
Sep 04	21	17	03.1	+07	06	32	2.414	3.340	152.3	9.3
Sep 09	21	14	02.9	+06	04	28	2.433	3.335	148.8	9.3

(1) Ceres

Date	RA (2000)			Dec.			Distance from Earth	Distance from Sun	Elongation from Sun	Mag.
	h	m	s	°	'	"	AU	AU	°	
Oct 06	02	14	48.5	-00	21	13	1.939	2.889	157.6	7.6
Oct 11	02	10	56.4	-00	40	05	1.919	2.886	162.1	7.6
Oct 16	02	06	46.3	-00	57	33	1.906	2.882	165.6	7.5
Oct 21	02	02	24.4	-01	12	58	1.900	2.879	167.2	7.4
Oct 26	01	57	57.2	-01	25	46	1.901	2.876	166.1	7.5
Oct 31	01	53	32.0	-01	35	23	1.909	2.873	162.8	7.5
Nov 05	01	49	16.1	-01	41	22	1.924	2.870	158.4	7.6

(4) Vesta

Date	RA (2000)			Dec.			Distance from Earth	Distance from Sun	Elongation from Sun	Mag.
	h	m	s	°	'	"	AU	AU	°	
Dec 01	08	31	48.9	+19	21	09	1.853	2.535	123.5	7.3
Dec 06	08	31	45.5	+19	34	57	1.797	2.532	128.7	7.2
Dec 11	08	30	56.3	+19	52	10	1.744	2.529	134.0	7.1
Dec 16	08	29	20.9	+20	12	46	1.696	2.527	139.5	7.0
Dec 21	08	26	59.4	+20	36	33	1.652	2.524	145.2	6.9
Dec 26	08	23	53.0	+21	03	09	1.614	2.521	151.1	6.8
Dec 31	08	20	04.9	+21	32	02	1.582	2.518	157.1	6.7

(9) Metis

Date	RA (2000)			Dec.			Distance from Earth	Distance from Sun	Elong-ation from Sun	Mag.
	h	m	s	°	'	"	AU	AU	°	
Dec 01	10	36	04.9	+14	49	12	1.892	2.192	93.9	10.5
Dec 06	10	41	24.8	+14	35	26	1.837	2.197	97.7	10.5
Dec 11	10	46	12.7	+14	25	14	1.783	2.203	101.6	10.4
Dec 16	10	50	26.1	+14	18	54	1.730	2.209	105.6	10.3
Dec 21	10	54	02.5	+14	16	45	1.678	2.215	109.8	10.2
Dec 26	10	56	58.8	+14	19	04	1.628	2.221	114.2	10.2
Dec 31	10	59	12.0	+14	26	04	1.579	2.227	118.8	10.1

NEAR-EARTH ASTEROID APPROACHES

A list of some of the most interesting numbered, named and 'provisionally designated' close-asteroid approaches during 2016 is presented in the table in this section. It should be borne in mind that the visibility of close-approach asteroids is highly dependent on whether they are close to the solar glare and from which hemisphere the observer is based, but by their very nature they move rapidly across the sky. The table gives the closest separation between the Earth and the asteroid in AU (1 AU = 149.6 million km) and the date of that closest approach and also the constellation in which the object can be found when closest. The brightest magnitude achieved and the corresponding date and constellation for that condition are also given. The reader might wonder why the dates when the Near-Earth asteroid is closest and when it is brightest are not the same. Well, sometimes the dates do coincide, but if an object is not behind the Earth, it may have a poor phase (like the crescent Moon) with most of it in shadow, regardless of its proximity to the Earth. Some of the objects listed are as faint as magnitude 17.8 at their best and so present a real challenge, even to advanced amateur astronomers using large amateur telescopes and CCDs. However, some are much brighter and the objects (1685) Toro, 2009 DL46, (154244) 2002 KL6, (2100) Ra-Shalom, (164121) 2003 YT1, (96590) 1998 XB, (5143) Heracles and (2102) Tantalus are currently predicted to be magnitude 14.0 or brighter at their peak and so are within visual range of large amateur telescopes. Comet 252P/LINEAR is, clearly, not an asteroid as it has volatile material on its surface which evaporates when warmed by the Sun and so it looks fuzzy, rather than being a star-like point. Nevertheless, I have included it in the table as it is passing as close to the Earth as a Near-Earth asteroid, which is a rare event.

Remarkably, the first object in the table, (85990) 1999 JV6, was the second object in last year's table, because it has an orbital period very close to 1.0 years, the same as the Earth's orbital period. The asteroid (2100) Ra-Shalom has an interesting history as it was the second asteroid discovered in the rare Aten class, after 2062 Aten. Both objects were discovered by Eleanor F. Helin, in 1976 and 1978 respectively.

Many of the best, brightest and closest encounters for 2016 are listed in the table. Scores of other tiny asteroids, with provisional designations, will come within advanced amateur CCD range during 2016, many of them undiscovered as this Yearbook goes to press.

In total there are some 250 known minor planets that will come within 0.2 AU (30 million kilometres) of Earth in 2016; some are tiny, car-sized boulders, and others are mountain-sized asteroids. Sixteen of these are classed as Potentially Hazardous Asteroids (PHAs), meaning they have an absolute magnitude of 22.0 or higher (corresponding to a size greater than 110–240 metres) and can pass within 0.05 AU (7.5 million kilometres) of the Earth. Fourteen of those sixteen are bright enough in 2016 to be included in the table, which shows all objects predicted to peak above magnitude 18.0. Five large, low-numbered and named asteroids return to the Earth's vicinity in 2016 and, not surprisingly, are among the brightest of the close-approach objects. These are (1685) Toro, (3103) Eger, (2100) Ra-Shalom, (5143) Heracles and (2102) Tantalus. None of these are Potentially Hazardous Asteroids (PHA) although they are relatively large, with diameters estimated to be as large as 4 kilometres, depending on their albedo.

The closest asteroid passage brighter than magnitude 18.0, currently predicted for 2016, is that of 2009 DL46, a PHA which will pass 0.01419 AU (2.1 million km) from Earth on 24 May, assuming its orbit is accurate. It is thought to be between 150 and 300 metres in diameter. The brightest asteroid passage in 2016, assuming no better candidates are discovered during the year, will be that of (164121) 2003 YT1. This object will pass 0.03477 AU (5.2 million kilometres) from Earth on 31 October, which will make it as bright as magnitude 10.8, due to its diameter of around 1 kilometre or more. Having said that, the aforementioned small comet 252P/LINEAR passes a similar distance from Earth on 21 March and will probably be of a similar magnitude, although it will appear fuzzy in appearance, rather than a star-like point.

Some interesting numbered, named and provisional designation close-asteroid approaches during 2016. All dates are for 2016. Objects with a '[PHA]' in the first column are considered to be Potentially Hazardous Asteroids. 252P/LINEAR is a close-approach Comet described earlier in the Comets section. Objects 2008 DL5, 2009 DL46 and 2008 SC had only been seen at one apparition at the time of writing so their orbits are far from precise. In keeping with Minor Planet Center protocol all asteroids with precise orbits are given with their number in parentheses, followed by the original year of discovery. Those asteroids for which a precise orbit has yet to be determined are given a provisional designation, e.g. (85990) 1999 JV6. Asteroids which have been named have their number in parentheses, followed by the name, e.g. (1685) Toro.

Asteroid (number)/ designation/name	Closest (AU)	2016 date when closest	Constell. closest	Peak mag.	Peak date	Const. brightest
(85990) 1999 JV6	0.03241	Jan 6.05	Carina	14.5	10 Jan	Canis Maj.
(1685) Toro	0.156550	Jan 22.30	Cetus	13.0	31 Jan	Eridanus
(141052) 2001 XR1	0.191117	Jan 23.80	Centaurus	16.3	18 Jan	Hydra
2008 DL5 [PHA]	0.04580	Feb 27.70	Puppis	17.2	1 Mar	Puppis
252P/LINEAR	0.03575	Mar 21.52	Apus	10	21 Mar	Aps
(7350) 1993 VA	0.153260	Mar 23.26	Camelop.	15.7	2 Apr	Ursa Major
1994 UG [PHA]	0.044405	Mar 25.31	Perseus	17.3	15 Mar	Lynx
(363599) 2004 FG11	0.049565	Apr 11.69	Cygnus	17.1	9 Apr	Hercules
2003 KO2 [PHA]	0.045974	Apr 25.01	Monoceros	17.1	28 Apr	Cancer
2006 UK [PHA]	0.045766	May 3.66	Pyxis	15.8	6 May	Hydra

Minor Planets in 2016

Asteroid (number)/ designation/name	Closest (AU)	2016 date when closest	Constell. closest	Peak mag.	Peak date	Const. brightest
2008 TZ3 [PHA]	0.033728	May 5.44	Sagittarius	14.8	3 May	Ophiuchus
2009 DL46 [PHA]	0.01419	May 24.12	Virgo	13.5	25 May	Libra
(382758) 2003 GY	0.079279	June 15.73	Columba	17.8	20 May	Vela
2010 NY65 [PHA]	0.027528	June 24.44	Leo Minor	16.9	27 June	Canes Vn.
(359369) 2009 YG	0.172539	July 21.78	Virgo	17.4	27 July	Libra
(154244) 2002 KL6	0.06838	July 22.01	Cygnus	13.8	18 July	Lyra
(3103) Eger	0.189414	Aug 2.56	Cetus	14.3	28 July	Cetus
(357024) 1999 YR14	0.055816	Sept 1.49	Eridanus	14.7	27 Aug	Cetus
2009 ES [PHA]	0.048383	Sept 5.81	Microscop.	15.6	7 Sept	Microscop.
(250458) 2004 BO41	0.099947	Sept 7.28	Coma B.	16.4	15 Sept	Hercules
2008 SC [PHA]	0.03945	Sept 24.23	Octans	17.4	22 Sept	Hydrus
2011 DU [PHA]	0.03907	Oct 6.84	Monoceros	16.7	13 Oct	Eridanus
(2100) Ra-Shalom	0.149907	Oct 9.59	Horolog.	14.0	4 Oct	Fornax
2003 TL4 [PHA]	0.025859	Oct 27.70	Hercules	15.1	25 Oct	Draco
(164121) 2003 YT1	0.03477	Oct 31.39	Auriga	10.8	31 Oct	Auriga
(138852) 2000 WN10	0.126707	Nov 10.53	Fornax	17.4	15 Nov	Eridanus
(96590) 1998 XB	0.119766	Nov 15.72	Lepus	13.2	23 Nov	Eridanus
(162911) 2001 LL5	0.106711	Nov 16.45	Boötes	17.2	3 Nov	Draco
1997 XR2 [PHA]	0.04756	Nov 18.63	Ursa Maj.	16.0	29 Nov	Auriga
(5143) Heracles	0.147017	Nov 28.98	Boötes	12.4	24 Nov	Draco
(369264) 2009 MS	0.069607	Nov 30.86	Virgo	14.9	27 Nov	Canes Vn.
2008 UL90 [PHA]	0.039143	Dec 12.62	Microscop.	16.7	18 Dec	Pegasus
2006 XD2 [PHA]	0.048455	Dec 21.33	Canes Vn.	16.8	18 Dec	Leo Min.
(2102) Tantalus	0.137484	Dec 30.61	Cetus	13.9	30 Dec	Eridanus

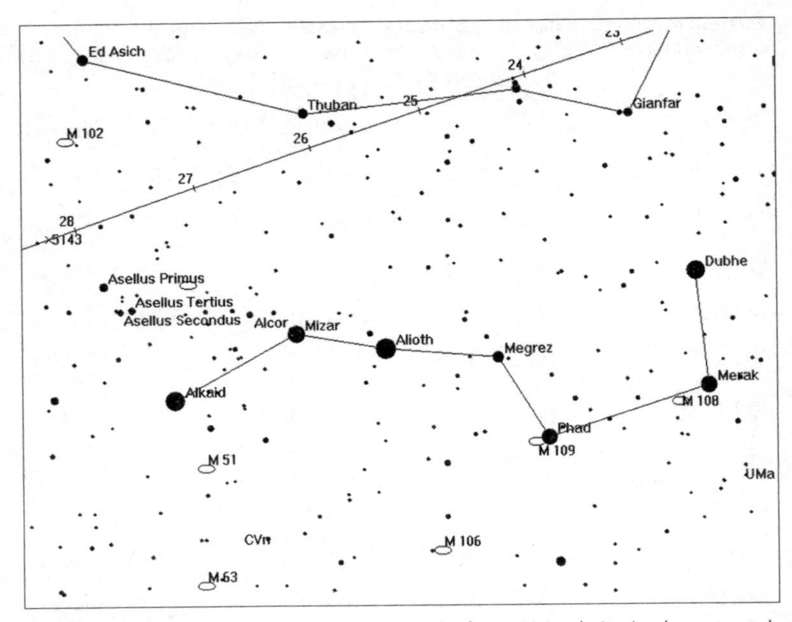

Figure 6. Minor planet 5143 Heracles tracks north of Ursa Major during its close approach to the Earth as it moves from Draco to Boötes from 24 November to 28 November. Its magnitude should peak at around 12.4.

Accurate minor planet ephemerides can be computed for your location on Earth using the MPC ephemeris service at: www.minorplanetcenter.org/iau/MPEph/MPEph.html. Using this facility can be important if an object is passing so close that the Earth's gravity affects the trajectory of the object. Even if an object is passing as far as 1 million kilometres away, there will be a significant parallax caused by the Earth's radius, which may be as much as twenty arc minutes or so, compared to the apparent position viewed from the Earth's centre, so it is important to set the software you use to your location on Earth. This means entering an Observatory Code on the Minor Planet Center site. The code for Greenwich, London, is 000.

Meteors in 2016

JOHN MASON

Meteors (popularly known as ‘shooting stars’) may be seen on any clear moonless night, but on certain nights of the year their number increases noticeably. This occurs when the Earth chances to intersect a concentration of meteoric dust moving in an orbit around the Sun. If the dust is well spread out in space, the resulting shower of meteors may last for several days. The word ‘shower’ must not be misinterpreted – only on very rare occasions have the meteors been so numerous as to resemble snowflakes falling.

If the meteor tracks are marked on a star map and traced backwards, a number of them will be found to intersect in a point (or within a small area of the sky) which marks the radiant of the shower. This gives the direction from which the meteors have come. Meteor radiants are not stationary because of the Earth’s motion around the Sun. They move eastwards about one degree of ecliptic longitude per day. A meteor shower is usually named after the constellation in which the radiant lies at the peak of the shower.

Bright moonlight has an adverse effect on meteor observing, and for about five days to either side of full Moon, lunar glare swamps all but the brighter meteors. Visual observers may, however, minimize the effects of moonlight by positioning themselves so the Moon is behind them and hidden behind a wall or other suitable obstruction. In 2016, the maxima of the Lyrids, Northern Taurids, Leonids and Geminids will all suffer interference by moonlight. Lyrid maximum on 21 April coincides with a virtually full Moon in Virgo. In autumn, the broad peak of the Northern Taurids around 12 November occurs just before full Moon, although the Southern Taurids, which peak five days earlier, will be less affected. Observations of the Leonids later in November will be hampered by a Moon only a couple of days past full on the Taurus/Gemini border, which is a pity as the peak is near 04h, a good time for UK observers. The Geminids, which are now the richest of the annual meteor showers, also coincide with a virtually full Moon in Taurus.

There are many excellent observing opportunities in 2016. The peak of the Quadrantids occurs just after dawn in the UK on 4 January, but good rates should be seen in the early morning hours, weather permitting. The Eta Aquarids in early May coincide with new Moon, and shower members may be observed in the eastern sky as the radiant rises before dawn. The complex of showers which peak in late July and early August, e.g. the Capricornids, Delta Aquarids, Piscis Australids, Alpha Capricornids and Iota Aquarids, are very well placed with respect to the Moon this year, and the combined activity of all these showers should provide good observed rates at this time. Perseid maximum occurs during daylight hours from the UK, but high activity should be seen on the nights of 11/12 and 12/13 August. The Moon will be a waxing gibbous in Ophiuchus, setting at around midnight. Of the autumn showers, conditions are quite favourable for the later parts of the broad Orionid peak, although maximum itself will be affected by a waning last quarter Moon in Cancer. Taurid activity in late October/early November should also be relatively unaffected by moonlight. The Ursids in late December are also quite favourable this year and, although rates are low, this is a shower badly in need of observation.

As always, observations away from the major shower maxima, and of year-round sporadic activity, are every bit as important as those obtained when high rates are anticipated.

The following table gives some of the more easily observed showers with their radiants; interference by moonlight is shown by the letter M.

			Radiant			
Limiting Dates	**Shower**	**Maximum**	**RA**		**Dec.**	
			h	m	°	
1–6 Jan	Quadrantids	4 Jan, 09h	15	20	+50	
19–25 April	Lyrids	21 Apr, 22h	18	08	+33	M
24 Apr–20 May	Eta Aquarids	5–6 May	22	20	−1	
17–26 June	Ophiuchids	10, 20 June	17	20	−20	M
July–Aug	Capricornids	7, 15, 25 July	20	44	−15	
			21	00	−15	
15 July–20 Aug	Delta Aquarids	28 July, 6 Aug	22	36	−17	
			23	04	+02	
15 July–20 Aug	Piscis Australids	31 July	22	40	−30	
15 July–20 Aug	Alpha Capricornids	2–3 Aug	20	36	−10	

Meteors in 2016

Limiting Dates	Shower	Maximum	Radiant			
			RA		Dec.	
			h	m	°	
July–Aug	Iota Aquarids	6 Aug	22	10	–15	
			22	04	–06	
23 July–20 Aug	Perseids	12 Aug, 12h	3	11	+58	
16–30 Oct	Orionids	21–24 Oct	6	24	+16	
20 Oct–30 Nov	Taurids	5, 12 Nov	3	54	+22	M
			3	33	+13	
15–20 Nov	Leonids	17 Nov, 04h	10	16	+22	M
Nov–Jan	Puppid-Velids	early Dec	9	00	–48	
7–16 Dec	Geminids	13 Dec, 20h	7	33	+32	M
17–25 Dec	Ursids	22–23 Dec	14	28	+78	

Some Events in 2017

ECLIPSES

There will be four eclipses, two of the Sun and two of the Moon.

11 February: Penumbral eclipse of the Moon – North America, South America, Atlantic Ocean, Europe, Africa, Indian Ocean and Asia (except the far eastern part).

26 February: Annular eclipse of the Sun – South America (Chile and Argentina), South Atlantic Ocean and South-West Africa (Angola and Zambia).

7 August: Partial eclipse of the Moon – Europe, Africa, Indian Ocean, Asia and Australasia.

21 August: Total eclipse of the Sun – 'The Great American Eclipse'. North-East Pacific Ocean, Oregon, Idaho, Wyoming, Nebraska, Kansas, Missouri, Kentucky, Tennessee, Georgia, South Carolina and North Atlantic Ocean.

THE PLANETS

Mercury may be seen more easily from northern latitudes in the evenings about the time of greatest eastern elongation (1 April) and in the mornings about the time of greatest western elongation (12 September). In the Southern Hemisphere the corresponding most favourable dates are 30 July and 24 November (evenings) and 19 January and 17 May (mornings).

Venus is visible in the evenings from January until early March. It reaches greatest eastern elongation (47°) on 12 January and attains its greatest brilliancy on 18 February. Venus passes through inferior conjunction on 25 March and will then be visible in the mornings from early April until November. It attains its greatest brilliancy on 26 April and reaches greatest western elongation (46°) on 3 June.

Mars does not come to opposition in 2017. The planet is visible in the evening sky from January until May, is in conjunction with the Sun on 27 July, and becomes visible in the morning sky from September until the end of the year. Mars will be next at opposition on 27 July 2018, when the planet will be in Capricornus.

Jupiter is at opposition on 7 April in Virgo.

Saturn is at opposition on 15 June in Ophiuchus.

Uranus is at opposition on 19 October in Pisces.

Neptune is at opposition on 5 September in Aquarius.

Pluto is at opposition on 10 July in Sagittarius.

Part Two

Article Section

Comet Chasing: The Rosetta Story

NATALIE STARKEY

INTRODUCTION

Some twenty-two years have passed since the Rosetta mission was approved as part of the European Space Agency's Horizons 2000 Science Programme. Since then, the Rosetta spacecraft, together with its Philae lander, have been on a decade-long journey into the Solar System, passing through the asteroid belt to plunge into deep space to catch up with comet 67P/Churyumov-Gerasimenko (67P/C-G).

Comets have long been mysterious objects in the night sky and, historically, the sight of them has been thought to indicate a threat or a bad omen. We now know that these mysterious, icy visitors which orbit our Sun originate in the cold, far outer reaches of the Solar System. Comets are known to contain all of the earliest ingredients that formed our planets – dust, gas, organic matter and water – preserving them in deep-freeze. We also have evidence that comets collided with Earth in the past, and it cannot be ruled out that this could happen again in the future, potentially leading to mass devastation. However, the fact that comets collided with our planet – potentially delivering water and even organic material – may be the reason Earth is now habitable.

Understanding more about comets, and how we would predict future cometary collisions, relies on us discovering more about the composition and behaviour of comets. The Rosetta spacecraft was packed off on a journey into the unknown, filled with laboratory instruments, to catch up with and land on a comet. It is set to answer some of the main questions about the behaviour of comets as they orbit the Sun, and to help us understand the role they played in the history of the Earth. Not only this, but Rosetta was also to become the first spacecraft to enter into orbit around a comet, and when the Philae lander was released from the orbiter to touch down on the surface of comet 67P/C-G, it became the first spacecraft in history to do so.

THE HISTORY OF THE ROSETTA MISSION

In March 2004, the three-tonne Rosetta spacecraft successfully lifted off from Kourou in French Guiana. The launch of the spacecraft was planned for 2002, but was delayed because of the failure of an Ariane-5 rocket, the planned launch vehicle, prior to the Rosetta launch. Rosetta's journey to catch up with a speeding comet was not set to be simple, as even the very powerful Ariane-5 rocket was not capable of sending the spacecraft direct to 67P/C-G. The European Space Agency (ESA) described Rosetta's journey as resembling something more like that of a 'cosmic billiard ball', involving three close fly-bys of the Earth (2005, 2007, 2009), and one of Mars (2007), in order to gain gravitational pushes to increase the spacecraft velocity, but it also made four orbits of the Sun and two transits through the asteroid belt.

The Rosetta spacecraft, formed of an orbiter and lander, with its enormous fourteen-metre-long solar-panel wings, was named after the famous Egyptian Rosetta Stone that can be seen today at the British Museum in London. The Rosetta Stone revolutionized our understanding of the ancient civilization of Egypt by providing a key to decipher the mysterious written language of hieroglyphs. Similarly, the Rosetta spacecraft was designed with the potential to unlock the ancient secrets of the Solar System by analysing a comet – an icy space rock formed 4.6 billion years ago before the planets were in existence. The orbiter itself, which transported the lander until its release on 12 November 2014, also contains the main propulsion system composed of two large propellant tanks, one containing fuel and the other containing an oxidiser, as well as twenty-four thrusters for trajectory and attitude control. In order for the orbiter to achieve its primary science objectives, it also contains several cameras, spectrometers, a number of sensors, and experiments that work at different wavelengths – infrared, ultraviolet, microwave and radio. These instruments provide high-resolution images of the comet and analyse the gases and dust grains in the coma as well as studying the interaction of the coma with the solar wind as Rosetta approaches the Sun.

The hundred-kilogramme box-shaped Philae lander was named after the Philae obelisks, one of which is on display in the gardens at Kingston Lacy in Dorset, which was used along with the Rosetta Stone to decipher the Egyptian hieroglyphs. As such, the Philae lander, along with its ten laboratory instruments, was designed to help the Rosetta orbiter unlock the secrets of the Solar System from analysing the chemical and physical properties of the surface of the comet. Philae includes a drilling system which was planned to be used to sample cometary material down to twenty-three centimetres below the surface, material which would then be transferred to the instruments on Philae for analysis. Philae was also designed to measure the density, texture, porosity, ice phases and thermal properties of the comet, as well as performing microscopic studies of individual grains.

WHY CHASE COMET 67P/ CHURYUMOV-GERASIMENKO?

Comet 67P/C-G, like most comets, is named after its discoverers, Soviet astronomers Klim Ivanovych Churyumov and Svetlana Ivanovna Gerasimenko, who first observed the comet in 1969. 67P/C-G originated in the Kuiper belt, which is a vast region of the Solar System past the orbit of Neptune and extending out to at least 50 AU (astronomical units where 1 AU is the distance from the Earth to the Sun). 67P/C-G has an orbital period of 6.45 years and a rotation of 12.4 hours. It previously passed through perihelion on 18 August 2002, and again on 13 August 2015. Prior to the Rosetta launch, 67P/C-G was observed with ground-based telescopes during almost all its appearances since its discovery in 1969.

The original target chosen for the Rosetta mission was comet 46P/Wirtanen, but the delay in launch meant that a new target was required, and so 67P/C-G was chosen.

Both comets are short-period Jupiter-family comets and were known to be active so that observations could be made during their active phase, and this meant that both comets had orbited the Sun a few times already. However, it was also important to find a comet that has not orbited the Sun too many times because each orbit causes changes to the comet as it is heated up and sheds material. Hence, comets evolve with time as they are affected by their interaction with the Sun. In fact, the Rosetta orbiter was designed to observe the interaction of comet 67P/C-G as it approached the Sun and to understand the processes that occur as the comet is heated up. 67P/C-G is not a sun-grazer, and so did not approach extremely close to the Sun at perihelion in August 2015, but instead remained further from the Sun than the Earth is at its closest point (186 million km). However, this is close enough for the comet nucleus to be heated and to produce substantial and observable activity. After all, it was not so long ago in 2011 that the closely observed Comet ISON disintegrated at perihelion, an event that scientists were unable to predict for certain because so little is known about comets and how they behave on their journeys into the inner Solar System.

ROSETTA'S LONG JOURNEY

Rosetta entered into deep-space hibernation from June 2011 to January 2014 in order to save energy because its solar panels would receive too little light during this time to fully power the spacecraft. With only a computer and several heaters staying active, the spacecraft remained in hibernation as it travelled from 660 million km from the Sun out to 790 million km, towards the orbit of Jupiter, and back.

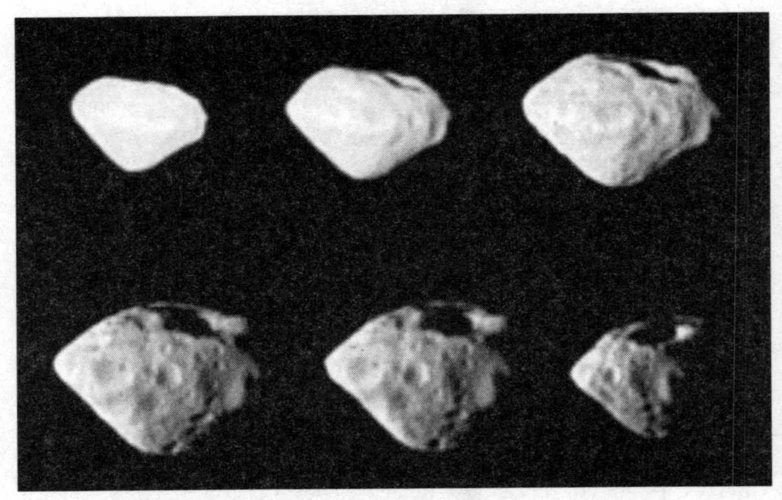

Figure 1. Asteroid Steins: A diamond in space. Taken from a distance of 800 km by OSIRIS, this image of Asteroid Steins with a diameter of 5 km shows a huge crater at the top of the asteroid that is approximately 1.5 km in size (as seen in this image). (Image, released 8 September 2008, courtesy of ESA ©2008 MPS for OSIRIS Team/MPS/UPD/LAM/IAA/RSSD/INTA/UPM/DASP/IDA.)

Rosetta did not relax on its long journey to catch up with 67P/C-G. In 2008 and 2010, it made close approaches to two asteroids: 2867 Steins and 21 Lutetia, respectively. Rosetta flew past Steins at a distance of 800 km and a relatively slow speed of 8.6 km per second. The encounter was only seven minutes long but the asteroid was illuminated on the side facing Rosetta, which meant it was possible to obtain very clear and bright images of Steins. Uwe Keller, Principal Investigator for the OSIRIS (Optical, Spectroscopic and Infrared Remote Imaging System), described Steins as a 'diamond in the sky' because of the shape of the asteroid and clarity of the images (see Figure 1).

Whereas Steins is a relatively small asteroid around 4.6 km in diameter, 21 Lutetia is much larger, at around 100 km in diameter, and Rosetta passed it at a distance of just over 3000 km. Despite this much greater distance, high-resolution images and spectra of Lutetia (see Figure 2) were obtained to reveal that the asteroid was probably very

Figure 2. Lutetia at closest approach. The image shows Asteroid 21 Lutetia at a distance of 3168 km in July 2010. Lutetia is 100 km wide and has numerous grooves and impact craters across its surface that might be related to shock impacts. (Image, released 10 July 2010, courtesy of ESA 2010 MPS for OSIRIS Team MPS/UPD/LAM/IAA/RSSD/INTA/UPM/DASP/IDA.)

old – a primordial planetesimal formed at the very beginning of the Solar System. Lutetia is heavily cratered, with crater-counting analyses revealing a complicated and tumultuous history for the asteroid, and also showing that it is formed of different geological regions spanning a wide range of ages. The combined results from Rosetta instruments indicate that the asteroid might have a metallic core surrounded by a more primitive outer crust; a differentiated structure which was unexpected for an asteroid of Lutetia's relatively small size.

ROSETTA WAKE-UP – HELLO, WORLD!

At 10.00 GMT on 20 January 2014, Rosetta's internal alarm clock went off, triggering a complex series of events to bring the spacecraft out of its long hibernation. This was a big step for the mission; if the alarm clock had not worked, the mission would have been over – there was no second chance and no way to trigger the wake-up sequence from Earth. It was a long, tense wait as, although the official wake-up signal was not expected until 17.30 GMT on 20 January 2014, the expected time window came and went with still only a background signal on the screens watched nervously at Mission Operations Centre ESOC (European Space Operations Centre) in Darmstadt, Germany. However, at 18.17 GMT, the Acquisition of Signal (AOS) finally arrived, tentatively at first, but then building up to a strong spike on the spectrum (see Figure 3). The wake-up procedure involved activation of Rosetta's core systems, such as heaters and avionics, to slowly bring the spacecraft back to life. Rosetta's thrusters were fired to stop its slow rotation and to perform a slight adjustment to its orientation to ensure that the solar arrays were facing directly towards the Sun. The star trackers also became active – an on-board navigation system similar to, but more sophisticated than, that used by eighteenth-century mariners. The purpose of this navigation system was to determine the spacecraft's attitude in order for Rosetta to turn directly towards Earth, before switching on its transmitter and pointing its high-gain antennae to send the signal to announce that it had awoken and which was relayed by Goldstone (US) and Canberra (Australia) antennae. With this, the famous 'Hello, World!' message was sent from the official @ESA_Rosetta social media account and the world was waking up to the excitement of this groundbreaking mission.

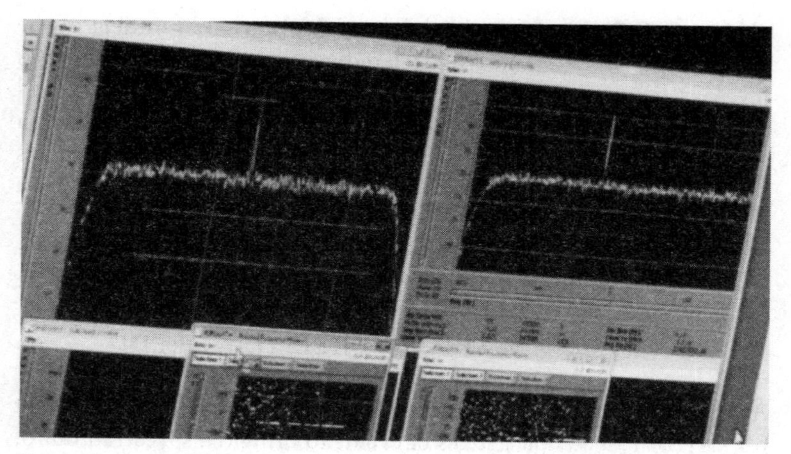

Figure 3. Rosetta Wake-up signal. This image shows the Acquisition of Signal (AOS) spectrum for the Rosetta spacecraft when it awoke on 20 January 2014. (Image, released 20 January 2014, courtesy of ESA.)

CATCHING UP WITH 67P/C-G – RENDEZVOUS AND ORBIT

The next big moment in Rosetta's journey was not until August 2014 when the spacecraft was set to rendezvous with 67P/C-G and enter into orbit above its dusty, icy surface to carefully map and characterize 100 per cent of the comet in order to locate a suitable landing location for Philae. Characterizing the comet meant determining its shape, rotation rate and orientation, gravity field, albedo (reflectivity), surface features and surface temperature. These features were all unconstrained but were vital for planning how to orbit and land on the comet. Rosetta's orbiter instruments showed that the comet had a mass of 10^{13} kg, a volume of 25 km^3 and a density of 0.4 g/cm^3 as calculated by RSI (Radio Science Investigation), OSIRIS and RSI/OSIRIS respectively. The surface temperature was estimated by the VIRTIS (Visible and Infrared Thermal Imaging Spectrometer) instrument at 205–230K in July/August 2014 and it was found that the surface was dark, porous and probably dry. The subsurface temperature was estimated by MIRO (Microwave Instrument for the Rosetta Orbiter) to be 30–160K in August 2014. ROSINA (Rosetta Orbiter Sensor for Ion and Neutral Analysis) managed to detect a number of gases whilst in orbit around the comet including water, carbon monoxide, carbon dioxide, ammonia, methane and methanol – the first three being the most abundant species. ROSINA also observed that the ratio of gases varied depending on location around the comet, with carbon monoxide being as abundant as water in some places and only 10 per cent as abundant in others.

Despite the low activity at the comet during the orbit and mapping phase of the mission, on 24 August 2014, scientists saw that COSIMA (Cometary Secondary Ion Mass Analyzer) had collected a number of large dust grains, around fifty to seventy micrometres in size, on the first of its twenty-four target collector plates. The results revealed that the dust grains were very fragile and loosely bound, having shattered on contact with the collector, despite only travelling at one to ten metres per second. COSIMA scientists concluded that the dust grains did not contain any ice and were found to be rich in sodium. It was concluded that the collected grains were similar to so-called Interplanetary Dust Particles – dust that arrives naturally in the Earth's stratosphere and is thought to originate from cometary meteor streams. The dust grains collected by COSIMA are thought to originate from the dusty outer surface of the comet, a layer that formed during the comet's last close approach of the Sun. It is thought that this dust layer disappeared around December 2014 (although this will be confirmed by future results from Rosetta) to reveal the layer below the protective dry dust – that of a mix of dust held in frozen gases. As Rosetta follows 67P/C-G towards perihelion it will be able to see what this mixed material looks like as it, too, will start to come off the comet surface with increased cometary activity towards perihelion. In turn, these future results will inform us about the interior of Comet 67P/C-G. After perihelion, this dust and frozen-gas layer is expected to become dehydrated to form, once again, a protective outer dust layer to the comet for the next few years until the next perihelion.

UNDERSTANDING THE SHAPE OF COMET 67P/C-G

Prior to Rosetta arriving at Comet 67P/C-G, scientists were able to build up an estimate of the comet's shape using ground-based telescopes and the Hubble Space Telescope, and by applying a technique called 'light curve inversion'. This led to the prediction that 67P/C-G was potato-like in shape, which was not unexpected for a comet. This allowed the orbit scientists to build up computer models to tell the spacecraft how to orbit the comet without crashing into the surface or flying so far away that the surface image resolution was too low. However, it soon became clear as Rosetta got closer to 67P/C-G that the simple 'potato-shaped' prediction was completely inaccurate. The first intriguing, blurry images, initially only a few pixels in size, from Rosetta's OSIRIS camera were beamed back to Earth and they showed, as Rosetta got closer and closer to 67P/C-G, that the comet was starting to look more like two comets than one (Figure 4). The media quickly started to refer to 67P/C-G as 'the rubber duck' because the comet appeared to have a small head, with dimensions 2.5 x 2.5 x 2.0 km, and a big body, with dimensions 4.1 x 3.2 x 1.3 km (see Figure 4). This unexpected shape raised the question as to whether the two lobes

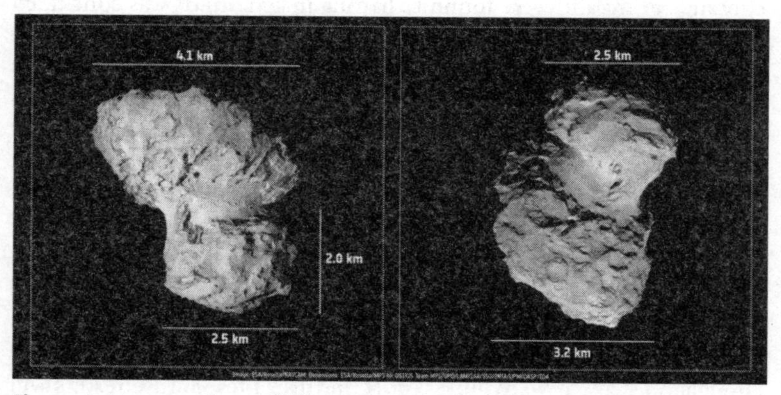

Figure 4. Measuring Comet 67P/C-G. The images here show the dimensions of Comet 67P/C-G as measured from images taken by Rosetta's OSIRIS imaging system on 19 August 2014. The larger lobe measures 4.1 x 3.2 x 1.3 km, while the smaller lobe is 2.5 x 2.5 x 2.0 km. (Image, released 3 October 2014, courtesy of ESA/Rosetta/NAVCAM; dimensions: ESA/Rosetta/MPS for OSIRIS Team MPS/UPD/LAM/IAA/SSO/INTA/UPM/DASP/IDA.)

represent a contact binary comet – formed 4.5 billion years ago as two comets became one on collision with each other – or a single body that has experienced mass loss over time to develop a gap at the 'neck' between the head and body.

The crisp images of this new world were quite a surprise for scientists and the comet's unexpected shape made it somewhat harder to plan a safe orbit programme. However, with some of the worlds' best scientists working on the mission, these problems were overcome and led to a successful mapping programme, and the most fantastic images being returned to Earth.

The level of detail obtained in the images of the comet's surface improved throughout August 2014 when the spacecraft reached a distance of just fifty kilometres from the comet. Going into September, when the spacecraft entered the global-mapping phase, Rosetta was as close as just twenty kilometres from the surface. The closer the spacecraft could get to the comet surface the better, as it allowed for higher-resolution imaging, which was crucial for understanding exactly how the surface looked; how uneven or smooth. In some images the resolution was so good that one image pixel was only seventy-five centimetres on the comet.

With the main mapping phase well underway, and with a good overview of the different terrains that made up the comet's surface by early September, it was time to choose a suitable landing location. It soon became clear that although there were some smoother surface areas, these were not necessarily deemed the most interesting scientifically – in terms of how active they were expected to become as the comet approached the Sun – or geologically. After waiting for so long to get to 67P/C-G, the choice was not easy: aim for somewhere flat, preferred by the landing engineers in order to ensure a safe landing, or aim for a riskier, more active, possibly boulder-ridden landing location, which might be scientifically more interesting as the mission progressed? In early September 2014, at a special meeting of the ESA teams in Toulouse, France, a compromise was made. At this stage there was also a good shape model of the comet that had a resolution of 3.6 m on the surface and 0.5 m in elevation, details of which were factored in to the landing site selection. The ten potential landing sites (A–J) were whittled down to five locations (A, C, B, I, J) and Site J (later named Agilkia after an island on the River Nile) was chosen as the primary landing spot on the so-called 'head' of the comet (Figures 5 and 6), with Site B, located on the body, as the back-up site. Site J was thought to offer interesting surface features and good illumination, which was deemed important for receiving enough power for continued lander operations after touchdown and completion of the primary science sequence. Both locations were viewed as scientifically interesting, with the potential for activity nearby, but although they contained some very large boulders and terracing (Figure 6), they were deemed not as

Figure 5. Philae candidate landing sites. An OSIRIS narrow-angle camera image taken on 16 August 2014 from a distance of about 100 km showing the five candidate landing sites during the Landing Site Selection Group meeting held 23 to 24 August 2014. (Image, released 25 August 2014, courtesy of ESA/Rosetta/MPS for OSIRIS Team MPS/UPD/LAM/IAA/SSO/INTA/UPM/DASP/IDA.)

Figure 6. Landing Site J. The image was taken with Rosetta's OSIRIS narrow-angle camera on 16 August 2014, from a distance of about 100 km, showing a zoom into landing Site J on the smaller lobe of the comet. The frame is about 1 km across and is centred on the mid-point of the landing ellipse and the image has a resolution of 1.85 m per pixel. (Image, released 25 August 2014, courtesy of ESA/Rosetta/MPS for OSIRIS Team MPS/UPD/LAM/IAA/SSO/INTA/UPM/DASP/IDA.)

risky as some of the other potential landing spots. However, mapping continued throughout October to better characterise Site J in more detail, with the spacecraft getting as close as 9.8 km from the cometary surface to improve image resolution.

The time of close approach for Rosetta was ideal for performing a range of scientific experiments to understand the comet before landing. These first findings have now been published, although there is much interpretation still to be done. Rosetta discovered that the comet's surface morphology is the result of the removal of not only small dust grains, as witnessed by COSIMA, but also larger volumes of material, possibly via explosive release of subsurface pressure or via the creation of overhangs by sublimation. These are processes that scientists hope to understand in more detail in order to shed light on how comets are affected by their interaction with the Sun, and, importantly, when they are likely to break up – a cometary behaviour of which, currently, we have little understanding. It was also found that outgassing of water from the comet appears to be localized to the neck region which has been observed to be particularly active. It is thought that this active neck region might be a zone of weakness, partly because of the higher levels of activity and partly because of the observation of a huge crack running along the length of the neck. These features hint at the possibility for the two cometary lobes to break apart at the neck in the future if activity preferentially continues to remove material from this region. In turn this might provide clues about the origin of the comet, potentially suggesting that it formed as one comet rather than two comets that came together as a contact binary. Another surprise from 67P/C-G is that the surface appears to be dehydrated, with no major ice-rich patches, and the comet's low thermal inertia indicates a thermally insulating powdered surface. VIRTIS revealed that the surface appears to contain opaque minerals associated with non-volatile organic materials. These are thought to be a mix of carbon-hydrogen and/or oxygen-hydrogen chemical groups, but with almost no nitrogen-hydrogen groups. Results from the lander and continued measurements by the orbiter will certainly add to this data set to build up a better picture of the organic components in and on the comet. Direct in-situ measurements of the D/H ratio, which is the ratio between two different isotopes of hydrogen found in water, in the coma reveal that it is approximately three times terrestrial value. This, along with previous findings from other comets, shows that Jupiter-family comets display a wide range of D/H ratio that may help scientists to work out where and when they formed in terms of their distance from the Sun and timing of formation relative to the planets, and how they might be related to the delivery of water on Earth.

LANDING ON A COMET

Once a landing location was selected, it was time to carefully plan the landing procedure, a complex manoeuvre that would get the spacecraft into the correct position, at the correct speed, to release the Philae lander so that it would fall to the landing spot simply under the force of gravity. Philae has no navigation system, so if the orbit and release calculations were wrong, there was a high chance that the lander could approach the comet too quickly and bounce off, or approach too slowly and miss it altogether. Towards the end of September 2014, a landing date of 12 November 2014 had been chosen, just one day later than originally planned – despite the shape of the comet being considerably more complex than had been first expected.

On 11 November 2014, the first of a series of 'Go/No Go' checks started. At any point the mission could be delayed if one of the spacecraft systems was not working. At 19.00 GMT, the first check confirmed that Rosetta was in the correct orbit for delivering Philae to the comet surface, with the next check coming at midnight GMT, confirming that the commands to control the separation and delivery of Philae were complete and ready to upload to Rosetta. At 02.35 GMT, the third GO was given after a final check that the lander was ready for touchdown. The final Rosetta manoeuvre was conducted at 07.35 GMT, which took the spacecraft to a point around 22.5 km from the comet's centre ready for the release of Philae. The only major problem found during this time was that Philae's on-board nitrogen-cooled gas propulsion system, designed to be fired during touchdown to prevent any bounce motion up off the comet, was found not to be working. As it was not possible to fix this remotely, it was decided to go ahead with the landing procedure regardless. A tense but ultimately successful evening meant that systems were in a good state by 08.35 GMT on 12 November 2014, when the landing sequence began for the release of the Philae lander. The lander's descent was imaged by Philae's ROLIS (Rosetta Lander Imaging System) camera as a continuous series of photos, and the OSIRIS narrow-angle camera on the orbiter captured the iconic shot (Figure 7) of the lander just after it was released and began its fall to the surface of the comet, with its legs deployed ready for landing. The lander completed more than half a rotation during its seven-hour descent, starting on the opposite side of the comet to the chosen landing site but falling to meet it as 67P/C-G also rotated during this time. Rosetta descended at around 2.5 km per hour, a slow walking pace, and used gyroscopic stabilization, in the form of a momentum wheel connected to a motor, to stabilize itself during descent. This wheel is spun up to a high rotational speed to provide an angular momentum bias to the lander and works in the same way as a spinning gyroscope.

There were also various pieces of technology designed to aid the crucial touchdown of the Philae lander onto the surface of 67P/C-G. With the gas-cooled thruster known not to be functioning, Philae now had to rely on its two harpoons, set to fire on touchdown, to anchor it onto the surface. However, it was thought that even the firing action of these powerful harpoons required a counter force, the defunct thruster on top, to help prevent Philae being flung off into space. The lander's three legs were capable of absorbing some of the shock of landing, with

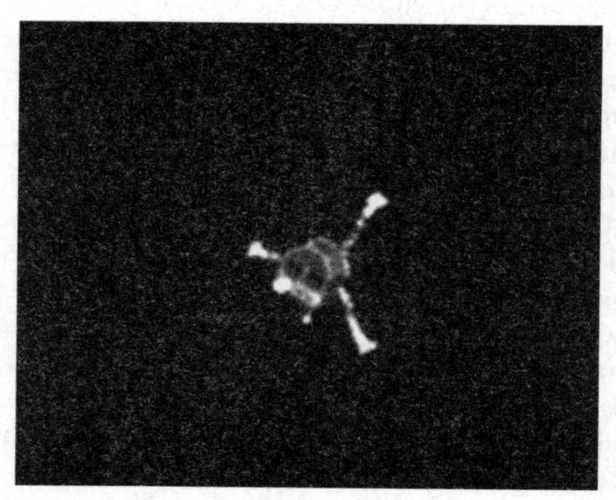

Figure 7. Lander departure. This image from Rosetta's OSIRIS narrow-angle camera shows the lander at 10.23 GMT (on-board spacecraft time) on 12 November 2014 after separation, including the deployment of the three legs and of the antennae. (Image, released 13 November 2014, courtesy of ESA/Rosetta/MPS for OSIRIS Team MPS/UPD/LAM/IAA/SSO/INTA/UPM/DASP/IDA.)

the impact energy being utilized to drive ice screws into the comet that were located on each of the legs. An already very complicated landing – possibly the most technically challenging in history – was made even more unpredictable without the thruster. History now tells us that without the thruster Philae probably had no chance of a bounce-free landing and what resulted was an unexpected and confusing time for the Rosetta scientists. Not only did Philae encounter an impenetrably hard surface, comparable to that of solid ice, as revealed by MUPUS (Multi-Purpose Sensors for Surface and Subsurface Science instrument package), but the harpoons also failed to fire properly in order to anchor the lander onto the surface. The first touchdown recorded by Philae, occurring at 15.34 GMT (with the signal arriving on Earth at 16.03 GMT), was almost precisely where intended in Site J (see Figure 8), showing that the flight dynamics teams had done a great job of setting up the release of Philae from the orbiter. However, the lander rebounded off the surface and travelled up to one kilometre above the comet because, it later transpired, the harpoons and ice screws did not deploy as planned. Philae then experienced two further touchdowns at 17.25 and 17.32 GMT (spacecraft time), and came to rest in a new region. Despite the lander having bounced around the surface of 67P/C-G, the primary science sequence was automatically set to begin straight after the first touchdown, and to last for around forty minutes. The Ptolemy instrument, a gas chromatograph mass spectrometer designed to measure the so-called light elements, such as carbon, nitrogen and oxygen, along with COSAC, a system designed to detect complex organic molecules, were fired up just after first touchdown to make the first measurements and upload data to the orbiter. However, with the lander already some distance above the surface when these measurements were made (see Figure 8), it is unknown exactly what information the instruments will have obtained until the data is inspected and interpreted in detail. It is likely the instruments will have measured material ejected from the surface on initial touchdown – which could tell us a lot about the composition of the gases and organic material composing the comet.

The first images of Philae's final resting place were beamed back to Earth on 13 November 2014 to reveal that the lander was sitting in a rather precarious, but as yet unidentified, location (see Figure 9). Furthermore, with most of Philae in the dark, it was soon apparent that the lander would receive only a fraction of the solar energy hoped for, and

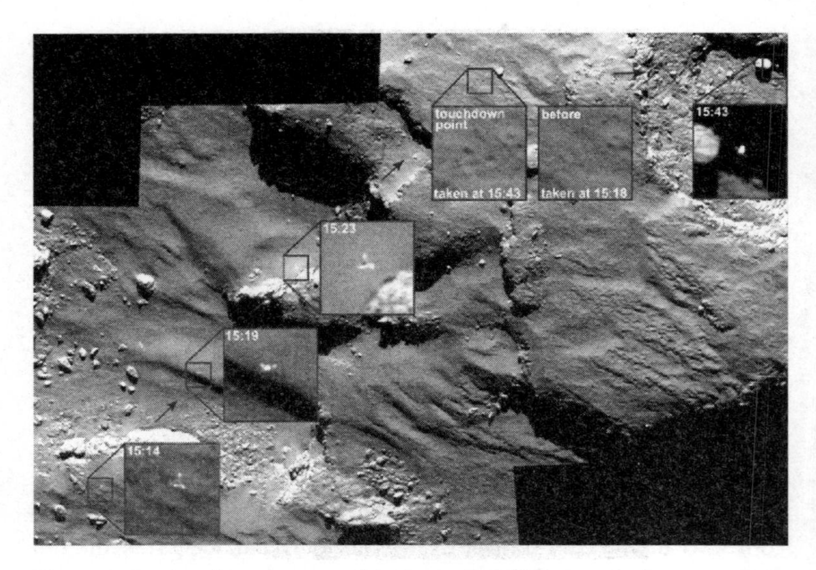

Figure 8. A series of images captured by Rosetta's OSIRIS narrow-angle camera over a 30-minute period spanning the first touchdown on Comet 67P/Churyumov-Gerasimenko on 12 November 2014. The time of each image is marked on the corresponding insets and is in GMT. Touchdown occurred at 15:34 GMT spacecraft time; the image marked 'touchdown point' was taken afterwards, at 15:43 GMT, but clearly shows evidence of the touchdown event when compared with an image taken previously. (Images courtesy of ESA/Rosetta/MPS for OSIRIS Team MPS/UPD/LAM/IAA/SSO/INTA/UPM/DASP/IDA.)

required for continued lander operations. Philae was designed to operate for sixty hours on its primary batteries, after which it was designed to switch to its main batteries which would be recharged from its solar arrays, needing six or seven hours of sunlight a day. As there wasn't enough sunlight for the main batteries, it became apparent that there was limited time to carry out experiments and that a new post-landing operations procedure was rapidly required to get the most out of the new situation – but at this stage it was still unknown which instruments would be capable of working. Furthermore, it was still unknown exactly where Philae had ended up and, with the lander not being safely anchored to the surface with harpoons and ice screws, it would be risky to drill into the comet because it might result in the lander lifting off the surface again and moving to an even riskier location.

With limited lander battery life, meetings were held between mis-

Figure 9. Comet lander panoramic. This unprocessed image, captured by the CIVA-P imaging system on the Philae lander, shows a 360-degree view around the point of final touchdown – the first panoramic image from the surface of the comet. The feet of Philae's landing gear can be seen in some of the frames and the sketch of the Philae lander is superimposed on top of the image showing the configuration of the lander as it was understood at the time. (Image, released 13 November 2014, courtesy of ESA/Rosetta/Philae/CIVA.)

sion control and the lander instrument teams to decide how best to proceed. The great news was that the lander was in one piece, sitting, albeit rather precariously, on the surface of the comet. The downside was the challenge of how to make best use of the power available to find out as much about the comet as possible. This was the first time in history that scientists had an opportunity to make real scientific measurements on the surface of a comet. It was an intense few days for all involved as they worked around the clock to make sure each instrument could attempt to make some measurements. Philae's Sampling, Drilling and Distribution (SD2) subsystem was activated towards the end of lander operations, despite fears that it might alter the position of the lander. However, it would not be known whether the drill contacted the surface successfully – the lander was thought to be sitting over a gaping hole – until the results from the other lander instruments had been carefully assessed. The other instruments were turned on and they made measurements, but with power rapidly declining, the operations had to come to a halt and the lander was safely powered down just after the last data was transmitted back to the orbiter to be beamed back to Earth. Philae was now, once again, asleep, but this time it was on the surface of a comet heading towards the Sun.

SINCE LANDING

It is expected that cometary activity, for any comet orbiting the Sun, will increase towards the time of perihelion as the comet nucleus gets warmer, allowing for frozen gases to sublimate and carry dust particles away from the surface. By March 2015, this activity was well documented, with dust jets at all times on the day side of the comet. In mid-March 2015, OSIRIS captured images of the exciting emergence of a new jet of dust bursting from the surface of the comet (see Figure 10). The sudden release of this new jet could be due to heat reaching ices trapped in a deeper layer beneath the surface, but it is unknown if this was a short-lived outburst or continuous jet. With this information, and further monitoring of the comet, scientists will be able to test dif-

Figure 10. Emergence of jet outburst. The scene at 67P/C-G at 07.13 (left) and 07.15 (right) on 12 March 2015, clearly showing the moment a jet of activity bursts into existence. With less than six months until perihelion, observing such activity allowed scientists to better understand cometary activity and predict how comets behave as they approach the Sun. (Image courtesy of ESA/Rosetta/MPS for OSIRIS Team MPS/UPD/LAM/IAA/SSO/INTA/UPM/DASP/IDA.)

ferent models of cometary activity to better understand how comets behave as they approach the Sun.

THE RE-AWAKENING OF PHILAE

In June 2015, after receiving enough sunlight on its solar panels during seven months of hibernation on the comet's surface, the Philae lander woke up once again. It was unclear how long Philae had been awake before the lander signal was received on 13 June 2015 because Philae's memory had already stored eight thousand data packets. With continued sunlight towards perihelion, the power situation was only set to improve, allowing for the possibility of continuous lander operations. Despite the precarious resting place that the lander ended up in after its tricky landing in November 2014, at an angle under a cliff, had it remained in the direct sunlight of Site J, the intended landing spot, then the lander almost certainly would have become too hot to operate that close to perihelion. With Philae awake, it meant that the lander instrument teams had to decide which experiments should run, but this had to be carefully balanced by the requirements of the orbiter instrument teams. If the orbiter came too close to the active comet, then it risked getting lost because its star-tracking system could no longer operate amongst all the confusion of dust particles streaming from the comet surface. However, the orbiter was required to collect and relay the lander's data back to Earth, which required it flying as close as possible to 67P/C-G to receive Philae's signal.

THE FUTURE FOR ROSETTA

What does the future hold for Rosetta? The fact that the lander awoke on the comet surface made for an even more exciting time for the Rosetta mission and comet science in general. If the lander was able to perform some more experiments up to perihelion, despite being perched rather precariously on the active comet as it approached the Sun, then the Rosetta scientists will gain further invaluable information about the comet's composition. Only time will tell what more Rosetta will reveal to us but 67P/C-G will continue its cometary journey after August 2015 regardless, journeying once again into deep space. The

Rosetta orbiter will also continue to track the progress of the comet during this time, performing experiments to understand how the comet's coma interacts with the Sun. There will be many new results released from the lander and orbiter teams over the next few years and scientists will continue to analyse and interpret the plethora of data obtained by the mission well into the future as it will inform our understanding of other cometary, and even asteroidal, bodies in space. There have been many firsts for the Rosetta mission, with the most obviously impressive being orbiting a comet and performing a controlled landing on its surface. The media buzz around the mission has continued even after the spectacular events of 2014 and 2015. After all, the rather audacious Rosetta mission, over twenty years in the making, arguably has to be one of the greatest of our lifetimes.

The Alpha Magnetic Spectrometer

STEPHEN WEBB

INTRODUCTION

The International Space Station (ISS) is the single most expensive object ever built: the total cost of assembling this structure, which is about the size of a soccer pitch and has about the same mass as two fully laden Boeing 787 Dreamliners, is estimated to be in excess of $150 billion. The ISS is unquestionably an impressive political and technological achievement. This collaborative venture of America, Brazil, Canada, Japan, Russia and eleven member states of the European Space Agency has led to the creation of the largest man-made object in space – many readers of this volume will have seen the ISS as a bright, naked-eye object as it speeds westerly across the sky at an altitude of between 330 and 435 kilometres. Furthermore, by offering a home for astronauts, it has permitted a continuous human presence in space since November 2000. But $150 billion is a lot of money for an enterprise supposedly devoted to science and technology. After all, the Large Hadron Collider recently provided the scientific community with knowledge of the essential make-up of the universe (see 'The Nature and Mass of the Higgs Boson', *Yearbook of Astronomy 2014*) and the machine that made this fundamental scientific breakthrough came with a price tag of about $10 billion. The eye-wateringly large expenditure involved with the space station has led several scientists to ask: is the ISS worth it?

Will future generations come to view the International Space Station as an orbiting white elephant? This is a difficult question to address. The ISS has produced some political and cultural benefits, and many would argue that the cost is a necessary down payment on our future as a space-faring species. On the other hand, it is difficult to defend the cost in terms of the station's capacity to make basic scientific discoveries: although some of the science that takes place on-board can only happen in a microgravity environment, much of the research could be carried out more cheaply and efficiently by robotic probes. Nevertheless, the ISS is home to an experiment – the Alpha Magnetic Spectrometer – that aims to shed light on one of the biggest enigmas in science: the nature of dark matter. If that experiment can discover dark matter, then the ISS will have cemented its place in scientific as well as technological history. The construction cost then might well be considered to be money wisely spent.

The Alpha Magnetic Spectrometer is a particle detector. The purpose of this chapter is to report on the early findings of this space-based experiment, and the current status of its hunt for dark matter, but also to describe the story behind its development: the deployment of a particle detector on the ISS is in large part due to the drive and ambition of a remarkable scientist – Samuel Chao Chung Ting. However, in order to understand the rationale behind the detector, and why Sam Ting devoted years of his life to building it, we first need to understand something about the particles it detects. The Alpha Magnetic Spectrometer looks for antimatter. So what is antimatter – and what is its relationship to dark matter?

ANTIMATTER AND DARK MATTER

The material world consists ultimately of various types of elementary particle – electrons, neutrinos, quarks, photons and so on – and every type of elementary particle comes with a so-called antiparticle. If certain conditions are met, and in particular, if a particle is electrically neutral, then it is possible for the particle to serve as its own antiparticle: the photon, for example, is its own antiparticle. However, if the particle carries an electrical charge, then it always has a distinct antiparticle, an object with precisely the same mass as itself but with opposite electrical charge. (Some particles carry more exotic types of charge in addition to electrical charge; an antiparticle always has the same mass as the particle but opposite values for all the charges.) The most familiar example of antimatter is the anti-electron, more commonly known as the positron, which possesses exactly the same mass as the electron but carries a positive unit of electric charge. Another example of an antimatter particle is the antiproton, which is not itself elementary but consists of constituents – quarks – that are elementary. The proton and the antiproton have precisely the same mass, but whereas the proton carries a positive unit of electrical charge the antiproton carries a negative unit of electrical charge.

One of the more dramatic consequences of the existence of antimatter is the phenomenon of mutual annihilation. If a particle encounters its antiparticle, then the pair can mutually annihilate and vanish into a flash of high-energy photons; subsequently, this radiation can convert back into particle/antiparticle pairs. The amount of radiation released in this process is vast: Einstein's famous equation $E = mc^2$ tells us not only that matter and energy can be interchanged but also, because the term c^2 is so large, that the annihilation of just a small amount of mass creates a huge amount of energy.

The phenomenon of mutual annihilation is routinely observed in particle accelerators, and it is extremely well understood by physicists, but it generates a thorny problem for astronomers. The problem is that all the 'stuff' we see out in space – planets, stars, clouds of gas and dust – is made of matter. In other words, everything we observe consists of negatively charged electrons and positively charged atomic nuclei. Although this fact might seem self-evident to the point of banality, it is in fact a profound puzzle. The laws of physics suggest that the Big Bang must have created equal amounts of matter and antimatter, and yet we see no signs of the tremendous release of energy that would occur if a galaxy encountered an anti-galaxy or a star collided with an anti-star. Where is all the antimatter that was created in the Big Bang? What could have caused the imbalance we see today between matter and antimatter?

One way to learn more about this question might be to study any antimatter that *does* exist. Although our telescopes observe only matter it turns out that we *do* encounter antimatter – in the cosmic radiation that rains constantly down upon us. As one would expect, given that the material in the universe appears to be made only out of matter, the cosmic ray flux consists mainly of particles: about 90 per cent of the cosmic rays that originate outside Earth's atmosphere are protons; about 9 per cent are alpha particles or the nuclei of even heavier elements; and about 1 per cent are electrons. But the observed flux of cosmic rays does also contain a small number of positrons; cosmic ray detectors occasionally even see antiprotons. (If we were ever to detect significant numbers of anti-helium nuclei, then this would be good evidence for the existence of primordial antimatter.) At this point, however, you might reasonably ask: if the universe consists of matter, then why should we expect to detect any antimatter at all?

The reason cosmic rays contain an antimatter component is that small amounts of antimatter are being created all the time. Those fast-moving protons that constitute about 90 per cent of cosmic rays occasionally smash into other particles in the interstellar medium, and the collision energies involved can be huge – far in excess of those available at the Large Hadron Collider, for example. And just as Einstein's equation $E = mc^2$ tells us that massive particle/antiparticle pairs can mutually annihilate into a burst of pure energy it also tells us that the reverse can happen: pure energy can convert into particle/antiparticle pairs. There are constraints on what can be created in such a process: various quantities such as momentum, angular momentum and electric charge must all be conserved, and the total mass of the particles produced cannot exceed the available energy, but within those constraints particle/antiparticle pairs can be brought into existence from pure energy. The process is called pair production (see Figure 1 for an illustration) and physicists routinely observe it in their particle accelerators. Since the electron is much less massive than the proton, it follows that electron/positron pairs are much more likely to be produced in collisions than proton/antiproton pairs, which themselves are much more likely to be produced than even heavier objects. (This is why we would not expect to see large numbers of anti-helium nuclei, unless they were relics of the Big Bang.) We can conclude, therefore, that the cosmic ray flux will contain a small amount of antimatter – not necessarily primordial antimatter from the Big Bang, but antiparticles created in particle collisions – and that it will be mainly in the form of positrons with only the occasional antiproton.

The observation of primordial antimatter in the cosmic ray flux would be tremendously exciting, but that possibility is not the main thrust of this article. Rather, we are concerned here with the possibility of a different addition to the antimatter component of the cosmic ray flux – an addition that might come from so-called dark matter.

Various lines of evidence suggest that the universe contains about six times more dark matter than regular matter. Unfortunately, although physicists have been very good at determining what dark matter *isn't*, they have signally failed to determine what it *is*. Whatever dark matter turns out to be, we know that it lies outside the scope of our

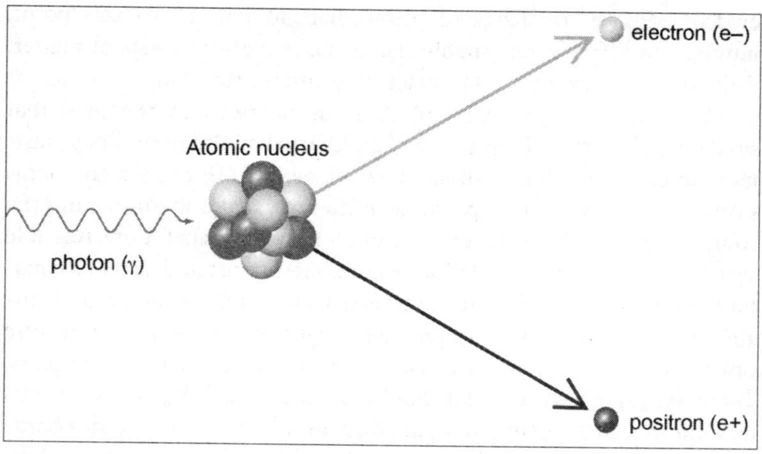

Figure 1. When a photon (represented by the wiggly curve) encounters an atomic nucleus, the photon's energy can be converted into matter, in the form of an electron and positron pair, through Einstein's equation $E = mc^2$. Conservation of energy means that the photon must have energy greater than the combined mass of the electron and positron. Furthermore, conservation of momentum means that the photon must be in the vicinity of a nucleus for the conversion to occur: the nucleus recoils when this pair production takes place. If the constraints are satisfied, however, this process converts energy into matter. It is an example of how antiparticles can be produced in the present-day universe. (Diagram courtesy of Stephen Webb.)

currently accepted theories and for that reason physicists desperately want to identify it.

Now, although science has yet to identify the constituents of dark matter, there is a longstanding suggestion that it might consist of weakly interacting massive particles – WIMPs, for short. Such particles would make their presence felt only via gravity and the weak force. This would explain why we don't see WIMPs and why it is so difficult to detect them – they simply don't interact readily with matter (or, indeed, with other WIMPs). Various hypothetical particles have been proposed as WIMP candidates, with many physicists favouring the neutralino – an entity predicted to exist by certain theories in high-energy physics. If neutralinos are indeed the source of dark matter – and, to reiterate, this is only one of several suggestions – then the Galaxy must be home to vast clouds of them. Occasionally, in those clouds of dark matter, a neutralino will by chance come very close to another neutralino. The result? Well, like the photon, the neutralino happens to be its own antiparticle. The result of the encounter is the same as when any particle encounters its antiparticle: mutual annihilation. The two dark-matter particles would vanish and produce high-energy radiation and particle/antiparticle pairs. If neutralinos exist, then, their occasional mutual annihilation will produce high-energy positrons, which would become part of the cosmic ray flux. Furthermore, their annihilation would produce positrons with a particular and distinctive spread of energies: in 1990, Mike Turner and the Nobel laureate Frank Wilczek showed that neutralino annihilation would cause the number of positrons to increase with energy, up to a maximum energy determined by the neutralino mass, and then drop off sharply. This distinctive dark-matter signal could in principle be detected.

Unfortunately, it would be very difficult to search for this distinctive spread of cosmic ray positron energies using a ground-based detector. When primary cosmic rays from the distant universe smash into the molecules in our atmosphere they generate a cascade of secondary particles. Teasing out the signal from those secondary particles would be extremely difficult. What's needed is a space-based experiment that can detect the primary cosmic rays directly. And that's where Sam Ting comes in.

SAM TING AND THE ALPHA MAGNETIC SPECTROMETER

Sam Ting is a daring, demanding and driven physicist with a reputation for leading large, complex experiments and carrying them out with exquisite precision and attention to detail. In 1974, for example, he directed a team at the US Brookhaven National Laboratory that discovered a new fundamental particle – the charm quark. (More precisely, his team discovered a new meson that is the bound state of a charm quark and an anticharm quark. A rival team from Stanford found the new meson at about the same time.) The discovery was of huge importance in the field of high-energy physics, and initiated a revolution in our understanding of fundamental particles. Just two years later, Ting was awarded the Nobel prize (along with Burton Richter, who led the rival team). This is one of the shortest intervals between discovery and award in the history of the physics prize. The Nobel committee compared the precision of Ting's experiment, his achievement in isolating just a few particles of interest from the millions of other particles passing through the detectors, to 'hearing a cricket close to a jumbo jet taking off'.

Ting's reputation grew and by 1983 he was leading the largest physics collaboration in the world: an experiment, called L3, which involved five hundred physicists at CERN's Large Electron Positron collider. L3 was housed in the tunnel that is now home to the Large Hadron Collider. Ting was able to persuade not just the 'usual suspects' to take part in his experiment; he got China and the then Soviet Union to participate. L3 eventually closed, however, and Ting's proposed upgrade to the experiment was turned down. (The huge magnet that was part of the L3 experiment has since been inherited by the ALICE experiment at the Large Hadron Collider.) Ting then led a coalition bidding to run the planned Superconducting Super Collider in Texas. This would have been a huge device, much larger even than the Large

Figure 2. A portrait of the physics Nobel laureate Sam Ting, taken in 2008 to celebrate the inauguration of the Large Hadron Collider. The photographer, Volker Steeger, asked Ting to make a drawing of his most important discovery; Ting drew a diagram showing how he discovered the charm quark. He signed his name in both Mandarin and English. (Image courtesy of CERN.)

Hadron Collider, and the size of Ting's proposed team matched the ambition of the project: a thousand scientists from more than ninety institutions situated in thirteen countries. Ting's proposal was declined, however, and plans for the Superconducting Super Collider were cancelled.

After the rejection of these two large-scale projects, Ting and his collaborators turned their thoughts to something smaller. They wondered whether a relatively small grouping of scientists – a hundred physicists, say – would be able to develop a table-sized experiment, for deployment in space, that could study the composition of cosmic rays. As described in the previous section, by studying the antimatter component in cosmic rays scientists might learn whether the universe truly does consist of matter or whether there might be hitherto undiscovered regions of the universe dominated by antimatter. They might make other discoveries, such as strangelets – hypothetical particles that are of interest to theoretical physicists. And, as already alluded to, they might detect a signal from dark matter. The lure of possibly making these discoveries convinced Ting to think deeply about how to construct such a space-based cosmic-ray detector.

At this point it's worth considering the first question faced by Ting: how would the detector distinguish between a particle and an antiparticle? The answer, as Ting of course knew, is to use a magnetic field. When an electrically charged particle moves in a magnetic field it experiences a force that causes its path to bend. This fact allows physicists to develop a device that can identify the charged particles passing through it. If a trajectory bends one way in a strong, uniform magnetic field, then we have observed a particle; if the trajectory bends in precisely the opposite way in the same magnetic field, then we have observed its antiparticle. By measuring the curvature of the trajectory we can determine the object's mass – and thus distinguish, for example, between the negatively charged electron and the negatively charged antiproton (which is 1836 times more massive than the electron). By measuring, in addition, quantities such as kinetic energy, we can build up a detailed picture of the particle or antiparticle. Such a device, making use of a strong magnetic field, would be an example of a magnetic spectrometer. This was straightforward in principle, if not in practice. Ting's real challenge, the next question he faced, was: how would it be possible to situate such an experiment where it needed to be – in space?

In 1994, Ting convinced Dan Goldin – the then boss of NASA – that a magnetic spectrometer on-board the ISS could do important science. Perhaps Ting did not need to apply all his legendary skills of persuasion: even two decades ago, Goldin was sensitive to criticisms regarding the scientific contributions that a space station could make. It must have helped, too, when Ting told him NASA wouldn't have to pay for the development of the spectrometer; NASA would only need to get the experiment onto the ISS. Goldin was sold on the idea. He agreed to a place for the spectrometer on the station and to a Space Shuttle flight for delivering it. (Other scientists, it must be said, who were obliged to go through a peer-review process to get their experiment on the ISS, were not pleased at this seeming favouritism.)

Ting's initial notion – that he would lead a small-scale collaboration to develop a device the size of a tabletop – soon crumbled. But then Ting never seems to work on a small scale. The delivery of his vision turned out to require input from five hundred scientists, from fifty-six institutions based in sixteen different countries – a complex task, but precisely the sort of large-scale endeavour at which Ting excels. Ting himself confesses to having worried about the details of the experiment every day for thirteen years.

In 1998, Space Shuttle *Discovery* carried a prototype of the experiment into space. The device was called the Alpha Magnetic Spectrometer, or AMS for short. Since it was only a prototype, a proof-of-concept for a more advanced experiment, this first device was also called AMS-01. It operated for ten days on the Mir space station. AMS-01 was a simple experiment, but it successfully demonstrated how data could be extracted from charged cosmic-ray particles passing through a magnetic field. Furthermore, from the fact that it looked for but did not detect anti-helium, it put limits on the presence of primordial antimatter. Not bad for a prototype. Work immediately began in earnest on the full-scale experiment: AMS-02. (In the rest of the article I shall simply write AMS; the reader can safely assume that the acronym refers to AMS-02.) Even with a large team working on AMS the development took years. Once complete (see Figure 3), the full AMS set-up underwent rigorous testing at CERN, including exposure to particle beams generated by the accelerators there. It was then shipped to an ESA facility in the Netherlands, where it underwent a variety of other tests to ensure it would function in the harsh environment of space; it was known that the thermal stresses on the experiment would be particularly acute. By any measure, this was a

Figure 3. AMS-02 standing in the clean room at CERN. The experiment was specifically designed to work as an external module on the ISS. It was assembled at CERN, and underwent thorough testing there (and later at the European Space Research and Technology Centre in the Netherlands), in order to be sure that it was ready for space. To give some idea of scale, the completed instrument has a height of 4 metres and a volume of 64 cubic metres. It is the largest scientific payload carried by the ISS. (Image courtesy of AMS-02 Collaboration.)

sophisticated operation. The intention was to send the experiment to the ISS in 2003.

On 1 February 2003, Space Shuttle *Columbia* disintegrated over Texas and Louisiana upon re-entry into Earth's atmosphere. All seven astronauts on-board died. Following this disaster, NASA decided to reduce the number of flights and to stop the shuttle programme completely by 2010. The flight that Goldin had promised Ting was one of those that were cancelled. Other options for delivering AMS to the ISS were discussed, but declined as being prohibitively expensive. For a long time it seemed to many people that AMS would never be launched – but those people underestimated Ting's political guile and sheer single-mindedness. He got the Italian foreign minister to argue his case to the then Secretary of State Condoleezza Rice. Ting himself pleaded with a Senate committee, and he later managed to convince a number of powerful senators of the value of AMS. In 2008, President Bush signed a bill authorizing NASA to add a shuttle flight carrying the Alpha Magnetic Spectrometer before the shuttle programme came to its end. On 16 May 2011, AMS was launched on what was the final flight of Space Shuttle *Endeavour* and the penultimate mission of the entire Space Shuttle Program. The experiment was installed on 19 May 2011 (see Figure 4). Sam Ting was perhaps the only physicist who could have made this happen.

Before discussing what AMS has seen since it began detecting cosmic rays in 2011, let's first take a brief look at the instrument itself.

Figure 4. The AMS-02 instrument as it was installed on the starboard truss of the ISS in May 2011 by the crew of the Space Shuttle *Endeavour.* (Image courtesy of NASA/AMS-02 Collaboration.)

THE ALPHA MAGNETIC SPECTROMETER

The Alpha Magnetic Spectrometer, as mentioned earlier, is essentially a particle detector. It is 8500 kg in mass and has a volume of 64 m^3. At its core is a large magnet, producing a magnetic field four thousand times greater than Earth's. The field configuration is carefully set so that it does not couple with Earth's magnetic field: if the two fields *did* couple, the motion of the ISS itself would be affected! The field bends the path of any electrically charged particles moving through it: electrons and antiprotons bend one way, positrons and protons bend in the opposite direction. And *lots* of particles pass through. If you visit the AMS web-page, you will see a count of how many cosmic rays the instrument has measured; as I write this, the AMS has measured 59,843,134,844 cosmic rays. That's roughly 45 million particles every day since it began operations in 2011.

Five different subsystems and detectors measure various properties – charge, velocity, trajectory and energy – of all those particles. A subsystem rejects cosmic rays that pass through the walls of the magnet; only particles that pass through from top to bottom are recorded. Two star trackers define the position and the orientation of the experiment, and there's a sophisticated on-board GPS system too. An advanced set of electronics transforms the signals generated by the various particle detectors into useable information and handles the torrent of data that the experiment generates – a CD's worth of data every second. And all this has to work in the unforgiving environment of Earth orbit. (See Figure 5 for more details.)

The full cost of developing all this equipment, and then installing AMS on the ISS, is estimated to be $1.5 billion. Although that figure represents only a fraction of the cost of the ISS itself, it *is* expensive for a scientific experiment. The question that was pointed at the ISS – namely, is it worth it? – has therefore been pointed at AMS. The answer, I presume, will depend on whether the experiment discovers something of interest. At the time of writing there have been no sightings of strangelets or primordial antimatter – two of the discoveries that its supporters hoped it might make. However, AMS *has* made a discovery that *might* tell us something about the nature of dark matter: the key lies in the exquisite precision with which it has been able to study the so-called positron fraction.

THE POSITRON FRACTION

We argued earlier that particle collisions taking place in the Galaxy will generate antiparticles, and therefore the cosmic ray flux should contain a small amount of antimatter in the form of positrons. But just how many positrons should we expect to see? Well, astrophysicists would

Figure 5. AMS-02. The detector module is a complicated piece of equipment! At the top of the instrument is the transition radiation detector (TRD), which measures the velocities of ultrarelativistic particles. The small tube visible in front of the TRD is a star tracker, which can tell where the module is pointing and allows the system to reconstruct the arrival direction of high-energy radiation. The corrugated vacuum case protects the heart of the experiment: the cooling system, the magnet, the silicon tracker (which measures the path of charged particles in the magnetic field), the anti-coincidence detector (which rejects particles entering through the side of the experiment), and so on. Directly above and below the vacuum case are time-of-flight counters, which measure the velocities of lower-energy particles. Below the main vacuum case is a Cerenkov counter, which measures the velocity of relativistic particles; and below that is an electromagnetic calorimeter, which measures the total energy of particles and can distinguish very accurately between positrons and protons. (Image courtesy of AMS-02 Collaboration.)

like to be able to calculate the positron flux – the number of positrons observed with a particular energy in a given period of time – but such calculations are extremely difficult. The galactic environment is a messy place, with twisting magnetic fields and winding threads of gas and tangled coils of plasma, so the mathematical models that attempt to describe the motion of positrons rapidly become complicated.

Figure 6. AMS-02 is clearly visible towards the top left of this photograph. Astronauts installed the experiment on the ISS in May 2011, and each and every second since then about a thousand cosmic ray particles have passed through the top of the instrument. (Image courtesy of AMS-02 Collaboration.)

However, the magnitude of all these complications should be the same for positrons and electrons because these particles are, in a sense, mirror images of each other. Apart from charge, they are identical. Therefore, even if astrophysicists are unable to calculate the positron *flux*, they should be able to use their models to at least estimate the positron *fraction* – the ratio of positrons to the total number of electrons plus positrons. When they carry out these calculations they find that the various models all make the same clear prediction: the positron fraction should decrease as the energy of the particles increases.

In the 1990s, balloon-borne cosmic-ray experiments saw tantalizing hints that the positron fraction did *not* decrease with increasing energy; in fact, it might even increase. If these hints were true, then there must be a new, as yet unknown, source of positrons that the models had not taken into account. And, as discussed earlier, the annihilation of dark-matter particles might provide just such a source of positrons. This possibility naturally caused a great deal of interest.

Unfortunately, because the balloon experiments came with large measurement uncertainties and covered only a small range of energies, astronomers were unable to reach any firm conclusions: it was quite possible that those hints of a positron excess were nothing more than experimental 'noise'. The situation changed in 2008, when data from an orbiting mission called PAMELA (Payload for Antimatter Exploration and Light-nuclei Astrophysics) suggested that the positron fraction initially fell with energy, as expected, but then increased at higher energies. Since PAMELA operated over a larger energy range than the balloon experiments, and made its measurements with more precision, this was strong evidence for a positron excess. One could still argue that this result required independent verification: PAMELA registered up to ten thousand times as many protons as positrons, so if the experiment misidentified just a small number of protons as being positrons, then its results would be misleading. In 2011, however, scientists using the Fermi Gamma-ray Space Telescope confirmed the PAMELA result. Fermi and PAMELA employed radically different approaches but both teams saw a positron excess. It was clear, then, that the positron fraction increases with energy. However, Fermi and PAMELA lack the capability of answering the questions that might help determine the source of the positron excess; questions such as: At what energy does the positron fraction begin to increase? At what rate does the positron fraction increase with energy? Are there any sharp structures in the graph of positron fraction against energy? Do the positrons come from particular directions or is the signal isotropic? At some energy the positron fraction will presumably cease to increase; at what energy does that occur? After that turning point, at what rate does the positron fraction decline? To answer those questions one requires an experiment such as the Alpha Magnetic Spectrometer – and, from just two years of data, the AMS collaboration has answered five of the questions posed above.

The answers to those questions can be seen in the diagrams below, which are based on the first 41 billion cosmic ray events registered by AMS and which were released in September 2014.

Figure 7 shows the behaviour of the positron fraction up to an energy of 30 GeV. (For those who haven't encountered the unit before, the GeV is simply a convenient unit for discussing particle energies. To give some idea of scale, if a proton and an antiproton were to meet slowly and annihilate into two photons, then those photons would each have an energy of 0.938 GeV. Making use of the equation $E = mc^2$, and employing appropriate units, physicists say that the proton has a mass of 0.938 GeV. For comparison, in these units the mass of a gold atom is about 182 GeV. The recently upgraded Large Hadron Collider will produce collision energies of 14000 GeV.) It is clear from Figure 7 that the positron fraction falls initially, but at 8 GeV it begins to increase. This answers the first question.

Other questions are answered by a consideration of Figure 8. AMS has discovered the location of the positron fraction maximum: the ratio increases up to an energy of 275±32 GeV, after which it begins to fall. The graph of positron fraction against energy is smooth: there are no structures that might indicate the existence of some particular process. Further observations indicate that the flux is isotropic: there is no preferred direction in space from which the positrons arrive. Positrons strike Earth equally from all directions.

AMS has made yet another contribution to the debate: it clearly confirms that the rise in positron fraction is due to an excess of positrons rather than a lack of electrons. So astrophysicists now know for certain that they have a puzzle to explain. They have to explain why Earth is being bombarded by positrons, a bombardment that seems to increase with increasing energy up to about 275 GeV and that comes equally from all directions.

Don't we have an answer to this puzzle? Isn't the answer that dark-matter particles are annihilating and producing this positron excess? Hasn't Ting found dark matter? Well, the AMS results represent a tantalizing hint – but unfortunately they aren't yet proof of dark matter.

DARK MATTER OR SOMETHING MORE MUNDANE?

A quarter of a century ago, Turner and Wilczek predicted what we might see if dark matter consists of neutralino particles that can annihilate and thereby produce electron/positron pairs. They argued that, if we studied the cosmic ray flux, we would see the fraction of positrons increase smoothly with increasing energy, and then at some maximum energy (determined by the mass of the particle that constitutes dark matter) the fraction of positrons would decrease sharply. The positron flux itself would be isotropic. The graphs shown in Figures 7 and 8 exhibit almost all the characteristics predicted by Turner and Wilczek, and are consistent with what one might expect from dark-matter particles with a mass of around 1000 GeV. One might therefore be forgiven for believing this to be 'smoking gun' evidence for dark matter. However, such a belief would be premature: the graphs don't show *every* characteristic predicted by Turner and Wilczek. A key signature for dark matter is that the positron fraction drops sharply after the maximum is reached; the graphs, quite noticeably, do not continue much beyond the maximum. We simply do not yet know whether the positron fraction drops quickly or slowly after the maximum. We will have to wait for the AMS collaboration to release more data.

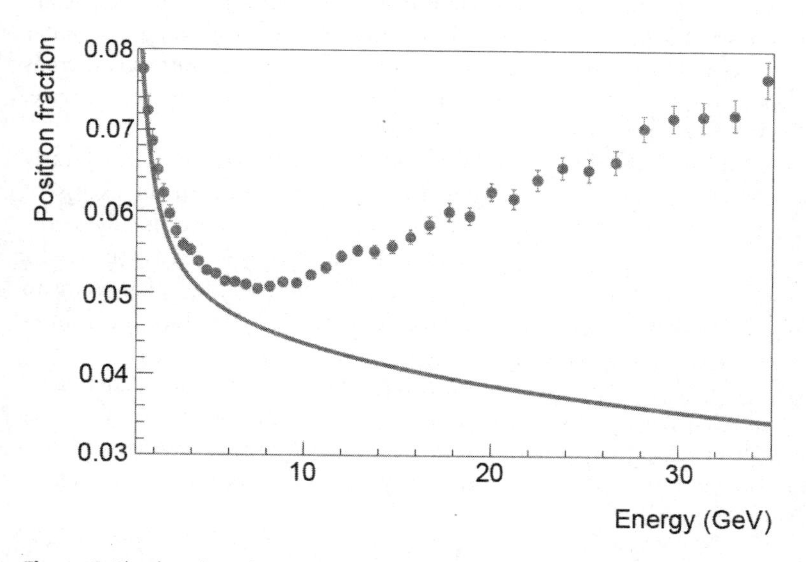

Figure 7. The dots show the positron fraction at different energies as measured by AMS. The solid curve shows the expected behaviour, assuming that positrons originate from the collision of cosmic rays. Above 8 GeV the positron fraction increases. Note the exquisite precision of the measurements. Because there are so many events, and because AMS subsystems are able to distinguish so clearly between positrons and protons (thanks in no small measure to Ting's fabled attention to detail and insistence on getting every minor detail correct), the error bars on the measurements are much smaller than Fermi or PAMELA could possibly achieve. (Image courtesy of AMS-02 Collaboration.)

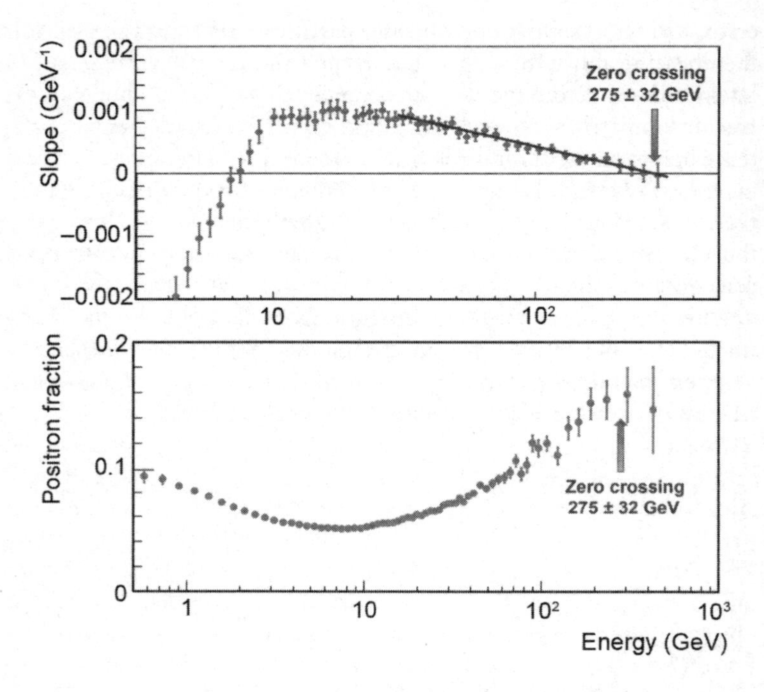

Figure 8. In the upper plot the dots represent the slope of the positron fraction at different energies, as measured by AMS. At high energies a straight line can be drawn through the dots, and where the line crosses zero is where the positron fraction ceases to increase. Measurements show that the positron fraction stops increasing somewhere in the energy range 243 GeV to 307 GeV. The lower plot shows the measured positron fraction over a large energy range. AMS sees no evidence for sharp structures in the graphs. (Image courtesy of AMS-02 Collaboration.)

At the time of writing it is possible that Sam Ting has an idea of how the positron fraction behaves after the maximum. Ting, however, has a reputation for not publishing a result until he is absolutely sure he has addressed all possible issues. Such a painstaking approach is surely admirable in a scientist, but on one occasion it was almost Ting's undoing. His team discovered the charm quark, the discovery for which he was awarded the Nobel prize, on 31 October 1974. Despite being told by at least one colleague that he was crazy not to publish immediately, Ting preferred to check his results and ensure that he wasn't in error. Richter's group found the same particle on 10 November 1974. It then became a race to publish and in the end the prize was shared. In attempting to determine the behaviour of the positron fraction Ting has no competitors: his AMS is currently the only experiment that can tell us precisely what happens to cosmic ray positrons as their energy increases. We just have to wait until Ting is certain enough to publish the results. With luck, the reports of those results will appear around the time this Yearbook is published: readers should keep their eyes peeled for news!

There remains an important question: If AMS observes the positron fraction to fall precipitously, in the manner outlined by Turner and Wilczek, would Ting then have observed dark matter? The somewhat frustrating answer probably would be: Maybe, maybe not.

When PAMELA and Fermi observed a rising positron fraction, astrophysicists rushed to consider whether there could be some explanation other than dark matter. They came up with another scenario: perhaps nearby pulsars are injecting high-energy positrons into space. The pulsar scenario at first glance seems plausible, and it has the advantage of being able to explain the positron excess in terms of known objects (pulsars) rather than unknown objects (dark-matter particles). Fortunately, experiments are taking place that will soon confirm this picture or else rule it out. Even if the pulsar picture is ruled out, it's possible that our surprise at the behaviour of the positron fraction is simply due to a poor understanding of how cosmic rays are accelerated to high energy and of how they propagate. Nevertheless, if AMS observes the positron fraction to fall sharply, you can expect to hear some commentators claim that dark matter has been discovered. You can also expect to hear other commentators argue that the observation is not explained by dark matter because the WIMP interpretation of AMS is in conflict with constraints provided by other experiments, such as that from the Planck satellite (see 'Planck's New View of the Universe', *Yearbook of Astronomy 2014* for more details). If nothing else, the debate will highlight just how important it is for science to understand dark matter – and what a fascinating area of astrophysics the search for dark matter has become!

CONCLUSION

The problem of dark matter is so long-standing and so puzzling that ever-increasing levels of resource are being diverted to solving it. Several underground experiments are trying to detect dark matter by looking for signs of it scattering off atomic nuclei; physicists at the Large Hadron Collider are hoping to produce dark-matter particles; and the AMS collaboration is attempting to detect the effects of dark-matter annihilation. Historically, these three approaches – *scattering* experiments, *production* experiments and *annihilation* experiments – have all led to exciting discoveries in particle physics. If future scientists look back and conclude that AMS was the experiment that first detected dark matter, then I believe they'll also conclude that its home, the International Space Station, was a scientific success. For this discovery alone, the ISS will have been worth the huge cost of its construction and maintenance. If AMS fails to find evidence for dark matter, primordial antimatter, or strangelets . . . well, I'll leave the reader to decide whether the non-scientific benefits of the ISS have been sufficient to justify the cost.

Broughty Ferry, 1882

RICHARD BAUM

It is widely understood in astronomy that while many look, few actually observe: the distinction, of course, marks the difference between rigorous, exact observation and the casual regard of the dilettante. While there is nothing incompatible in this it is important to remember that the enthusiastic but often untutored response of the latter occasionally results in misleading or spurious assumptions as we have so often seen in recent times. A good illustration is the so called 'Strange Phenomenon' seen on the 21 December 1882 from Broughty Ferry, a suburb on the east side of Dundee in Scotland, overlooking the Firth of Tay.

According to the *Dundee Advertiser* of 22 December 1882:

> Yesterday forenoon, between ten and eleven o'clock, the attention of several persons in Broughty Ferry was directed for a time to a somewhat unusual sight in the heavens. The sun at the time was shining brightly, being about due south, when a star was seen in close proximity to it. The star was a little above the sun's path, and the peculiar phenomenon was seen by various persons, who had their attention directed to it. Being daytime, the star did not have the brilliant luminous radiance stars exhibit at night, but was of a milky white appearance, and seemed, when seen through the glass, to be of a crescent shape. Being on a light-blue ground, and lying between two white clouds, it was seen to great advantage . . .

Three days later, a letter from a well-informed but anonymous writer in the Christmas Day issue of the paper quietly dismissed the alien character of the appearance, thus:

> The Broughty Ferry correspondent does not seem to be aware that the planet Venus can be seen with the naked eye at any time during the day when she is within a few weeks of her greatest brilliancy. The

> phenomenon which he records was easily witnessed in Dundee yesterday in bright sunshine and from this time till about the middle of February [1883] it can be seen daily if the sky is clear. The description which he gives of it makes it quite certain that it was the planet Venus and no other. It is at present a little to the west of the sun, and above it. It will continue to recede from the sun, going westwards from it, till the beginning of February. At present it is of a crescent shape; but as it continues to course round the sun it will exhibit more and more of the fully rounded till September 20th, when it will be in superior conjunction with the sun, and again appear as an evening star. The planet attains its greatest brilliancy on January 10th, [1883] when it will be quite easy for the most unskilled eye to see Venus in the day time.

And that, it might well be supposed, was the end of the affair. Yet a century later, in 1982, its repercussions were still being felt, if faintly.

It is famously told by the French astronomer Nicolas Camille Flammarion (1842–1925) that:

> In December 1797, the young General Bonaparte, after his wonderful conquest of Italy, returned to Paris to receive from the Directory the honours which presaged the consulate. Attended by a brilliant staff, he was going on horseback to the palace of the Luxembourg, where the Directory awaited him, when he was surprised to see, in the Rue de Tournon, all the people who stood ready to greet him turn round and look at a point in the sky instead of looking at him. His aide-de-camp told him that a star shone in the sky and that the French saw in it the star of the conqueror of Italy. It was Venus, then at maximum brightness. Political conditions have changed, but the star remains.

Venus is so resplendent at twilight that it is easy to overlook the fact that it can also be seen with the naked eye at midday. It is then visible as a tiny white speck on the pale-blue background of the sky, if one knows where to look, and providing it is not too close to the sun. Anyone unfamiliar with the calendar of celestial events can thus be excused for not connecting the glorious display at twilight with the muted spectacle of daytime. In 1897, British amateur astronomer P. M. Ryves (1880–1956), sometime Director of the Mars Section of the British Astronom-

Figure 1. Camille Flammarion (1842–1925) from *Astronomers of Today and Their Work* (Edinburgh, Gall & Inglis, 1905). (Image by Hector Macpherson, courtesy of Stebbing, Paris.)

ical Association, published a list of naked-eye sightings of Venus made in daylight between AD 398 and 1750, besides adding a series he himself had made in March 1897.

Naturally, the attempt at detection is not to be lightly undertaken. Yet if the sky is a deep blue and the planet is close to its greatest brilliancy, magnitude –4.6, and is about 40° east of the sun at an evening elongation or the same angular distance west of the Sun in the morning, it is entirely possible, and, as the record eminently shows, has been accomplished on numerous and diverse occasions. For instance, at the beginning of April 1881, the English astronomer E. E. Ledger (1841–1913) reported the experience of his brother-in-law who 'returned home one afternoon, said he had been much surprised to see, in the middle of the day, while walking along a street in the east of London, the Sun, the Moon and a star, all comparatively near together in the heavens. It was the young Moon and Venus which he thus perceived, without knowing where to look for the planet, or even being aware that it might be possible to see it.'

J. Henry Carleton (1814–1873), a lieutenant in the 1st US Dragoons, noted in his log of a mission to the Pawnee Indian Villages in Nebraska on the Great Plains of North America, that on 26 August 1844, about 11 a.m., he saw what he believed to be the planet Venus. It was very bright, and so distinctly visible that the unaided eye could catch it at once when directed to that part of the heavens. Venus was then at western elongation.

In 1882, Venus had been in transit on 6 December and having entered the western sky, was for a time a dazzling early morning object. No doubt this was known to the observers at Broughty Ferry, but it would seem they were unaware that at certain times it could also be seen in broad daylight, albeit with less ease. So for them to see a 'star' by day would not only be unusual but would also come as a surprise. Such a phenomenon would be sufficient to arouse widespread public emotion if literally reported. Resorting to optical aid they were further perplexed to find a crescent-shaped object the source of their interest, adding to the mystery. Hence without prior knowledge and understanding, the observers did not connect the splendour of the early morning display with the diminished spectacle of daytime, while magnification only added to their puzzlement.

As so often happens, in transmission, reports of this nature can undergo a subtle change of status, no doubt brought about in the present case by the use of the adjectives 'peculiar' and 'unusual' which, in the popular mind, would envelope the object in novelty, converting a local excitement into an enigma.

A bright object seen near the sun, proclaimed the French journal *l'Astronomie*. Was it Venus? How could it be?, was the astonishing response from the reputable astronomer and widely read author Richard Anthony Proctor (1837–1888), editor of *Knowledge*, the science weekly he had founded the year before.

He had reported on the sighting in the last issue of *Knowledge* for 1882; now, in the succeeding number (5 January, 1883), he was obliged to fall back on pedantry in defence of his opinion, simply to placate an adherent of the Venus hypothesis:

Figure 2. Richard Anthony Proctor (1837–1888). (Image by William S. Warren, Boston, available at www.picturehistory.com/product/id/18521.)

> A correspondent, writing from Scarborough, makes some fun out of our 'Gossip' note about a star-like body seen on December 21, entertaining, apparently, the belief that the object was the planet Venus. The idea is one which is naturally suggested by the description we quoted from a letter addressed to the *Dundee Advertiser*, 'the milky-white appearance' and the 'crescent shape' when seen with the telescope. But a little consideration would have shown our humorous correspondent that as Venus was in transit on 6 December, she could not have been in close proximity to the sun at noon on December 21. As a matter of fact, she was then 1h. 33m. in RA west of the sun (corresponding to an arc distance of 23° 15′ in RA), in declination 5° 2′ north, making an actual arc distance of about 24°. . .

To Proctor, that could not possibly imply 'close proximity' as stated in the account.

The Irish astronomer and author John Ellard Gore (1845–1910), well known for his translation of Camille Flammarion's celebrated *Popular Astronomy* (1895), nodded assent and, in the same issue of *Knowledge*, plausibly suggested:

> If the object was not a comet near perihelion, it was probably a temporary star [nova]. I find there is no bright star near the place, the nearest being ε and σ Sagittarii (of the 2nd mag.), each about 12° distant; but these would be south of the sun, and not 'a little above the sun's path', as the description states, besides being quite too faint to be visible in the daytime (to the naked eye) under any circumstances. The object could not have been the planet Venus, which was situated about 23° west of the sun on the day in question. It seems worthy of remark that the place of Kepler's celebrated 'Nova', of 1604, was – at the time of observation – only about 8½° to the west (and a little north) of the sun's place. So that, if the object was really a star, it seems possible that it may have been another outburst of Kepler's star.

It is, of course, true that some historical supernovae, such as those of AD 185, 1006, 1054 and 1572, were visible to the unaided eye in full daylight, and with that in mind, Gore proposed: 'The morning sky should be examined, as, if the object be still visible, of the same brilliancy, it should now be a conspicuous object before sunrise.'

And so it was. There, in the jade light of dawn, shone a star so splendid, so beautiful, as to invite the admiration of even the most tired eye. Then who has not stood at daybreak utterly enchanted by the moon-like radiance of Cicero's Lucifer – Venus?

Confident he was correct, Proctor distanced himself from any further part in the discussion. Why should he retract? He had checked

Figure 3. Venus: The star of evening and morning twilight. Drawing by Scriven Bolton. (Image courtesy of *Hutchinson's Splendour of the Heavens*, London, Hutchinson & Co Ltd, 1923.)

the ephemeris and established the facts beyond doubt, so what reason was there to deny truth?

Replaced by a question mark, Broughty Ferry passed from currency and, for several decades, nothing more was heard of the appearance; it had, apparently, slipped unnoticed through the Ivory Gate. Eventually, it did reassert itself. For, sometime early in the twentieth century, that indomitable collector of all things unexplained, Charles Foy Fort (1874–1932), stumbled upon the original account and cited it in *New Lands* (1923), his second iconoclastic book, thus enhancing the air of mystery which already surrounded it.

And so it might have remained, like so many other things, a forgotten oddity, washed up on the shores of the known. But in 1965, the year made memorable by the advent of Ikeya-Seki 1965 VIII, one of the most brilliant sungrazing comets on record, the British historian of astronomy Cicely M. Botley (1902–1992), having read of the Broughty Ferry object as reported in the 29 December 1882 issue of *Knowledge*, was prompted to ask, in a letter to the American magazine *Sky & Telescope*, whether the phenomenon could refer to a long-overlooked member of the same comet group?

Joseph Ashbrook, the editor, took the suggestion seriously, but since the visual data was so sketchy, before committing anything to print, he sought advice from the British astronomer Brian Geoffrey Marsden (1937–2010), director of the Minor Planet Center (MPC) at the Harvard-Smithsonian Center for Astrophysics (1978–2006), an expert in celestial mechanics and astrometry. He pointed out that the account, in fact, was compatible with a sungrazer moving in the same orbit as, say, comet 1882 II, with a perihelion passage late in December. Ashbrook was satisfied and, with that assurance, published the letter.

Jonathan Shanklin, long-time Director of the Comet Section of the British Astronomical Association, further commented that members of the Kreutz sungrazing group of comets would approach the sun from the south, curve round the north and then head south once more. Any comet would normally only be visible from the Southern Hemisphere. The rate of brightening would be rapid with a peak of short duration, barely a few hours. It is just possible a medium-sized fragment might be visible no more than a degree or two from the sun but would only be well placed for observers in the Northern Hemisphere just before it disintegrated.

Even so, on due consideration, he concluded that, given the circumstances of the Broughty Ferry phenomenon, the Venus hypothesis was in all likelihood the most likely explanation.

Proctor was undoubtedly correct when he said that 'an actual arc distance of about 24 degrees, roughly forty-four sun breadths, could scarcely be described as "close proximity"'. But then he was commenting technically as an astronomer. So what exactly does that phrase mean?

Consider, for example, the optical discovery of Neptune. This took place at Berlin on the night of 23 September 1846. It marked the close of a long period of speculation and mathematical inquiry in direct interpretation of Newtonian gravitational theory, and is rightly acclaimed as one of the great triumphs of theoretical astronomy.

Working on improved tables of Jupiter, Saturn and Uranus, the French astronomer Alexis Bouvard (1767–1845), at Paris, found that although Jupiter and Saturn moved in obedience to the gravitational principle, Uranus did not. Even after taking into account the perturbations of all the known planets, he failed to reconcile the prediscovery observations with those made in the early part of the nineteenth century. Rejecting the old observations as unreliable, he based his tables on the modern positions, and left to posterity the awkward decision as to whether the discrepancy was due to inaccurate observation on the part of the older observers, or to the action of some external agency.

Only a few years elapsed before Uranus again began to deviate from its calculated orbit. The ancient observers were completely exonerated. By 1830, the anomaly amounted to some twenty arc seconds. Ten years later, it had risen to ninety arc seconds, and continued to increase until by 1845 it had reached the so-called 'intolerable quantity' of nearly two arc minutes.

To the uninitiated such minute quantities may appear insubstantial, almost to the point of insignificance. If translated into the actuality of unaided visual observation, they may well feel justified in this belief. If we suppose two bodies, one in the place of the real planet, the other in that of the fictitious or theoretical one, then it would require an eye of exceptional acuity to distinguish them apart. Magnified by the telescope, the ostensibly small becomes measurably large, and is not for a moment to be neglected. In exact science, such unexplained, though unquestionably true, residuals are more often than not the very essence of discovery. In the case of Uranus, the residuals supplied the data from which the existence of another majestic world was disclosed – Neptune.

Against that background it is easy to understand Proctor's position. To him the phrase implied an exact measurement but in the context of the account, it expressed no more than a relationship, a concept rather than a quantitative statement. The Moon is near compared to, say, Mars or Jupiter, while a meteor is closer still. How long is a piece of string? we might ask. In the case of Broughty Ferry, all but the most pedantic would consider an angular separation of about 24° well represented by the phrase 'close proximity'. After all, it is only one fifteenth of the total sweep of 360°. Hence, by attaching an unwarranted degree of precision to a casually described naked-eye sighting, Proctor automatically precluded Venus from the equation and misleadingly encircled the Broughty Ferry object with a halo of mystery creating an enigma where none existed.

In response to the public alarm caused by the daylight brilliance of the planet in 1716, Edmond Halley (1656–1742) was motivated to investigate the phenomenon for:

> It may justly be reckoned one of the principal uses of the Mathematical Sciences, that they are in many cases able to prevent the superstition of the unskilful Vulgar; and by shewing the genuine causes of rare appearances, to deliver them from the vain apprehensions they are apt to entertain of what they call prodigies; which sometimes, by the artifices of designing men, have been made use of to very evil purposes. Of this kind was the late appearance of *Venus* in the day time, generally taken notice of about *London* and elsewhere; and by some reckoned to be prodigious.

'It was Homer who, 3,000 years ago, saluted Venus as the most beautiful of stars,' says Flammarion in his book *Dreams of an Astronomer* (1923), and he proceeds to tell of 'The Luminous Phenomenon of Cherbourg', a remarkable public event which excited a great deal of attention in the spring of 1905. It resulted in:

> . . . a series of the oddest and most contradictory descriptions. They spoke of an oval disc describing curves in the sky; the appearance of an electric meteor; a halo due to the deviation of the sun; of an illuminated captive balloon, of a new kind of maritime signal, of an unknown star, of a comet, even of a 'constellation.'

There was more to come, he further says:

> On the eleventh day of observation, April 11 (the strange apparition had commenced on April 1, and mariners might have thought of an April hoax), the maritime prefect of Cherbourg ordered the commander of the *Chasseloup-Laubat* to study the luminous phenomenon. A vessel was sent to look for Venus! The naval officers could not explain the mystery; one of them, however, wrote that it might be the planet Jupiter. Other commanders, having heard of the comet discovered at the Nice Observatory by M. Giacobini, announced that the 'unexplained light' might well be that comet! They did not know that that comet was a telescopic one, invisible to the naked eye. In the night of April 10–11 a meteorite was seen at Tunis. The question arose whether it was not this meteorite which had first been seen every evening in Cherbourg! And so on. Every sort of absurdity was given out on the subject . . .

To which Flammarion added a footnote: 'Humanity prefers its ignorance and its illusions.'

Resplendent in the western sky Venus reigned high in the western sky for three months in the spring of 1905. Yet few seemed to realize it was Venus. Its visibility in daylight, we need scarcely repeat, has become a public event on numerous occasions, as in the dark mornings of December 1887, when it was widely thought that the 'Star of Bethlehem' had returned. Its brilliance renewed sensation in 1913, especially in England, where many confounded its radiance for a mysterious German instrument of espionage, whilst, in our time, this same apprehension incorporated Venus into Unidentified Flying Object (UFO) mythology. So it seems the Broughty Ferry phenomenon was just another stirring of the public imagination by a rare, but not infrequent appearance.

On the clear morning of 31 December 1994, the British amateur astronomer Edward L. Ellis (1931–2006), at St Albans, England, followed Venus with the naked eye from 09h 00m until 11h 20m UT. He used markers and checked its position every five minutes before losing it at 11h 20m; it was then quite low in the south-west. What makes this observation significant, however, is the fact that eight sidereal revolutions of the Earth are approximately equal to thirteen revolutions of

Venus, which means that after an interval of eight years both planets occupy nearly the same positions in their orbits with respect to each other. Working backwards, Ellis found the circumstances of his observation were not too dissimilar from those of December 1882, the only difference being that Venus was at inferior conjunction on 2 November 1994, being further west of the Sun and a much easier naked-eye object.

As if to underscore the fact, the Revd T. E. R. Phillips (1868–1942), the well-known British amateur planetary observer, speaking at the December 1925 meeting of the British Astronomical Association, said that, in his experience, 'Venus . . . is not quite the same easy naked-eye object at noonday that it was years ago, I remember . . . once seeing it at midday . . . only twenty degrees from the sun.' Note the emphasis on that phrase, an interesting reflection on the opinion of Proctor, who submitted that 'an actual arc distance of about twenty-four degrees, about forty-four sun breadths, would scarcely be described as "close proximity"'.

The story of the Broughty Ferry phenomenon is one of water muddied by different perceptions of the same object or assertion. Moreover, in 1882, the transit of Venus on 6 December, the discovery of a small comet close to the Sun during the total solar eclipse of 17 May, and the advent of an even more spectacular visitor, sungrazer 1882 II in September, all provided a background that stirred the popular imagination. Could it then be that Proctor's abrogation of the obvious was an involuntary reflex to these wonders?

Yet like a fragment of cobweb clinging vicariously to some dusty relic, a vestige of mystery remains. The observers at Broughty Ferry say the object was a little above the sun's path but no mention is made of whether it was preceding or following the sun. So while a daylight sighting of Venus almost certainly explains what was seen, the case is still marginally open, allowing us room to speculate a little more. Indeed, in as recently as 1982, the Australian comet expert David A. J. Seargent listed the Broughty Ferry object as a possible sungrazing comet in his book *Comets: Vagabonds of Space* (1982).

In many respects, the matter reminds us of the question as to the visibility of the shadow cast by Venus. The elder Pliny in his *Natural History* says that, alone among the stars, 'she shines with such brilliance that her rays cast a shadow'. Its detection, however, is one of such delicacy that most attempts to catch it will fail. But, in 1899, the French astronomer E. Touchet did manage to register a shadow photographically. He attached a camera, minus its lens, to the ocular of a small telescope, and then, on exposure to the light of Venus, recorded the shadow and diffraction bands cast by an object occupying the position of the camera lens.

In this respect, the following extract from a letter dated 26 December 1912, from Dr Warde Fowler to the astronomer and seismologist Professor H. H. Turner (1861–1930), at Cambridge, is not without interest:

> As Venus is so fine just now, I went out about 5.30 this evening, after a day of torrential rain, to have a look at her. The Moon was not up, and hardly any other star was visible. The light of Venus seemed to be intensified by a huge bank of black cloud immediately below her. She shone on the open road down which I was walking, and which was wet and shiny with rain. I could not see any shadow of myself, but a shadow became plainly visible when anyone passed me: at about twenty yards away it was most obvious. I take it that the circumstances were unusual and combined to produce an unusual effect.

Figure 4. The Great Comet of 1882 as seen by Camille Flammarion on 9 October 1882 at 4 a.m. The comet was seen at noonday near the Sun on 16, 18 and 19 September. It passed in front of the solar disk on 17 September, but, like a transparent flame, was invisible. (Engraving by P. Fouché in *Popular Astronomy: A General Description of the Heavens*, translated by J. Ellard Gore, London, Chatto & Windus, 1894, p. 507, Fig. 216.)

Since the light of Venus is roughly ten times greater than that of Sirius, a search for both Venus in daylight and its fugitive shadow ought not to be without hope.

The Broughty Ferry episode is Shakespearean in character, reminding us of the bard's comedy *Much Ado About Nothing*. And yet, if one reflects on its elements, it will be seen to offer serious meditations on the subjective nature of reportage, of reliability and interpretation, essentially of the truth of observation – a subject to which Professor Hermann Bondi drew attention in his controversial but groundbreaking paper 'Fact and Inference in Theory and in Observation' in 1955.

REFERENCES

H. Bondi, 'Fact and Inference in Theory and in Observation', *Vistas in Astronomy*, Vol. 1, pp. 155–162 (1955).

J. Henry Carleton, *The Prairie Logbooks* (Chicago, Caxton Club, 1943; reprinted 1983).

Camille Flammarion, *Dreams of an Astronomer* (trans. from the French by Fournier D'Albe) (Fisher and Unwin Ltd, 1923).

Charles Fort, *New Lands* (New York, Boni and Liveright, 1923); reprinted in *The Complete Books of Charles Fort* (New York, Dover, 1974).

SERIAL PUBLICATIONS

Knowledge, An illustrated magazine of science plainly worded – exactly described (1882, 1883 and 1897).

Philosophical Transactions, No. 349, Vol. 29 (1716).

L'Astronomie Revue D'Astronomie Populaire de Météorologie et de Physique du Globe (Deuxième Année, 1883).

The Chaldean, No. 13, Vol. IV (Winter 1922).

The Dundee Advertiser (December 1882).

The Journal of the British Astronomical Association, Vol. 66 (1955–1956).

Sky & Telescope, Vol. 33, p. 84 (1967).

Choosing an Astrograph

MARTIN MOBBERLEY

In the last decade there has been an explosion of shiny new astronomical equipment on offer to amateur astronomers and, without doubt, the biggest growth area has been in the imaging sector. Telescope systems designed for imaging the night sky, historically referred to as 'astrographs', are everywhere. But a few decades ago, in the film era, things were very different indeed.

A TRIP DOWN ASTROPHOTOGRAPHY MEMORY LANE

Back in the 1970s, 1980s and early 1990s, most amateur astronomers were visual observers, with maybe a third being astrophotographers. The photography part of the hobby was quite challenging, because film was relatively insensitive stuff, and so if you had no darkroom, you quickly became a regular customer of the 'one-hour turnaround' at the local high-street camera shop, where the staff would often declare that there was: 'Nuffink on yer film, mate, 'cept a load of white dots and fuzzy things. So we didn't print 'em [sniff]!'

I was first hooked on amateur astronomy as a schoolboy, in the late 1960s, and the changes that have affected me in the last forty-eight years have been extraordinary; the ways I enjoy (and endure) the hobby, too, have changed totally since I first read Patrick's *Observer's Book of Astronomy* in 1968 and then joined the British Astronomical Association (BAA) in 1969. In my formative years, during the 1970s, I was simply a casual visual observer, learning my way around the sky with a 60 mm Dixons Prinz refractor and a homemade 8.5-inch reflector. These back-garden telescope experiences were supplemented with occasional visits to use the Victorian 4-inch refractor in the dome of the roof of the Bury St Edmunds Athenaeum, during the exciting Apollo Moon landing era. From 1980 onwards, I owned a splendid British-made 14-inch reflector, and from 1980 to the early 1990s, it was a rare week when I did not take some astrophotos and spend one or two nights a week developing films and printing them in the darkroom that my bathroom had rapidly become (after my patience with the high-street photographers and chemists had run out). Forget bath oil and rubber ducks, my bathroom was soon filled from wall to wall with the enlarger, the developer, stop-bath and fixer trays and some deep red darkroom lights. Indeed, when I was choosing both my first home (a first-floor flat) and then my second (a bungalow), my number-one criterion was that the bathroom must be indoors with no windows to let stray light in.

Those days are long gone now and while I still feel much nostalgia for the photography era, I certainly do not miss all my late-night sandwiches tasting of fixer. Perhaps my best memories of that time were chatting to the astrophotographic legends of my youth who were still alive in the seventies, eighties and later years. These men from a different era, such as Horace Dall, Harold Ridley, and even Commander Henry Hatfield, were people I had looked up to as a schoolboy and here I was in their company! The film-emulsion era was not all bliss, though. I recall an entertaining tale told to me by Harold Ridley (1919–1995) regarding a long, guided exposure of comet Halley he took in 1986, on a rare clear night. On turning the white darkroom light on, he realized that he had accidentally developed and fixed the ground-glass screen on his refractor astrograph, which had felt the same to his fingers, in the dark, as the glass film plate. The photographic plate itself was still on the worktop and was now totally ruined by the white darkroom light, which Harold had turned on when the image of Halley did not seem to be emerging in the dull red light. 'It took me a few weeks to get over that one,' he said, at a 1987 BAA meeting, lighting up yet another cigarette to 'settle down' his smoker's cough! Harold was a true BAA gentleman and a character from the old school, and he used a dedicated astrograph, designed to photograph comets. He employed the offset guiding technique, whereby one guided visually on a star using a guide telescope for the whole duration of the exposure, but slowly offset the star, sliding it along an illuminated micrometer wire in the eyepiece, frantically pressing the hand-controller buttons, to compensate for the slow cometary motion against the stars. It took a lot of preparation and experience to get that right. While other people of that era produced comet photographs with wiggly star trails, Harold's were ruler straight and the comet's nucleus was a frozen point: he was a perfectionist.

In fact, I think it was in connection with Harold that I first heard the term 'astrograph', which essentially means a telescope designed for the sole purpose of astronomical photography. Harold was a 'lens man'; no, not a character from the E. E. 'Doc' Smith 'Lensman' Science Fiction books, but a man who only used refractors and lenses for astrophotography. Reflectors never appealed to Harold as they easily lost collimation (precise optical alignment) and the mirror surfaces (silvered in his era) soon tarnished when a telescope was used every clear night. His final system consisted of an 18-cm (7-inch) aperture Zeiss lens with a focal length of some 1.2 metres (so, slightly faster than f/7). Used with medium-format film, covering a 6 x 6-cm frame, this resulted in a field of view of almost 3 degrees, so it was ideal for the brightest comets. The camera was guided by a splendid 6-inch refractor, where the guiding eyepiece was often at a convenient height to endure long exposure visual guiding. Prior to the CCD era, Harold's system was ideal for photographing comets as it could go fairly deep,

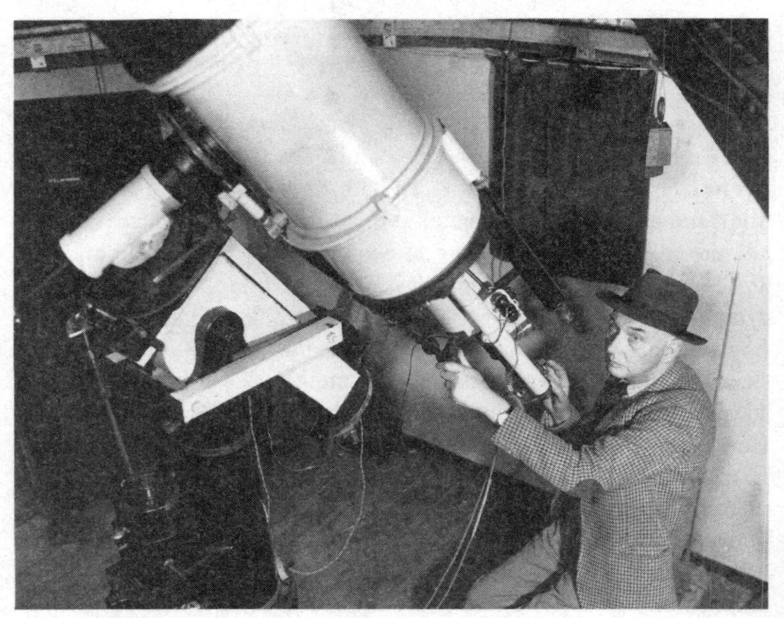

Figure 1a. Reggie Waterfield in his observatory at Woolston, Somerset during the 1950s. The 6-inch f/4.5 Dennis Taylor Cooke Triplet astrograph is shown. (Image courtesy of Denis Buczynski. Original photographer unknown.)

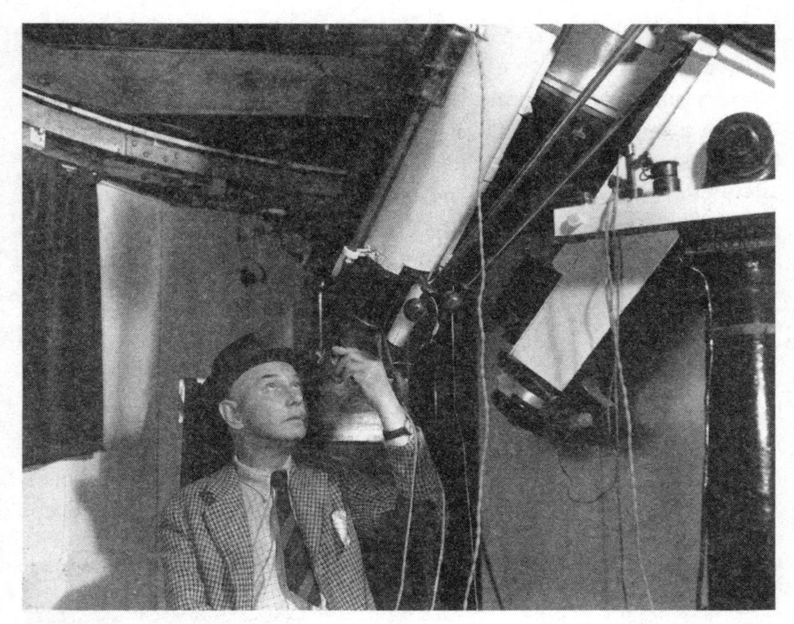

Figure 1b. Reggie Waterfield positioned at the 6-inch Cooke refractor guiding eyepiece of his astrograph at Woolston, Somerset during the 1950s. (Image courtesy of Denis Buczynski. Original photographer unknown.)

but it still had a nice wide field. Best of all, it was on a permanent, hefty, German Equatorial Mounting that would last for decades, installed in an observatory and ready to go every night. You would just switch the drive on, use the setting circles and find a guide star. There were no fragile plastic drive components whirring at 40,000 rpm under 'Go To' slewing that would fall apart after a few months of use and there was no software or menu structure to worry about and constantly upgrade. Here was a dedicated system to last a real astronomer a lifetime. Harold's astrographic system was very similar to that used by his own mentor, the remarkable Reginald Lawson Waterfield (1900–1986), known as Reggie to his friends. Despite being confined to a wheelchair from 1950, Reggie exposed and measured hundreds of comet photographs right up to his death in 1986. Once again, his astrograph used lenses. It was a 6-inch aperture f/4.5 Cooke triplet design, hand figured by Dennis Taylor, and could accept photographic plates up to 216 x 165 mm in size! Again, the system was guided visually with a 6-inch refractor (a Cooke) and served Reggie for more than fifty years of comet photography.

Such massive, dependable systems gave me an early appreciation of the most important aspects of having an astrophotography system. It had to, above all, be a total package that you could rely upon every single night, in summer or winter, to deliver the goods, and the actual imaging system was only part of the consideration. A reliable equatorial mount was just as important, but in this chapter I will just describe the Optical Tube Assemblies (OTAs) available to the modern amateur.

EARLY CCD REQUIREMENTS

Of course, the digital era changed the game for many astrophotographers and, increasingly from 1992, I (and many others) switched to mainly using cooled CCDs. My last big photographic effort was for comet Hale-Bopp in 1997. Unfortunately, early CCD chips were tiny in size and so, used with the typical long-focus telescopes of the 1990s, delivered microscopic fields of view. Also, imagine a world with no Internet, e-mail speeds of 1200 baud, no printers that could handle graphics and everything else belonging to the dark ages. Yes, CCDs went two or three magnitudes fainter than film with the same exposure, but, in the early 1990s, you then had to photograph the monitor screen, or send someone a floppy disk in the post, for anyone else to see what you had imaged! Early e-mail would not even let you send images without encoding the binary data into letters and numbers: the worst of all possible worlds! As well as that nightmare, home computers in the early 1990s were pitifully slow, the early hard disks soon filled up with images, and image-processing software could manage a contrast stretch, but little else. For telescope manufacturers the tiny sizes of the early CCDs did not mean that they had to rethink their basic systems; they just sold shorter focal-length optics. Therefore, one instrument that became very popular as an astrograph at that time was Celestron's diminutive 5.5-inch f/3.6 Schmidt-Newtonian Comet Catcher. On the positive side, with traditional Newtonian designs any optical aberrations at the edges of image fields were problems that affected the standard 36 x 24-mm film frames in the early 1990s, but were not an issue when using small CCD chips of less than 10 mm across. So, the standard telescope designs were still OK, in short-focus form, until CCD detectors became much bigger, which they have become, dramatically, in the last few years.

BACK TO THE FUTURE!

It is the recent increase in digital detector size, along with the realization that astrophotography does not have to be an outdoor battle with the elements, or an indoor battle with photographic chemicals, that has fuelled the modern explosion in astronomical imaging gear. However, because big digital detectors are now back to the size that film frames were in the 1980s, in many ways the ideal astrograph has been subjected to a retro effect. It's a case of 'Back to the Future' as, once again, focal lengths of two metres or so are not totally incompatible with deep-sky imaging, because the modern CCDs are big enough to deliver a useful field.

At this point I would like to involve a bit of mathematics. I know many people hate formulae, but I will try to keep things simple. Fundamental to the understanding of choosing or designing an astrograph for deep-sky or comet work is calculating the field of view. There are only two numbers required to achieve this, namely the focal length of the instrument and the size of the detector. Let us imagine that we want a system with a field of view one degree across, in other words, twice the diameter of the full Moon. The tangent function is invaluable here and tells us that the tangent of one degree is 0.017455 – in other words, over a distance of x, an angle of one degree will span a height of 0.017455 times x. Or, translated into astrograph language, over a focal length of 1,000 mm, or one metre, one degree will need a CCD chip width of 17.455 mm to capture it, or 17.5 mm to keep things simple. From this, and the fact that we are only talking about small angles, we can derive a useful formula or three, as follows:

CCD width required in millimetres = Focal length in metres x angle in degrees x 17.5

or

Focal length required in metres = CCD width in millimetres/(angle in degrees x 17.5)

or

Angle covered in degrees = CCD width in millimetres/(focal length in metres x 17.5)

There is one other calculation that needs to be borne in mind if you are keen to get the maximum signal-to-noise ratio out of a deep-sky image and that is the sky-sampling ratio. Bright stars look very big in astrophotos, simply because optical diffraction spreads the light out over many arc seconds from the actual centre of the point source that is the star. [NB 1 arc second = 1/60th of an arc minute and 1/3600th of a degree.] With very faint stars all you see is a much smaller disc, spread around by diffraction and by the tracking errors of the telescope drive. Typically, the faintest stars you see in good amateur deep-sky images will have diameters of around four arc seconds or so. Planetary imagers can achieve far tinier resolutions because their imaging systems work at up to a hundred frames per second and so can 'freeze' the turbulent atmospheric seeing, but for long exposures four arc second-diameter stars are about as good as you ever get. In fact, when objects are low down, and comets are often low down in the twilight murk, the faintest stars can be much larger, maybe five or six arc seconds in diameter.

Clearly, in any deep-sky image the aim is to record the faintest stars (or features) possible and for this you need to match your detector's pixel size to the atmospheric seeing by choosing compatible detectors and instruments. If each pixel is far smaller than the faintest (and smallest) star diameters, you will not collect much flux from such a small star sample. Inevitably, small CCDs used with a very long focal length might also leave you with a very narrow field of view too. Conversely, if each pixel is huge compared to the faintest stars, then each pixel will collect all the star's light in one go (or several stars together) along with a significant amount of the sky background (light pollution) and the star will tend to get engulfed by the background sky and look very blocky too, filling a single square pixel. However, this approach may well give you a nice, wide field of view, useful for big objects like the Andromeda galaxy. Clearly, between these two extremes there is an optimum sampling ratio and, although opinions differ, a good all-round figure is roughly two arc seconds per pixel. So, to get the most signal-to-noise from an astrograph you may wish to have another formula to tell you how many arc seconds per pixel your desired system is working at. The easiest formula to use here is:

Arc seconds per CCD pixel = 206 x pixel size in microns/focal length in millimetres. (I might add that one micron is a thousandth of a millimetre.) An example may help here. Let us imagine an astrograph with a focal length of 1000 mm, using a CCD with 9 micron pixels (a popular size). In this case we get: Arc seconds per CCD pixel = 206 x 9/1000 = 1.85.

Of course, even with these formulae there are still many decisions to be made by the prospective imager of the night sky. If a garden-based observatory is not an option, or if the budget is small, a 'grab and go' lightweight astrographic refractor system may be required. Also, what are the observer's main interests? If searching for, or imaging, supernovae is the prime interest, then a decent aperture Schmidt-Cassegrain telescope (SCT), with a modest-size CCD detector might work best. Galaxies are small in angular size, but supernovae are faint and often close to the nucleus, so aperture and a fine-image scale (one to two arc seconds per pixel) will be more important than the field of view. On the other hand, if you want a portable system for imaging those naked-eye comets that come along every few years, and you want to capture a degree or two of the tail, then a short focal-length instrument, combined with, for example, a digital SLR (DSLR), might be your best bet. Decisions, decisions, decisions, but where do you start? Well, the best thing to do is to familiarize yourself with all the optical tube options out there and I will attempt to do this next, starting with the simple reflector.

BASIC NEWTONIAN ASTROGRAPHS

The humble Newtonian reflector, for decades the mainstay of amateur astronomy, still gives you the most aperture for your money and prices have never been cheaper, because so many telescopes are now mass-produced in the Far East. If I was choosing a low-cost astrograph, the first instruments that would attract my attention probably would be the Skywatcher Quattro 8-inch and 10-inch f/4 Newtonians; in the UK, these cost in the region of £400 or £500 respectively for the metal tube versions and £600 or £800 for the carbon fibre versions. In terms of aperture inches per pound sterling they are hard to beat! The carbon fibre versions are lighter, expand or contract far less with temperature changes and, well, just look smarter!

When used with imaging sensors more than 10 mm in diameter, stars will look like seagulls at the edges of a frame at f/4, unless a coma-corrector is also purchased. Fortunately, such items are not too

Figure 2. The 8-inch f/4 Skywatcher Quattro with carbon fibre tube. (Image courtesy of Skywatcher.)

expensive and budget coma-correctors only cost around £100, although high-end versions (such as the TeleVue Paracorr, which increases the f-ratio by 15 per cent) can cost a lot more. As with any fast Newtonian, collimating the optics (lining up the mirrors precisely) is essential to get the best performance across the field and it is true that many beginners find this process tricky and quite scary too, because fiddling about with a screwdriver near to a brand new telescope's secondary mirror support is never a relaxing pursuit.

Most amateur astrophotographers quickly find out that the standard commercial telescope drives (with or without periodic error training) rarely track well enough to allow individual exposures of more than thirty to sixty seconds to be attempted, when using large telescopes, without some form of manual guiding or auto-guiding being employed. Overcoming these issues is a subject in itself, which I do not intend going into here, but a beginner will invariably just use a digital SLR with his or her first astrograph and fire off dozens of thirty-second exposures, stacking the sharpest shots later using appropriate software. Because comets themselves move against the background stars, and bright comets close to the Earth move very rapidly, this short-exposure strategy works very well when imaging these enigmatic fuzzy visitors to the inner Solar System. All the image processing can be done indoors the next day (rather than battling with long-exposure auto-guiding in the cold and damp outside) and you just stack the frames centred on the comet's inner coma. With an 8-inch (203 mm) f/4 reflector delivering a focal length of some 800 mm, a standard APS-C sensor (typically 25 mm x 17 mm) would yield a field of view of 1.8 x 1.2 degrees, which is very handy for the brightest and biggest comets, or for large objects like the Pleiades, or the Orion Nebula.

ULTIMATE NEWTONIAN ASTROGRAPHS

In recent years the design and fabrication of sophisticated corrector and reducer lenses for telescopes has given astrophotographers more options in the search for ultra wide fields, populated to the corners with faint, pinpoint stars. As I mentioned in the previous section, coma is the dominant Newtonian aberration and causes stars to rapidly look like tadpoles, or seagulls, in the corners of deep-sky images taken with Newtonians. Coma is also the reason why you need to precisely collimate a Newtonian for planetary work, which is an order of magnitude more sensitive to distorted star images. The faster the f-ratio the more coma distorts those stars outside the centre of the field and it worsens roughly with the inverse cube of that ratio. In the 1980s and 1990s, I used a 14-inch f/5 Newtonian telescope from Astronomical Equipment Ltd (AE), with 36 x 24 mm-format photographic film. At f/5 the stars on long exposures were just acceptable out to about 10 mm from the centre of the film. Some years later, I acquired a massive 19.3-inch f/4.5 Newtonian from AE and initially used film with that as well. The stars were only acceptable out to a radius of about 7 mm at f/4.5, but as the early CCDs were smaller than that, this was fine. Many years later I used an Orion Optics 10-inch f/6.3 Newtonian for planetary work, but did a few deep-sky tests on stars with it too. At f/6.3 this was far superior with regards to coma. Stars were virtually perfect out to the very corners of a Canon 300D sensor (23 x 15 mm) and the imperfections were caused by tracking errors, not coma. These experiences taught me a lot and hammered home something I already knew: namely that it is the modern craving for fast, portable systems that has created a need for sophisticated correctors to produce pinpoint star images across the biggest sensors. If Newtonian astrographs were all f/7, no correctors would be needed at all.

I am reminded here of a dedicated comet-photography astrograph used in the film era by the expert British comet photographer Michael Hendrie. This was a 10-inch (254 mm) f/7 folded reflector, engineered so that it resembled a Cassegrain, but a very large flat secondary mirror merely folded the light back down the tube and through the mirror hole, so that it looked more like an f/3 system in terms of its tube length. At f/7, the stars, even at the edge of medium-format film, looked OK, giving a useful field of view almost two degrees across. You might say that f/7 is too slow for deep-sky work, but it is not. After all, modern corrected Dall-Kirkhams and Ritchey-Chrétiens work at these f-ratios, and if you want more light per pixel, you can electronically 'bin' the pixels 2 x 2 or 3 x 3 anyway. However, folded f/7 Newtonian astrographs are not commercially produced and so, unless you enjoy making your own gear, we have to choose from the current commercial products, bringing us back to the subject of the ultimate Newtonian Astrographs.

Two major telescope companies specialize in large and medium aperture-corrected Newtonians for amateur deep-sky work: these are Orion Optics in the UK and Astro Systeme Austria (ASA). The Orion Optics AG system is based around an f/4 Newtonian design and features a huge four-lens Wynne corrector in the drawtube (180 mm

Figure 3. The Orion Optics AG12 astrograph. (Image courtesy of Orion Optics.)

long x 76 mm diameter) which reduces the f-ratio to f/3.8 and corrects the field across a 60-mm circle. With such a long optical corrector in the drawtube, and to accommodate the focus needing to reach the CCD detector (after passing through accessories like filter wheels), the focal plane sits some 260 mm from the far end of the corrector. This means the substantial focuser unit is mounted very high on the carbon fibre tube. This design results in a very compact system, as a lot of the light path occurs after the Newtonian secondary mirror.

Of course, the acid test of any astrographic system is the final result obtained and two individuals in particular have demonstrated that the Orion Optics 12-inch Astrograph, the AG12, can produce spectacular images in the hands of dedicated amateurs. Those individuals are Peter Shah, in the UK, and 'Strongman' Mike Sidonio in Australia. Many of their images can be found on the Orion Optics Gallery pages at: www.orionoptics.co.uk/imagegallery.html.

Admittedly, it is not just about the telescope here, but about the individuals, their patience, their high-quality focusers and their equatorial mountings. The importance of dark skies and a high-quality CCD detector are also crucial. The Orion Optics AG corrected Newtonians come in various apertures, from eight to sixteen inches (two hundred to four hundred millimetres).

The similar ASA corrected Newtonian astrographs also have expert users, most notably the well-known Austrian amateurs Michael Jaeger and Gerald Rhemann (who works for ASA). These ASA corrected Newtonians come in various configurations. The basic 'N' series Astrographs are f/3.6, with reducer and Barlow correctors available to provide f/2.75 or f/6.8. These come in eight- to twenty-inch apertures (two hundred to five hundred millimetres). There is also a specific 'H' series two hundred-millimetre astrograph working at f/2.8.

None of these dedicated astrographs are cheap, partly because the corrector lenses alone cost around £1,000, which is enough to buy an entire budget twelve-inch f/4 Newtonian, equipped with a coma corrector! You end up paying about seven times more so that you can fully illuminate the largest popular CCD detectors (36 x 36 mm) without light fall-off (called vignetting) and with pin-sharp stars, even in the corners. Of course, giant three-inch diameter focusers, large secondary mirrors and carbon fibre tubes are not cheap either. With a budget twelve-inch f/4 Newtonian and a basic two-inch coma corrector you would typically only be able to fully illuminate the popular KAF-8300 based CCDs, with dimensions of 18 x 13.5 mm, with fairly sharp stars in the corners, over a 0.85 x 0.65-degree field. So, to illuminate more than four times that area at the focal plane, you pay a very hefty price indeed. Even so, it is a price that the keenest and richest deep-sky imagers will be happy to pay, especially when you consider that the large Kodak KAF-16803-based CCD cameras alone will also cost more than £7,000 in the UK! Of course, if you downsize from a 12-inch f/4 system to an 8-inch f/4 system you get a 50 per cent wider field anyway for the same sensor size! You also get a better sampling ratio of 1.4, rather than 0.9 arc seconds per pixel, using the KAF-8300's tiny 5.4 micron pixels. Sometimes smaller can actually be better!

I have been in this hobby a very long time, and since the 1990s I have answered thousands of equipment-choice e-mails (and postal snail-mail letters) from beginners in astrophotography. Incredibly, yet another such e-mail arrived while I was actually typing the above paragraph! If there is one thing I have learned it is that the main ingredient required to take good astrophotos is the sheer, cussed determination of the individual involved. Yes, good equipment is very nice, but, clearly, if the telescope remains unused, or rarely used, you will not get those great results you dreamed of. It is easy, as a complete beginner, to imagine that a pricey piece of shiny new kit will do the work for you, but it never will. Only a determined (verging on manically obsessive) individual, who is a natural problem solver, will get great images and overcome the equipment niggles, cloud frustrations and software bugs that kill others' enthusiasm off. Every year millions of pounds are spent on expensive items of astro-kit, but, sadly, a lot of it is rarely used! For that reason, if you are a beginner to the hobby of astrophotography, it is better to start by spending modest amounts of money before you know whether you are going to really use the best kit to the max. For this reason, buying something like an 8-inch f/4 Skywatcher Quattro, and using it with a digital SLR, will be far less painful (if you lose interest in cloud-dodging and late-night problem solving) than spending £10,000 on the best kit. A large-aperture system that will have pinpoint stars across the fifty-millimetre diagonal of a top-grade CCD is the dream of many astrophotographers, but a huge amount of pleasure can be derived from getting decent results from budget kit that has not broken the bank, or taken a relationship to the divorce courts!

MAKSUTOV-NEWTONIANS

The Maksutov-Newtonian design has been around for a long time and various manufacturers have produced this form of astrograph over the years. A very similar design is that of the Schmidt-Newtonian, which features a full-aperture Schmidt corrector at the normally open Newtonian end, rather than a Maksutov meniscus corrector. Schmidt-Newtonians (like the 1990s Celestron Comet Catcher) tend to be a fast f/4, whereas Maksutov-Newtonians are a bit slower, typically f/5. Maksutov-Newtonians have been produced in far greater numbers and the Russian company Intes Micro has manufactured many versions, featuring a sealed tube and a small secondary, for excellent planetary views.

However, in recent years, the Maksutov-Newtonian has been aimed more at wide-field deep-sky imagers. The secondary mirror has been enlarged and the coma reduced substantially across a wide field by the full aperture Maksutov corrector. This glass window also ensures that the tube is sealed against the elements, like a Schmidt-Cassegrain, so that the internal mirror surfaces rarely become damp. In many respects you could query whether the Maksutov-Newtonian has any major advantages over a standard Newtonian equipped with a coma corrector, although they do tend to keep collimation better than a basic reflector where the secondary flat mirror is held in place by flexible 'twangy' vanes. Without doubt the most popular Maksutov-Newtonian astrograph currently in production is the Skywatcher MN190 which has a 190-mm aperture and a focal ratio of 5.3. It sells for around £1,000 in the UK. The potential buyer will look at the aperture and design and might ask if it is really worth paying twice the price of a budget two hundred-millimetre Newtonian with a coma corrector? Ultimately, it all boils down to what is more important to you: the price or a sealed tube? You could argue that the diffraction properties of the 190-mm Maksutov-Newtonian, with no secondary supports, are great for planetary viewing, but planetary observers will often prefer more aperture anyway, and deep-sky imagers quite like those diffraction spikes from the secondary mirror holder. Indeed, bizarre though it seems, some imagers actually 'pretty up' their deep-sky stars with software-generated diffraction spikes!

The other serious contender in this field is the Explore Scientific

'David Levy Comet Hunter' Maksutov-Newtonian. At 152 mm (6 inches) in aperture and with an f-ratio of 4.8 (surprisingly fast for a Mak-Newt), this is a smaller and faster alternative to the Skywatcher MN190 and it weighs in at only 7 kg, compared to 10 kg for the MN190. In addition, it includes a carbon fibre tube and a very handy carrying handle, a feature that is invaluable in the dark, but which other manufacturers do not seem to have latched on to!

At 7 kilograms, the David Levy Comet Hunter is easily carried on all

Figure 4. The Explore Scientific David Levy Comet Hunter. (Image courtesy of Explore Scientific.)

decent equatorial mountings and its short focal length of 730 mm will deliver a 1.1 x 1.4-degree field when one of the popular KAF-8300-format CCDs is used with the telescope. This is a very useful field of view to employ when imaging bright comets and wishing to record the first degree or so of tail. Using an 8300 sensor with its 5.4 micron pixels, the sampling ratio comes out at 1.5 arc seconds per pixel. This portable Maksutov-Newtonian is priced at a similar amount to the Skywatcher MN190 by UK dealers. In many ways it is a vastly improved twenty-first-century version of the 1990s Celestron 'Comet Catcher'.

HYPERBOLIC REFLECTORS

During the mid-1980s, the highly regarded Japanese Takahashi company introduced a range of superb little astrographs called the Epsilon series. These telescopes looked like short-focus Newtonians but the primary mirror was hyperbolic, not parabolic, and the company designed a special corrector lens in the drawtube to flatten the field and make stars pin sharp across a fifty-millimetre-diameter circle. I purchased one of these telescopes (a 160-mm model) in the early 1990s and still own it now and, like all Takahashi gear, it has lasted a long time and always delivered the goods. Various Epsilon series reflector astrographs came and went over the years and all operated at fast f-ratios of about f/3.6 or so. Modern coma correctors and short-focus Newtonians have tended to threaten the survival of the Epsilon line, which is a shame, but Takahashi currently still sell two outstanding Epsilon astrographs. The largest is the ultra-fast E-180, an astrograph with an effective aperture of 180 mm (primary mirror 190 mm) working at an impressive f/2.8, with a focal length of 500 mm. With the popular KAF-8300 series CCDs such a system will deliver an optimum 2.2 arc seconds per pixel and a 2.0 x 1.5-degree field, combined with a very handy aperture, a light weight of 10 kg and a tube length of just 570 mm. The current UK

Figure 5. The compact Takahashi E-180 f/2.8 astrograph. (Image courtesy of Takahashi.)

price is around £3,800. If that is way above your budget, there is also a smaller f/3.3 Takahashi E-130 astrograph which has an effective aperture of 130 mm, a focal length of 430 mm and a price of £2,300.

SCHMIDT-CASSEGRAINS

On the face of it the basic Schmidt-Cassegrain telescope (SCT) design is far from ideal for imaging any star fields more than a quarter of a degree across. Indeed, the long f-ratios of, typically, f/10 are really more suitable for planetary imaging when a Barlow lens is added. This is borne out in practice by the number of extraordinary high-resolution images of the planets, especially Jupiter, that are taken by amateur astronomers using the Celestron 8-, 9.25-, 11- and 14-inch models. Yet many amateur astronomers, including myself, also use Schmidt-Cassegrains for deep-sky astrophotography of galaxies and other objects. The simple fact is that a big Schmidt-Cassegrain is a very versatile and affordable instrument which combines a decent aperture with a short tube length and a relatively low weight. In addition, the position of the eyepiece, at the opposite end to where the telescope is pointing (as with a refractor), delivers a very user-friendly experience. The fact that the tube is sealed with a glass corrector plate has a number of advantages too. The obvious one is that, as mentioned earlier, it prevents moisture settling on the delicate mirror surfaces. It also keeps the compact tube very rigid, although it does tend to trap heat inside the tube after a warm day. Schmidt-Cassegrain mirrors are of the conical design, where a mirror does not sit inside a traditional cell, but has a thick centre and a thin edge, so that a single central support can hold the mirror without it distorting. This design works well up to about 16-inch apertures, but in the original Schmidt-Cassegrains the mirrors did tilt, or flop, at certain sky angles, causing a loss of collimation.

I have used a Celestron 14 Schmidt-Cassegrain for both deep-sky and planetary imaging since 2003 and it performs well, but I do not use it for deep-sky targets larger than a quarter of a degree across as I would need a very large CCD and it is not really a wide-field instrument; rather, it is more of a galaxy and variable star-imaging system. The UK supernova hunter Tom Boles discovered an incredible 155 supernovae with his three Celestron 14 Schmidt-Cassegrain telescopes, but, again, only with narrow-field imaging.

A few years ago, it became clear that affordable sensor chips were becoming larger, and when amateur astronomers started attaching digital SLRs to their SCTs, they didn't much like the tadpole-shaped stars at the field edges, or the vignetting. Other manufacturers, such as RC Optical Systems, were offering Ritchey-Chrétien telescopes (I will deal with these later) with wider, flatter fields, but with a very hefty price tag too. It was clear that the traditional Schmidt-Cassegrains would need to adapt to be able to cope with full-frame sensors and CCDs as big as old 35-mm film frames (36 x 24 mm) and the two big SCT manufacturers, Celestron and Meade, responded with new products. As reported by *Sky & Telescope* in 2007 and 2008, Star Instruments and RC Optical Systems filed successful lawsuits against Meade for calling their reduced coma design an 'advanced Ritchey-Chrétien design', which it was definitely not. It was simply an aplanatic version of the Schmidt-Cassegrain, so Meade eventually had to rename their system as ACF (Advanced Coma Free). No such legal battle affected Celestron's answer to the coma problem, which resulted in their splendid range of EdgeHD telescopes which have since proved very popular. Essentially, the Celestron EdgeHD product line incorporates a complex lens in the central baffle tube which provides a flat, virtually aberration-free field, over a diameter of forty-two millimetres at the focal plane. This is enough to easily cover APS-C format CCD sensors with a diagonal dimension below that size and even full-frame (36 x 24 mm) CCD sensors show stars that

Figure 6. The 9.25-inch Celestron EdgeHD telescope. (Image courtesy of Celestron.)

are almost acceptable near the field edge. The Celestron EdgeHD range also includes a set of optional dedicated reducers that can be used to tame the long f-ratios of each of these instruments by a factor of 0.7. This means that you can use a CCD sensor some 1.4 times smaller and it will cover the same area as a bigger sensor would do without the reducer.

However, whichever way you look at things, SCTs, coma free or otherwise, are mainly suitable for imaging deep-sky objects like galaxies, globular clusters or planetary nebulae and not objects that span a degree or more across the sky. Indeed, the C14 EdgeHD has a maximum practical field of about half a degree. Even so, you can take great images with a half-degree field and 14 inches of aperture, but you may well want to use a focal reducer, a CCD with big pixels, and 2 x 2 or 3 x 3 pixel binning, to get a sensible sampling ratio on the sky. I should add that third-party companies offer a variety of focal reducers for Schmidt-Cassegrains. For example, the Optec company market the Optec Lepus 0.62 x reducer which caters for a modest 22-mm diagonal CCD and tames the focal length of the Celestron EdgeHD instruments by 62 per cent.

HYPERSTAR AND THE ROWE-ACKERMAN SCHMIDT ASTROGRAPH

There is another way to drastically tame the focal length of a Schmidt-Cassegrain and make it into a dedicated astrograph and that is by using the Hyperstar lens system developed by the Arizona company Starizona. The basic design of a Schmidt-Cassegrain is not that dissimilar to the historic photographic Schmidt Camera, the first example of which was designed by Bernhard Schmidt in 1930. However, instead of a piece of photographic film being suspended on a curved surface inside the Schmidt Camera at f/2, in the modern amateur's SCT a secondary mirror is positioned inside the tube to shoot the focal plane through a hole in the primary mirror and out to a Cassegrain focus at f/10. But what if you remove that secondary mirror entirely and insert a CCD at the f/2 focus? Of course, you can't curve a CCD sensor, but you can insert a lens to correct the field and allow the f/2 light cone from the primary mirror to stay at f/2 but focus on the detector. In many ways it is returning the original Schmidt design to its roots, but with the focal plane flat and outside the tube. Starizona's Hyperstar system achieves this. They perfected the lenses required and their system requires the owner of a commercial Schmidt-Cassegrain to carefully remove the secondary mirror, insert a lens, and attach a suitable CCD camera, or a digital SLR, to the front of the telescope. It's a very clever system, and a very fast one at f/2, but it is certainly not for everyone.

First, most amateurs quite like to have a system that can be used both as an imaging system and as a visual telescope at the same time. No amateur in a sane state of mind would swap the secondary mirror of a Schmidt-Cassegrain with the Hyperstar lens system in the middle of the night, though! Indeed, despite many spectacular images being taken with Hyperstar-converted Schmidt-Cassegrains, most amateurs I know feel distinctly uneasy about swapping the secondary mirror of their SCTs for a Hyperstar lens and camera, even once, in daylight! Just the sight of a DSLR or CCD supported by the glass-corrector plate hole makes many amateurs cringe. Even so, the Australian amateur astronomer Terry Lovejoy has discovered several comets using a Hyperstar-converted 8-inch Celestron, which is proof that the system works very well indeed, for its brave users.

However, there is now another very similar option, engineered by Celestron itself, in the form of their Rowe-Ackerman Schmidt Astrograph: a telescope designed from the start to work like a Schmidt Camera at f/2, but purely as an imaging system, as it cannot be used visually. What Celestron have done is to take the body of their 11-inch f/2 Celestron, make the primary mirror fractionally slower (f/2.2), with a focal length longer by sixty millimetres, and add a built-in four-element lens group just before the corrector plate. With a CCD camera (or DSLR) attached, the imaging gear sits close in to the corrector plate and looks less fragile than a Hyperstar system. For wide-field imaging, the fact that a DSLR's rectangular shape obstructs the 11-inch aperture is not a big issue. In addition, the camera interface ring built into the corrector plate on the new Celestron astrograph is, reassuringly, a lot more solid than just using the old secondary mirror hole with the Hyperstar system.

Even so, for anyone who already owns a Schmidt-Cassegrain, and is keen on ultrafast imaging, converting their SCT using the Hyperstar system will probably be the preferred option, if only from a financial viewpoint. In addition, both Hyperstar and the Rowe-Ackerman Schmidt Astrograph work best using DSLRs or one-shot colour CCD cameras. There is simply not enough back-focus to consider using filter wheels for filtered work with monochrome cameras, and some narrow-band filters would not work perfectly at f/2 anyway. With a focal length of 620 mm the Celestron Rowe-Ackerman Schmidt Astrograph delivers a wide 3.3 x 2.2-degree field onto a 36 x 24 mm full-frame DSLR sensor and with a sampling ratio (with the 4 to 7 micron pixels typical in DSLRs) of roughly 1.5 to 2 arc seconds per pixel, it is close to the optimum value for deep-sky imaging. This exciting newcomer to the astrograph market will set you back about £3,000 in the UK.

Figure 7. Celestron's Rowe-Ackerman 11-inch astrograph front end, with DSLR attached. (Image courtesy of Celestron.)

REFRACTOR ASTROGRAPHS

Despite their relatively modest apertures, astrographs based on short-focus refractors have become very popular in the past twenty years. Indeed, even for observers in the film era, ex-Second World War reconnaissance lenses were very popular when used in photographic systems. I mentioned earlier the comet photographer Harold Ridley and his 18-centimetre Zeiss lens of 1.2-metres focal length and his mentor Reggie Waterfield. The development of triplet, quadruplet and even quintuplet lenses and, in recent times, low dispersion 'exotic' lens materials, along with sophisticated computer modelling, has meant that many manufacturers now offer relatively fast refractors suitable for wide-field imaging applications. As with the priciest camera lenses offered by Canon or Nikon, these top-range systems have maximum affordable apertures of around 150 mm due to the difficulty of manufacturing flawless glass discs of larger apertures in these low-dispersion glasses, often based on fluorite compounds. The best refractors, with very low colour dispersion across the CCD range, are frequently referred to as 'apos', short for apochromats, although in practice such a description is rarely justified and the term 'semi-apo' would be a better description.

From a financial and aperture perspective it would, at first glance, seem that short-focus semi-apo refractors make little sense, but many splendid images have been taken with such equipment. While it is true that the cost per inch of aperture is ludicrous compared to, say, a simple 8-inch f/4 Newtonian equipped with a coma corrector, when you compare side-by-side images taken with a hundred-millimetre 'apo' and a two hundred-millimetre reflector, the difference in limiting magnitude is far less than you might expect and the wider field of view with a shorter focal length, combined with rugged portability, can more than compensate if you are imaging a long tail on a bright comet. Why do these small semi-apo refractors seem to perform so well? There are a number of factors. First, the sampling ratio issue. Many large astrographs are seriously over-sampling the sky, with CCD pixels covering less than one arc second in some cases. While this will dim serious light pollution considerably, it will also dim the faint stars to the point where they may be too ghostly to be visible. Then there is the fact that many equatorial mountings in use by amateurs cannot track accurately enough to keep a faint star from drifting outside its own three- or four-arc second-diameter radius. In this case, having a fine sampling ratio of one to two arc seconds per pixel is impossible to justify. Large instruments often take hours to cool down too, so star images can be very bloated after a hot summer's day and small low-mass instruments tend to keep focused and collimated at all angles in the sky. All this conspires to mean that in less-than-perfect observing conditions the optimum sampling ratio may well be three or four arc seconds per pixel. The user-friendly 'grab and go' nature of these small refractors also means that they tend to be used more, especially if a permanent observatory is not an option.

Without doubt the most revered apochromatic astrograph in modern times is the almost-legendary Takahashi FSQ-106 astrograph, now called the FSQ-106ED. This compact and lightweight 106-mm aperture f/5.3 system, combined with a full-frame 36 x 24 mm sensor, like the Kodak KAF 11000, has delivered some of the most stunning wide-field images of the modern era. When you calculate the sampling ratio, it comes out at precisely 3.5 arc seconds per pixel with a 9-micron pixel size. The sky coverage with a full-frame sensor is 3.9 x 2.6 degrees, which is perfect for capturing the intricate details in those first few degrees of the tails of the best comets. Unlike many of the relatively new telescope companies that have emerged with products manufactured in the Far East during the 1990s and in the twenty-first century, Takahashi has a very long history of producing astrographs that can deliver pin-sharp stars across very wide fields. In late 2015, Takahashi announced a new big brother to the FSQ-106ED, namely the FSQ-130ED, an almost identical design, but with 23 per cent more aperture. It will be interesting to see if this new apo refractor gains such a legendary status in the astronomy world.

Perhaps the only other astrographic refractors treated with such reverence are those made by Roland Christen's Astro-Physics company. Many amateurs, especially in the US, regard an Astro-Physics 'Starfire Apochromat' as the most desirable telescope an amateur astronomer can own, whether new or second-hand. There is only one real problem here, though. Such is the desirability of these instruments, and the attention given to strict quality control throughout the manufacture, that there is a very long waiting list for brand new Astro-Physics products. Unlike any other telescope, if you order direct from Astro-Physics, you have to put your name down for a place in the queue and you can wait for eternity! But, even so, many amateur astronomers are prepared to wait many years to have a chance of owning such a beautiful instrument, desired by so many. The most recent flagship Astro-Physics Starfire apochromat was the 175-mm aperture f/8 Starfire EDF, which would set you back more than $20,000, but such was the demand for that model and the smaller 160-mm aperture Starfire apochromat that, unless you were lucky, you would never actually receive it. Indeed, at the time of writing, applicants for the last eleven manufactured had been entered into a lucky lottery, such was the demand!

Indeed, Astro-Physics say on their website: 'Please understand that it is more important to us that we produce a few excellent instruments than a lot of mediocre ones'. So, if you really crave a legendary Astro-Physics refractor astrograph, going onto the second-hand market may be your best option. Like a Fabergé egg, some things are worth dreaming of acquiring, even if you never actually succeed!

Of course, the Takahashi FSQ-106ED and Astro-Physics Starfire apochromats are not the only astrographic refractor systems. In recent years a flood of Chinese-manufactured instruments have entered the market at remarkable prices for the optical technology involved. Admittedly, they do not have the kudos of an Astro-Physics apochromat, and the quality control means that there is a small risk that what you get will not be 100 per cent perfect (as with most things in life), but there has never been so much choice for the refractor purchaser. Foremost amongst these competitively priced short-focus refractor astrographs are the Skywatcher Esprit Pro series (80- to 150-mm apertures ranging from under £1,000 to over £4,000 in price) along with a bewildering array of options from William Optics, Borg, TeleVue, Celestron, Vixen and others. Competition is fierce, but the Takahashi FSQ-106ED remains the small apo astrograph to beat and it is only one of many high-quality Takahashi refractor options.

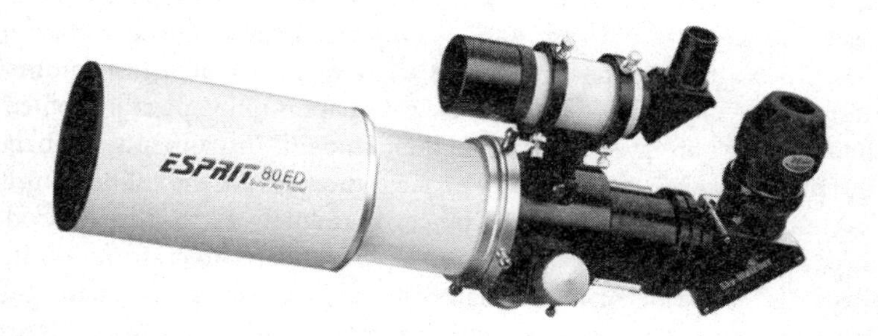

Figure 8. Skywatcher's smallest Esprit refractor, the Esprit 80, which makes an affordable astrograph. (Image courtesy of Skywatcher.)

RITCHEY CHRÉTIENS AND CORRECTED DALL-KIRKHAMS

As I mentioned when discussing Schmidt-Cassegrains, in the traditional SCT design the stars rapidly become tadpole shaped as you move outside the central portion of a traditional full-frame camera sensor. With an old style Celestron 14, for example, imaging fields of view wider than twenty arc minutes or so, and recording sharp stars in the corners, becomes impossible, even with standard and affordable reducer/corrector lenses. As we have seen, this problem has been partially alleviated by the modern range of coma-free SCTs such as the Celestron EdgeHD, although, even now, a Celestron 14 cannot image a perfect field of view larger than about half a degree across. Professional astronomers, outside of small university departments, have rarely used Schmidt-Cassegrains, mainly because their budgets have allowed them to go for much larger astrographic instruments, often employing the Ritchey-Chrétien (RC) design. In this system, invented a hundred years ago by the American George Willis Ritchey and the Frenchman Henri Chrétien, both the primary and secondary mirrors have hyperbolic curves and no lenses or corrector plates are involved. This design is free of the severe off-axis coma that affects fast Newtonian telescopes and easily outperforms old-style Schmidt-Cassegrain systems and even the newer coma-free designs too. Up to a few years ago, very few companies sold Ritchey-Chrétien systems in the apertures typically used by amateur astronomers. In the US, the companies RCOS (RC Optical Systems) and OGS (Optical Guidance Systems) did, and still do, sell Ritchey-Chrétien systems in apertures as small as 12.5 and 10 inches respectively, but at eye-watering prices compared to standard Schmidt-Cassegrain systems of a similar aperture. However, in the past few years, various European and Far Eastern optics producers have found ways to mass-produce the complex hyperbolic mirrors needed for affordable Ritchey-Chrétien telescopes, thus giving the Schmidt-Cassegrain manufacturers like Celestron some additional competition. The companies and dealers now offering Ritchey-Chrétien systems for amateur astronomers include Officina Stellare, Altair Astro, GSO (Guan Sheng Optical of Taiwan), Astrosib and the aforementioned top of the range RCOS and OGS manufacturers. An examination of the less-expensive models reveals that the GSO 8-inch models are roughly comparable in price with that of a classic Schmidt-Cassegrain of the same aperture and rather cheaper than a Coma-free/EdgeHD model. The Ritchey-Chrétiens work at around f/8, compared to f/10 for the SCTs and the secondary mirror in the RC telescopes is, typically, 42 per cent, compared to 34 per cent for the SCTs, making the SCT much more versatile if both deep-sky and planetary imaging are being considered. Move up to 10-inch RC models, or SCTs in the range 9.25 to 11 inch, and the newer Coma-free/EdgeHD SCT models are similarly priced to the 10-inch RC offerings, with prices, depending on precise model and specification, spanning the £2,500 to £4,000 range. What is most interesting, however, is the huge difference in the prices of classic Schmidt-Cassegrains and the newer coma-free models, especially here in the rip-off UK, where the cost is far higher anyway (even allowing for import taxes) than in the US. At the time of writing the standard Celestron SCT models were approximately 40 per cent of the cost of the EdgeHD models for the 9.25- and 11-inch apertures, making them an absolute bargain if you are happy to image with fields smaller than, say, 20 arc minutes! Move up to 12-inch Ritchey-Chrétiens, and the cheapest models compete in price with the Celestron 14 SCT and Celestron 14 EdgeHD models, which do not exhibit this remarkable factor of 2.5 x variation in the UK price. The more high end, 12-inch and larger, Ritchey-Chrétiens, such as those made by RCOS, OGS and Officina Stellare, are in a higher-price bracket than even the Celestron 14 EdgeHD at its UK price.

So, it is clear that the traditional 8- to 14-inch SCT range telescopes are now in competition with the new wave of 'budget' Ritchey-Chrétien telescopes, but most amateur astronomers that I know still have a huge preference for Schmidt-Cassegrains, despite the optical purity of a system without lenses, where no chromatic (colour) aberration is being introduced by the RC design. Why this continued preference for both old-style and coma-free SCTs? Well, there are a number of reasons. First, Celestron, despite almost disappearing in 2001 (when its relatively new parent company Tasco faced bankruptcy), has been manufacturing traditional Schmidt-Cassegrains since 1970. In fact, the original owner of Celestron, Tom Johnson, founded Celestron Pacific in 1964, but it was in 1970 that the now famous, orange tubed 8-inch f/10 Celestron SCT first appeared. With a name that has now been familiar to astronomers for forty-five years, and substantial brand loyalty, especially in the US, Celestron SCTs have a solid following.

Admittedly, Meade have also made Schmidt-Cassegrains since 1980, and were threatening Celestron's very survival in the 1990s, but the Celestron range is so well known that it is hard for other manufacturers to gain such a following.

In addition, the sealed tube design of Schmidt-Cassegrains is exceptionally attractive to amateur astronomers who invariably keep their biggest telescopes in damp outdoor observatories, crawling with enormous spiders and other insects. Amateur astronomers want to keep their aluminized mirror surfaces clean, which is easily achievable inside a sealed tube. As well as this, the Ritchey-Chrétien design features a much larger secondary obstruction, which is detrimental to high-resolution planetary work where details close to the telescope limit are being sought.

There is another strong rival to the coma-free Schmidt-Cassegrain and Ritchey-Chrétien designs, though, and that comes in the form of the Corrected Dall-Kirkham (CDK) design. In last year's Yearbook I related the life story of Horace Dall (1901–1986), the Luton amateur astronomer and optical genius who designed and built the first Dall-Kirkham Cassegrain telescope. Alan Kirkham, an American, shares the design name because he resolved the optical formulae that defined these instruments a few years after Dall built the first example.

The original Dall-Kirkham (DK) design is a Cassegrain system that is easier to make than a traditional Cassegrain but has one major shortfall, namely that it is only good for narrow-field work and not for wide-field observing or imaging. With the advent of modern ray-tracing software and exotic glass materials it has, in recent years, been possible to insert a correcting lens into the Cassegrain baffle tube of a Dall-Kirkham, to make it work very well over wide CCD fields, as much as seventy millimetres in diameter. The insertion of a correcting lens prior to the Cassegrain focus is, of course, very similar to the strategy adopted by the coma-free Schmidt-Cassegrains produced by Meade and Celestron. However, the resulting DK design allows much wider aberration-free fields and at faster f-ratios, too, typically around f/6.8 or so. The downside is that, as with the Ritchey-Chrétien design, the secondary mirror obstructs a substantial amount of the aperture, making the instruments less desirable for planetary imaging, especially with the smaller apertures.

There are three major manufacturers currently involved in the production of corrected (CDK) or optimized (ODK) Dall-Kirkhams for use by wealthier amateurs. These are the US Planewave company (founded by two former Celestron directors), the UK company Orion Optics, and the Italian Company Officina Stellare. A fourth corrected DK option is available from the Japanese manufacturer Takahashi whose 'Mewlon' range of Dall-Kirkhams has long been considered a high-quality option for the planetary observer. In this case Takahashi have manufactured a separate corrector/reducer lens that can be retro-fitted to all existing Mewlon models going back twenty years; it reduces the system f/ratio from f/10 to about f/7.3. The Takahashi Mewlon Dall-Kirkhams, corrected or not, have a smaller secondary mirror obstruction than the Planewave or Orion Optics offerings so, in many ways, compete with the coma-free Schmidt-Cassegrains, such as the Celestron EdgeHD. The main difference here is that the Mewlon models feature an open tube and are expensive relative to SCTs. Indeed, if you are a diehard Takahashi astrograph fan and have very deep pockets, the company now sells a dedicated 'Corrected Cassegrain Astrograph' of 250-mm aperture called the Takahashi CCA-250. It is a 250-mm f/5 Cassegrain system with pin-sharp stars out to a remarkable diameter of 88 mm. Unfortunately, the current cost in the UK works out at around £13,000, which, for all but total perfectionists, will be a step too far for a 250-mm aperture!

The Planewave CDK range consists of serious astrographic instruments that are outside the range of all but the most dedicated and wealthiest amateur astronomers. The smallest instrument in the line-up is a 12.5-inch f/8 system which sells for $12,000 in the US and almost £10,000 in the UK; this is far higher than the price of a Celestron 14 EdgeHD and more than twice the price of a Celestron 11 EdgeHD! But, be under no illusion, you are paying for serious instrumentation here. The most expensive Planewave tube assembly in the range is the 24-inch model which will set you back $50,000 and if you want the 27-inch model with alt-azimuth mounting, you can wave goodbye to $200,000. Essentially, to buy that beast you would need to sell your house to raise the funds, although a Premiership footballer could presumably buy a new one every single week! The 12.5-inch Planewave model is a solid tube system, but all the larger models, including the 14-inch are open-tube truss designs.

At the time of writing, the specialist Italian optics company, Officina Stellare, who make Ritchey-Chrétiens and a unique Riccardi-Honders f/3 Astrograph, also offer one 300 mm Corrected Dall-Kirkham, called

Figure 9. Planewave's 14-inch Corrected Dall-Kirkham Astrograph. (Image courtesy of Planewave.)

the RiDK 300 (Riccardi 300 Dall-Kirkham). It costs almost £10,000 in the UK, so is competing with Planewave at this aperture, but I know of no one who has hands-on experience with that particular model.

Fortunately, there is a far more affordable option in the form of the aforementioned Orion Optics ODK line, which, as they are British, means that there is no mysterious, baffling and enormous disparity between the UK and US prices. The Orion Optics ODK pricing is far more attractive to the British amateur astronomer, with the 10-inch model selling for under £4,500 in the UK, the 12-inch for around £6,000 and the 14-inch for £7,500 or so. The company also offers weighty 16- and 20-inch ODK models for around £9,000 and £26,000 respectively. Unlike the Planewave CDK range, the Orion Optics ODK range employ solid tubes, apart from in the largest 20-inch model. Clearly, for anyone contemplating purchasing, for example, a Celestron 14 EdgeHD optical tube in the UK, the Orion Optics 12- and 14-inch ODK models are wider-field alternatives in a similar price range. With the 14-inch ODK working at f/6.8 and claiming a flat field greater than 52 mm in diameter, it has a significant real-field advantage to the Celestron 14 EdgeHD working at f/11 with a 42-mm field. In fact, it is a two-fold increase in the diameter of the pin-sharp star field, roughly 37 arc minutes for the C14 versus 74 arc minutes for the ODK 14. It should be added that (as mentioned earlier) the C14 does have an optional dedicated 0.7 x focal reducer available. Indeed, it is worth mentioning that the Planewave CDKs all have dedicated corrector/

reducers available at great cost, reducing their focal ratios even further, down to an impressive f/4.5 in the case of the f/6.8 models. At this point, as focal reducers/correctors are often used in astrophotography, I think it is worth saying a bit more about these accessories.

FOCAL REDUCERS AND CORRECTORS

Prior to the 1990s, using lenses to reduce the f-ratios of telescopes was virtually unheard of. Plenty of amateur astronomers used 2 x or 3 x Barlow lenses to increase the focal length for planetary work, but not to decrease it. Those using Schmidt-Cassegrains at f/10 were mainly still using photographic film in the early 1990s and even with a 14-inch Celestron at f/11, a standard 36 x 24-mm film frame would still capture a 32 x 21-arc minute field: big enough for all Northern Hemisphere galaxies except M31 and M33. A few amateurs were even willing to guide for two hours, visually, at f/11! With the advent of tiny CCDs, there became an urgent need to shrink the focal length of SCTs. One early strategy devised by Celestron was to market a range of Schmidt-Cassegrains that worked at f/6.3, without a reducer, but this quickly died a death as it was clear that the resulting aberrations were not acceptable. So, as a result, various companies, namely Celestron, Meade and Optec, started producing focal reducers to tame SCTs from f/10 to f/6.3 and even f/3.3! In this way a tiny CCD could cover a field up to three times wider than it could at f/10, with a more sensible sky-sampling ratio too. With CCD chips smaller than ten millimetres in width this worked very well.

However, CCDs are now much larger and so one has a choice of not using a focal reducer at all, or using one but accepting that the field might not be perfect in the corners. SCTs are not the only telescopes for which focal reducers are now available. Short-focus refractor manufacturers routinely offer focal reducers to make their refractors even faster, or to sharpen the field. These devices are labelled in various ways, such as field correctors, field flatteners, focal reducers and reducer/correctors. The choice is truly bewildering, even to the dealers! To get an overall view of just how many such accessories are on offer to the telescope user I would recommend going to the website of the giant German/European distributor Teleskop Service at www.teleskop-express.de and perusing their Astrophotography and Photography Reducers, Converters and

Correctors section. The last time I checked, there were more than seventy options in that section. Please stop the world, I want to get off!

As mentioned earlier, I receive huge numbers of e-mails each month regarding telescope purchases and, increasingly, many of these messages are about focal reducers and correctors that do not work as expected. In some cases, the reducers have simply been incorrectly assembled. The correct lenses have been inserted in a factory somewhere in the Far East, but not necessarily in the right order, or the right way around! In other cases, the field is highly vignetted, or distorted (or both), at the corners because the purchaser did not appreciate that the reducer was only intended for use with a relatively small CCD. Another problem frequently encountered is ensuring that the precise distance from the final lens of the corrector to the CCD surface is exactly as recommended by the manufacturer. This is absolutely critical, as these optical systems are highly sensitive to the lens-CCD distance and if this spacing is incorrect, the resulting star images can be appalling. The correct distance can be adjusted by buying spacer rings, but if filter wheels and other accessories are in the light path, it can be tricky to get things right. My advice here would be to initially use every astrographic system without any optional reducers at all, and if a reducer is purchased, an inexpensive one that does not work will not be as painful to bear as a bank-bursting one that does not work either! Online astronomy forums such as Stargazers Lounge, Cloudy Nights, the Progressive Astro Imaging Group and various Yahoo! user groups can be invaluable if unbiased help is required with specific accessories like these.

IMAGING SENSORS

Anyone contemplating carrying out astrophotography and purchasing an astrographic system would be well advised to design the system around the CCD camera or DSLR sensor that they intend using. One could even find that the actual telescope choice is the least-expensive decision compared to the choice of the sensor and the equatorial mounting, at least if a simple Newtonian reflector is chosen as the main instrument. If you want a big sensor and want to do simple colour imaging on a budget, then a DSLR with a full frame or APS-C-sized sensor will surely be the best choice. Such a camera can be used for daytime photography as well, which means that even if the endless cloud, rain, night-time dew, freezing cold nights and kamikaze moths turn you away from the hobby, you still have a decent camera. DSLRs do not record well the deep red Hydrogen-alpha line prominent in nebulae, but if comets or globular clusters, or even galaxies, are your main interest, then this is not a big problem. Canon do produce an astronomy-tweaked DSLR that overcomes this problem, the 60Da, which costs around £700, but unless you are seriously into imaging nebulae you will probably prefer a cheaper standard model with perfect colour balance on daytime objects.

As I mentioned earlier, in deep-sky astrophotography an important criterion is the sampling ratio, with two arc seconds per pixel often being considered optimum. If you end up sampling at finer resolutions (with long focal lengths and/or small pixels), this is not a total disaster, as using most cooled CCD cameras in binning mode effectively increases the pixel size, thus binning 2 x 2 is like having pixels twice as large, or even like halving the f-ratio, from a 'photons per effective pixel' perspective. Basically, in the early planning stages, thinking about the sampling scale that your proposed sensor and telescope will deliver is important, as are the likely targets for your astrograph.

Arguably the most popular small pixel CCD sensor systems now in use by keen astrophotographers are based around the 8.3 megapixel Kodak KAF-8300 sensor which employs an array of 3326 x 2504 pixels of 5.4 microns in size. Clearly, unbinned, this tiny pixel size is best employed using short focal lengths of around 500 mm or so, delivering 2.2 arc seconds per pixel. With chip dimensions of 18.0 x 13.5 mm, such a system would cover roughly 2.1 x 1.5 degrees. By comparison, the pixels in DSLRs are around 4 to 7 microns in size and the sensors are roughly 30 per cent bigger in the APS-C DSLR models than in the KAF-8300 sensor.

Moving up in pixel size, 9 microns is the next most common to be found in astronomical imaging systems. Examples of such CCDs include the Kodak KAF 6303E, KAI 11002M, and KAF 16803 sensors. These chips are all different sizes, namely 28 x 18 mm, 36 x 25 mm and 37 x 37 mm, and so will span different-sized fields on the sky. With a focal length of 1000 mm, a 9 micron pixel will cover 1.9 arc seconds and the chips in question will cover fields of 1.6 x 1.0 degrees, 2.1 x 1.4 degrees and 2.1 x 2.1 degrees respectively. There are specialist CCDs out there with bigger pixels, up to twenty-four microns, in fact, but I have covered the most popular pixel sizes above, along with four of the most popular Kodak CCDs. Sony also make some splendid CCD chips which are used by astro-camera manufacturers such as Starlight Xpress. The Kodak KAF 6303E mentioned above is a so-called 'NABG' sensor which means that it has no 'anti-blooming gates', so excess charge from bright stars will overfill the pixel 'bucket' and 'bleed' across the nearby columns. Although unsightly on very bright stars, these CCDs are better for purely scientific purposes such as photometry, because the measurement of charge in the pixels is very linear, right up to the saturation point.

REMOTE IMAGING ONLINE

In recent years, a new way of imaging astronomical objects has become popular and it is a method I have fully enjoyed using. Traditional outdoor amateur astrophotography can be nothing short of an energy-sapping battle and it is hardly surprising that amongst the hundreds of newcomers to this masochistic aspect of astronomy each year, only a few are still persevering a few years later. Most adults in the twenty-first century lead complex and stressful lives with little spare energy left in the tank for frequent nights battling with equipment under the stars – until they retire from the dreaded day job, that is! British nights are usually cloudy and often very cold, and despite the initial enthusiasm for purchasing that gleaming astro-kit seen in the glossy magazine ads, the reality of the cost, complexity and hassle of astronomical imaging means that, for many, just one year after the financial investment is made the kit is gathering dust in the spare bedroom or shed, or appearing for sale in the second-hand columns!

It quickly becomes obvious that those who persevere with this hobby to the bitter end are either dangerously addicted to it, heading for an imminent divorce, happy to take grotty images a few times a

Table 1 (opposite). Twenty-two of the telescopes mentioned in this article and how they perform using three popular Canon DSLR models and three Kodak KAF sensors, used in popular cooled CCD cameras. For each telescope/sensor combination the first number gives the sky-sampling ratio in arc seconds per pixel and the second number gives the field coverage in arc minutes. So, for the top left box, a Skywatcher 8-inch f/4 Quattro, used with a Canon EOS 700D has a sampling ratio of 1.1 arc seconds per pixel and covers 94 x 63 arc minutes on the sky.

Astrog./CCD	EOS700D		EOS6D		EOS60Da		KAF8300		KAF11002		KAF16803s	
Quattro 8 f/4	1.1	94x63	1.7	152x101	1.1	94x63	1.4	76x57	2.3	152x101	2.3	152x152
Quattro 10 f/4	0.9	75x50	1.3	122x81	0.9	75x50	1.1	61x46	1.8	122x81	1.8	122x122
OOAG12 f/3.8	0.8	66x44	1.2	107x71	0.8	66x44	1.0	53x40	1.6	107x71	1.6	107x107
MN190 f/5.3	0.9	76x51	1.4	123x82	0.9	76x51	1.1	61x46	1.9	123x82	1.9	123x123
ES Levy f/4.8	1.2	104x70	1.9	169x113	1.2	104x70	1.5	85x64	2.6	169x113	2.6	169x169
Tak E180 f/2.8	1.8	152x102	2.7	245x163	1.8	152x102	2.2	122x92	3.7	245x163	3.7	245x245
Tak E130 f/3.3	2.1	179x119	3.2	288x192	2.1	179x119	2.6	144x108	4.4	288x192	4.4	288x288
C925 Edge f/10	0.4	33x22	0.6	53x35	0.4	33x22	0.5	26x20	0.8	53x35	0.8	53x53
C11 Edge f/10	0.3	27x18	0.5	44x30	0.3	27x18	0.4	22x17	0.7	44x30	0.7	44x44
C14 Edge f/11	0.2	20x13	0.3	32x21	0.2	20x13	0.3	16x12	0.4	32x21	0.4	32x32
Row-Ack f/2.2	1.4	124x83	2.2	199x133	1.4	124x83	1.8	100x75	3.0	199x133	3.0	199x199
TakFSQ106 f/5	1.7	145x97	2.6	233x155	1.7	145x97	2.1	116x88	3.5	233x155	3.5	233x233
TakFSQ130 f/5	1.4	118x79	2.1	190x127	1.4	118x79	1.7	95x71	2.9	190x127	2.9	190x190
Esprit 80 f/5	2.2	191x128	3.4	309x205	2.2	191x128	2.8	154x116	4.6	309x205	4.6	309x309
Esprit 150 f/7	0.8	73x49	1.3	118x79	0.8	73x49	1.0	58x44	1.8	118x79	1.8	118x118
CCA 250 f/5	0.7	61x41	1.1	99x66	0.7	61x41	0.9	49x37	1.5	99x66	1.5	99x99
ODK12 f/6.8	0.4	38x25	0.7	60x40	0.4	38x25	0.5	30x22	1.0	60x40	1.0	60x60
ODK14 f/6.8	0.4	32x21	0.6	51x34	0.4	32x21	0.5	26x19	0.8	51x34	0.8	51x51
ODK16 f/6.8	0.3	28x19	0.5	45x30	0.3	28x19	0.4	22x17	0.7	45x30	0.7	45x45
Pwave 12.5 f/8	0.3	30x20	0.5	49x32	0.3	30x20	0.5	24x18	0.7	49x32	0.7	49x49
Pwave 14 f/7.2	0.3	30x20	0.5	49x32	0.3	30x20	0.5	24x18	0.7	49x32	0.7	49x49
Pwave 17 f/6.8	0.3	26x17	0.5	42x28	0.3	26x17	0.4	21x16	0.7	42x28	0.7	42x42

year, or are clinically insane! However, in the twenty-first century, things do not have to be quite so gruelling, sleep depriving, or so unsociable. For a monthly outlay, ranging from about £25 to £500 (for the richest addicts), telescope time can be rented out online so that you can just sit indoors in a warm room in UK daytime, imaging objects while it is the middle of the night in places like New Mexico or Australia. Sheer bliss!

Like many amateur astronomers who have experienced the traditional astronomical methods, in my case since 1968, I felt uneasy at this remote imaging approach at first. Was it 'real' observing at all? Maybe it was even a form of cheating? Surely a real observer has to stay outside until the bitter end, battling with cloud, rain, frostbite, rivers of dew, near-death experiences and equipment failures, pretending it is all great fun, even if it is total, never-ending misery? Surely remote CCD imaging is for wimps, I thought? Well, my views have changed, if only because the older I get the more I hate those freezing cold winter nights. Also, the images I can capture from locations such as Mayhill in New Mexico and Siding Spring in Australia, using the itelescope.net system, make the images I can take from the bright and mainly cloudy UK skies look, well, rather sad!

The telescope systems on offer to the remote observer at facilities such as itelescope.net are beyond the reach of most amateur astronomers. Systems comprising 20-inch Planewave telescopes, atop the best equatorial mountings, fitted with absolute encoders, with large format CCDs, will set you back more than $60,000! Even if you owned such equipment, you would, in the UK, be living under mainly cloudy and light-polluted skies too, and I can guarantee that when a bright supernova or comet appears you would experience endless gales and lashing rain. The skies at these remote locations are, in contrast, usually clear and very dark. For the pleasure of using these astrographs you pay a rate which varies from approximately 200 to 100 Australian dollars per hour, or roughly £2 per minute on the budget tariffs. In practice, even exposures of a few minutes' duration, with such equipment at such locations, can blow away any image you can ever grab from the UK. So, unless you are heavily into ultra-long deep-sky exposures, renting time on an astrograph based abroad makes perfect economic sense, even if it is not the same as having your own observatory. For the vast majority of amateur astronomers it is the only chance they will get to obtain spectacular images with such pricey equipment and it comes without the aforementioned physical battle, unsociable hours and sleep deprivation usually associated with astrophotography. Personally, I could not ever live without my own back-garden telescopes, but in the last few years I have thoroughly enjoyed using remote telescopes more, although, as yet, they cannot be used to take high-quality planetary images and most cannot image objects very low down, as is the case with many comets in twilight.

FINAL THOUGHTS

When all is said and done, deciding what astrograph suits your astrophotography needs best is a highly complex and highly personal decision, greatly influenced by the spare time, family commitments, back-garden size and financial situation of the individual. I have not even touched upon the subject of choosing a suitable equatorial mounting for an astrograph here, as that would take another article, just as long, and the price range of such equipment is enormous. The cheapest equatorial mountings will only track acceptably for thirty seconds, whereas the best will track perfectly for ten minutes, with no auto-guiding. If there is one thing I have learned in more than forty-five years as an amateur astronomer, it is that, if in doubt, stay with a simple system and a small financial outlay. Also, it is the long-term enthusiasm and problem-solving ability of the individual that results in the best images, not the price of the shiny new neglected toy, gathering cobwebs in the shed. I will say one thing about equatorial mountings, though: with the exception of high-end mountings such as the Paramount, Planewave and Astro-Physics systems, do not believe the manufacturer's payload figures on the standard mountings. Add up the weight of your telescope, camera gear and accessories and then look for a mounting where the blurb says it will handle TWICE that total payload. I must have had scores of e-mails over the years from amateur astronomers who have bought a mounting advertised as suitable for, say, a Celestron 14 SCT or a 12-inch Meade, only to find it will just about track and slew with half that payload. You have been warned! Budget equatorial mountings, used regularly at their weight limit, and their slew-speed limit, will often break down every few months.

Many budding astro-imagers reading this article will not have thousands of pounds at their disposal and my recommendation to them would be that purchasing something like a Skywatcher Quattro 8-inch f/4 Newtonian on, say, an EQ5-class mounting, with a coma corrector, is just achievable for £1,000. With a bit more cash a portable Skywatcher MN 190 or Explore Scientific David Levy Comet Hunter are worth serious consideration too. An entry-level DSLR will work well with all these systems and many amateur astronomers already own such versatile cameras. Stacking multiple thirty-second exposures with such equipment, on the popular objects like the Andromeda Galaxy and the Orion Nebula, will prove highly rewarding and such a modest system is portable and can be used visually too. If you crave rugged versatility (planetary and deep-sky imaging and visual observing), then an SCT such as a Celestron 8 or 9.25 (classic or EdgeHD) may be the wisest choice for a reasonable price.

If the astrophotography bug really bites after the initial financial outlay, that's great; if not, you still have a decent aperture telescope for visual use. Alternatively, for a modest initial outlay, and if the British weather has sapped your enthusiasm, you could try online remote imaging using a quality astrograph based in New Mexico, Siding Spring or at dark European sites.

Whatever your decision, Good Luck!

SOME USEFUL ONLINE WEBSITES TO VISIT

Orion Optics AG astrographs: http://www.orionoptics.co.uk/AG/agastrographs.html
Orion Optics ODK: http://www.orionoptics.co.uk/ODK/odkoptimiseddall.html
Skywatcher: http://www.skywatcher.com/products.php
Explore Scientific: http://www.explorescientific.co.uk/en/home/
Takahashi: http://www.takahashi-europe.com/en/index.php
Celestron: http://www.celestron.com/browse-shop/astronomy
Planewave: http://planewave.com/
ASA Astrographs: http://www.astrosysteme.at/eng/astrographs.html
Ian King Imaging: http://www.iankingimaging.com/_
Teleskop Service English page: http://www.teleskop-express.de/shop/index.php/language/en

The Square Kilometre Array

LISA HARVEY-SMITH

It is eighty-five years since Karl Jansky, a young engineer working at the Bell Telephone Laboratories in the United States, discovered that the regular and persistent static interference heard in long-distance telephone communications originated from outer space. From this simple discovery emerged a whole new field of astronomical enquiry – radio astronomy – in which scientists study the universe using invisible millimetre- to metre-wavelength electromagnetic radiation.

Radio waves are generated in space by a range of natural physical processes. The brightest and most widespread sources of radio waves in our night sky are active radio galaxies that cultivate super-massive black holes. These galaxies, dragon-like, spew out jets of hot plasma from their centres.

Another source of radio emission from space is hydrogen gas, the raw material that makes up stars and is distributed in clouds and clumps throughout the spiral arms of our own Galaxy. As hydrogen is the most widespread (baryonic) substance in our universe, radio

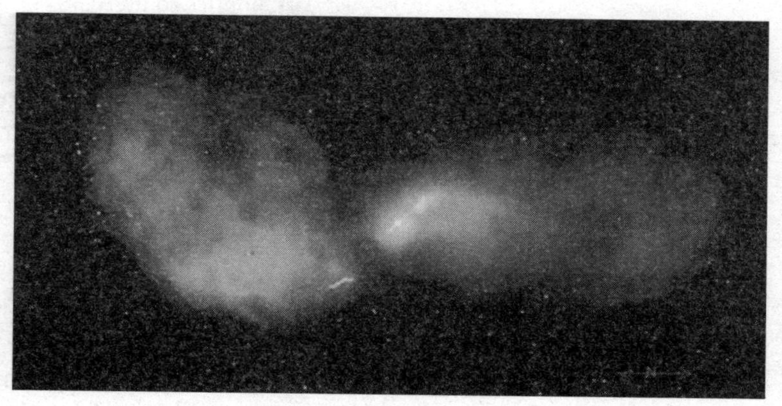

Figure 1. Radio emission from the nearby radio galaxy Centaurus A, imaged with the Australia Telescope Compact Array. (Image courtesy of Ilana Feain et al. (2011).)

astronomy is a powerful probe of the potential for building stars throughout the universe.

A small number of rare and mysterious objects also emit what seems like more than their fair share of radio waves. These include pulsars, which are rapidly spinning stars composed entirely of neutrons. Also prolific radio emitters are cosmic masers; regions of agitated gas that amplify background radio signals millions of times to produce a dazzling laser-like beam. To study the origin and properties of these radio waves, astronomers use radio telescopes.

UNDERSTANDING RADIO TELESCOPES

A radio telescope is a device that collects radio waves emanating from the sky and detects, amplifies and records the signals for further scientific study. Using sensitive electronics and modern computing techniques, astronomers can make images of the radio sky and measure the wavelength-dependent (spectral) properties of radio emissions from cosmic sources. This information can tell us about the physics and chemistry of the regions that are emitting the radiation.

The most common radio telescope design features a large collecting surface (known as a 'dish'). The surface area of the dish determines the

Figure 2. The Lovell 76-metre Telescope at Jodrell Bank Observatory. (Image courtesy of Dr Anthony Holloway, the University of Manchester.)

sensitivity of the telescope, so the bigger the dish, the fainter the signals we can detect.

The dish reflects radio waves that fall upon the metal surface and directs them to a radio receiver mounted at the focal point. The receiver contains an electrical conducting material that converts the electromagnetic waves into a tiny electrical signal, rather like the waves on an oscilloscope. This incredibly weak signal is then electronically amplified and digitized, with a system designed to minimize the unwanted system noise (static) introduced by the receiver system. Finally, the digital signal from the telescope is analysed in a powerful computer to produce a radio image and, in many cases, a range of additional scientific information. It is certainly not a straightforward process, but one that can yield amazing scientific results!

NOT JUST SINGLE-DISH TELESCOPES

For eighty years astronomers have built ever larger and more sophisticated radio telescopes, enabling us to see fainter objects and study more distant regions of the universe.

The Arecibo radio telescope in Puerto Rico is, at the time of writing, the world's largest single-dish radio telescope with a diameter of 305 metres. Most radio telescopes are mounted on tracks with gears and giant metal cogs that enable them to point at almost any object in the sky. The vast scale of the Arecibo telescope means that the dish cannot move at all and the target position can only be partially 'steered' by moving the receiver with respect to the dish. This trick is rather like lining up two mirrors to see around a corner. The largest fully steerable dish is the Green Bank telescope in the US, with a diameter of a hundred metres.

These giant eyes on the heavens are extremely powerful scientific instruments and for decades have been used to carry out cutting-edge astronomical research. But even these vast telescopes have their limitations. The long wavelengths of radio waves put us at a serious disadvantage relative to our optical, infra-red and X-ray colleagues.

The challenge is to achieve a high *angular resolution*, which means the level of fine detail that a telescope is able to distinguish. With low angular resolution, images of the sky resemble clumsy finger paintings, with images smudged and adjacent objects merged into one. Higher angular-resolution images are more detailed and are able to yield greater volumes of scientific information.

The number of waves that fit inside a telescope determine its angular resolution. For a 10-metre optical telescope studying green light with a wavelength of 550 nanometres, the number of wavelengths that fit side by side in the telescope aperture is 18 million. A typical wavelength that radio astronomers might study is 21 cm. In order to fit 18 million waves side by side (to achieve the same level of image detail), the equivalent radio telescope would need to be 3800 km across. Not very practical to build, nor for gaining planning permission!

RADIO TELESCOPE NETWORKS

To circumvent this problem, astronomers use a clever trick called 'aperture synthesis' which enables us to simulate a very large telescope by combining the signals from several smaller telescopes spread out across large distances. These networks of radio telescopes now stretch across the globe, maintained by scientific collaboration between many different countries.

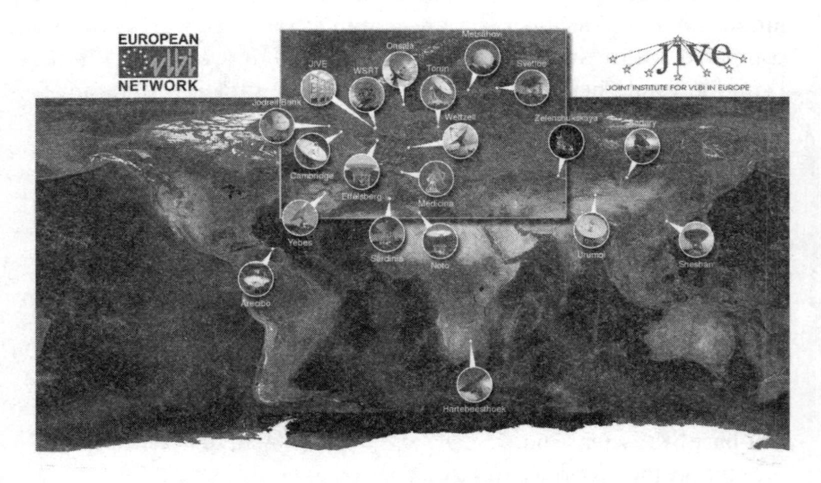

Figure 3. The European Very Long Baseline Interferometry Network. (Image courtesy of Paul Boven, JIVE.)

THE SKA: THE IDEA IS BORN

In its first five decades, radio astronomy made many transformational discoveries, leading to the award of five Nobel Prizes in Physics. Radio telescopes precipitated the discovery of pulsars, interstellar molecules, neutral hydrogen gas in galaxies and the ubiquitous cosmic background radiation that supports the Big Bang theory and helps us to map out structure in the early universe. After all this success, scientists began to consider the next steps in this evolution.

In the late 1980s and early 1990s, facilitated by discussions at several international scientific conferences, the concept of a highly sensitive next-generation radio telescope emerged. Driven by a desire to keep up the pace of discovery, astronomers recognized that the next big step in radio astronomy instrumentation would require an exponential increase in collecting area. This led to the idea of a telescope array with a total collecting area of 1 million square metres (one square kilometre).

Such a vast collecting area – fourteen times greater than Arecibo – would enable this next-generation radio telescope to detect millions of new galaxies in a far greater volume of space than is currently possible.

This unprecedented sensitivity combined with a high angular resolution would also enable radio images of distant objects to be directly compared with existing optical and infrared data for the first time. Doing so would help us to diagnose properties such as the mass, age, distance to galaxies and the amount and distribution of the mysterious 'Dark Matter' therein.

A number of specific proposals for a next-generation radio telescope design emerged from India, the Netherlands, the UK, the US, Canada, Australia and the Soviet Union around the turn of the 1990s.

These concepts quickly gained support and in 1993, an international working group began a worldwide effort to develop a more detailed design and science case for the project. The international research community was mobilized and by 2004, the scientific ideas for the project – now dubbed the 'Square Kilometre Array' (SKA) – were published in a 600-page SKA science book.

Support for the project is now greater than ever. The SKA is now an international collaboration of ten member countries (Australia, Canada, China, India, Italy, New Zealand, South Africa, Sweden, the

Netherlands and the United Kingdom), with a further nine countries participating in the detailed design of the telescope.

The SKA Organisation is a not-for-profit company responsible for coordinating the activities of the SKA project including science, engineering, site evaluation, operations and public outreach. The international project office is located at the Jodrell Bank Observatory at the University of Manchester and is currently staffed by more than forty people.

DESIGN AND LOCATION

The SKA baseline design has two key components and the design of these is driven by the science goals of the SKA project. The SKA low-frequency array (a sensitive telescope operating at frequencies between 50 and 350 MHz) is required to detect radio waves from the Cosmic Dawn – the period before the first stars and galaxies were formed after the Big Bang – and to study the evolution of galaxies in the early universe. The SKA mid-frequency array, operating at approximately between 350 MHz and 14 GHz, is required to study our own Galaxy, the evolution of galaxies and star formation over several billion years, and the formation of planets and the evolution of life in the universe.

CHOOSING A SITE

Along with the design, one of the most important factors in the success of the SKA telescope is its location. Just as optical telescopes require clear, dark skies (many of the first great optical observatories became obsolete due to light pollution from nearby cities), the major radio observatories also suffer from 'radio frequency interference' from modern human activities. The fact is that most of the planet's landmass is covered by radio, television, communication, navigation, GPS and other radio transmissions. Fortunately, astronomers have negotiated through the International Telecommunications Union a small number of special protected 'windows' that are avoided by other spectrum users in favour of passive astronomy observations.

One of these is centred at 1420 MHz, the frequency emitted by the most common radio emission from hydrogen atoms. This arrangement works well for studying hydrogen gas in the Milky Way and in nearby galaxies. When we come to more extensive studies of hydrogen gas throughout astronomical history, however, the expansion of the universe ensures that the frequency of hydrogen emission is shifted to a lower frequency outside the protected band. Astronomers then find themselves struggling to detect these extremely faint signals from distant galaxies against a cacophony of background noise.

Another important factor in choosing a location was the atmospheric stability of the site. At low frequencies this depends on the high-altitude charged particles that make up the ionosphere and at higher frequencies the water vapour in the troposphere has the most effect on radio waves. The right region should also not be prone to flooding, extreme weather or earthquakes.

The SKA Organisation conducted an exhaustive search for the best location for the SKA, which lasted several years. After a bidding process in 2006, a list of four candidate sites was reduced to two; South Africa and Australia. The decision was made by the International SKA Steering Committee, following advice from a committee of seven scientists from five different countries that examined the bids. Both these candidate sites (with Australia newly partnered with New Zealand) prepared detailed cases to demonstrate their suitability in time for a final site-decision process held in 2013.

The criteria were many, but the key factors considered were: radio quietness, climactic and geological stability, security, legal and taxation considerations, political stability and cost.

After considering thousands of pages of scientific and technical submissions, in May 2013, the SKA Organisation announced that it had decided upon a dual-site implementation, with the majority of the telescope comprising a vast low-frequency radio telescope in mid-west Western Australia and a huge array of radio dishes to be located in the Karoo region of South Africa.

Although the dual-site decision had some scratching their heads, this solution did have some clear scientific and practical benefits. The foremost advantage of multiple instruments is to increase the available observing time for science projects. As well as increasing the available telescope time, the dual-site implementation makes use of existing radio astronomy infrastructure that has already been built up in both countries through the SKA precursor projects.

THE SKA DESIGN

The concept design for Phase 1 of the project is expected to be operational by the mid-2020s, and we will look at this, together with the likely upgrade path for SKA Phase 2, below.

The Australian-based component of the SKA will be a low-frequency array comprising approximately one hundred and thirty thousand stationary antennas spread over 65 kilometres of the outback. Together, this vast array of antennas will scan the sky between 50 and 250 MHz, working as one giant radio eye. The array will have a survey speed more than 130 times that of the LOFAR telescope, which is currently the most powerful telescope in the world in this frequency range. The low-frequency SKA will search for gravitational waves using very precise measurements of radio signals from pulsars. It will also measure the distribution of neutral hydrogen from the Cosmic Dawn, which will tell us how gravity, dark matter and dark energy influenced the distribution of matter in the very early history of the universe.

Figure 4. Artist's impression of the Phase 1 SKA low-frequency array comprising hundreds of thousands of dipole antennas. (Image courtesy of SKA Organisation.)

The other component of Phase 1 of the SKA, located in South Africa, will be an army of approximately two hundred radio dishes – each towering three storeys high – and spread across 150 km of the Karoo. SKA dishes will be capable of tuning into a range of radio frequencies from 350 MHz to 14 GHz, centred on the 1420 MHz radio frequency commonly emitted by hydrogen atoms. The SKA dish array will enable astronomers to study the structure of our Milky Way Galaxy and other galaxies in unprecedented detail. It will carry out vast surveys of the sky and study the history of star formation and galaxy evolution over billions of years. The higher-frequency capability will be used to study the formation of planets in solar systems around other stars, and the role of complex biological molecules and their possible role in the genesis of life in the universe.

Figure 5. Artist's impression of the SKA dish array, including an existing MeerKAT antenna, in South Africa. (Image courtesy of the SKA Organisation.)

A second phase of the project will increase the collecting area towards a full square kilometre, enabling scientists to see even deeper into almost completely uncharted regions of space and to probe new, exciting and as-yet-undiscovered astrophysical objects. Smaller numbers of outlier telescopes will be located between tens and thousands of kilometres from the core of each array. This will provide a zoom capability, enabling astronomers to make high angular-resolution images and thereby classify galaxies at roughly half the distance to the edge of the observable universe.

SKA PRECURSORS

A large global scientific endeavour such as the SKA requires an army of highly skilled and experienced professionals, including engineers, technicians, tradespeople, computer programmers and astronomers. In order to assemble such a team, the SKA family has spent much of the past decade developing new technologies, establishing new observatories and research centres, and broadening collaborations and, in doing so, has prepared the ground for the SKA.

SOUTH AFRICA

South Africa is relatively new to radio astronomy, but, despite this, the country is now becoming a major international player.

In 1975, the Hartebeesthoek Radio Telescope near Johannesburg became the first radio astronomical observatory in Africa after it retired from spacecraft-tracking duties with NASA and was converted into a dedicated radio astronomy facility. In 2001, Hartebeesthoek joined the European VLBI Network, an international consortium of radio telescopes that join forces for three coordinated observing campaigns per year. This network specializes in making very high-resolution pictures of distant objects, including the central regions of galaxies surrounding their central black holes, and gravitational lenses; distorted images of background galaxies made by the gravitational pull of dark matter.

Since 2005 the SKA South Africa team has been steadily building the capacity and infrastructure required to host the SKA. They have established a new observatory at a remote site in the Northern Cape of South Africa, approximately ninety-five kilometres from the nearest town of Carnarvon. Not only will the Karoo site host the SKA, but the observatory will first be home to the South African SKA pathfinder MeerKAT – a very powerful radio telescope made up of sixty-four radio dishes spread across eight kilometres of the Karoo.

As one of the official SKA pathfinder projects, MeerKAT will develop, test and prototype new technologies for radio astronomy and will be a world-class research facility in its own right. The project is being rolled out in stages to enable research and development to be carried out throughout the project and to allow the best technologies to be incorporated into the final MeerKAT design.

As the first stage of this project, the KAT-7 (Karoo Array Telescope) array, a prototype radio array comprising seven composite twelve-metre diameter fiberglass dishes, came online in 2010. Each dish was made using an experimental fabrication technique that eliminates the need for assembly and adjustment of dish panels. This technique was an experimental solution for fabricating large numbers of dishes for the SKA.

Figure 6. Three of the South African KAT-7 precursor dishes. (Image courtesy of SKA South Africa.)

The early scientific results from KAT-7 were extremely promising. An international team of astronomers working with data from KAT-7 discovered an extremely faint and rather rare halo of radio emission surrounding a gigantic cluster of galaxies. Another team used KAT-7 to measure the distribution of mass and dark matter in a nearby spiral galaxy and measured variable radio emission from an 'X-ray binary',

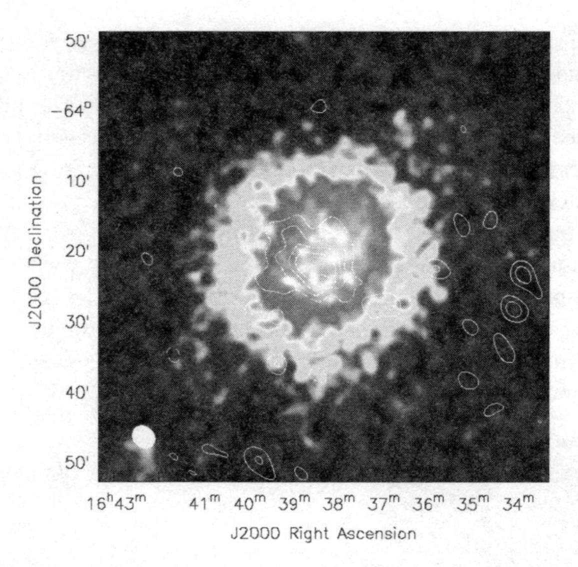

Figure 7. A vast radio halo surrounding Triangulum Australis, a merging cluster of galaxies, discovered using KAT-7. The contours show the KAT-7 results superimposed on the ROSAT X-ray image. A clear coincidence between the radio halo and the hot X-ray plasma is apparent. (Image courtesy of Anna Scaife and Nadeem Oozeer, KAT-7/SKA South Africa.)

where material from a star is pulled onto the surface of a partner neutron star or black hole.

As well as building a new observatory and telescope infrastructure in the Karoo, the SKA South Africa project has increased the national skills base in readiness for the SKA. The South African government has provided over six hundred student bursaries in astronomy and engineering in a push to increase the number of astronomers, physicists, mathematicians, engineers and technicians. Universities in South Africa have also broadened collaborations with research institutes across the world, attracted international academics to work in Africa and, recently, for the first time, provided astronomy courses to students in the SKA Africa partner nations of Kenya, Mozambique, Madagascar and Mauritius. This boost to the development of science and technology in Africa will surely prove to be one of the great legacies of the SKA project.

The next stage in the project is to build MeerKAT. The MeerKAT dishes are shaped to reflect the radio waves from the sky into a secondary reflector and a radio receiver that are both offset from the centre of the dish. This design has the advantage that the dish has an unblocked view of the sky, although its asymmetrical-shaped design may introduce other challenges, including structural strain on the dish and complex signal reflection paths.

Once built, MeerKAT will be the most sensitive centimetre-wave radio telescope in the Southern Hemisphere. Ten international teams of scientists are ready to use MeerKAT for a range of planned experiments that cover the search for gravitational waves, investigation of the dynamic radio sky and the evolution of gas in galaxies through billions of years of cosmic history.

Figure 8. The first of sixty-four dishes that will make up the MeerKAT array. (Image courtesy of SKA South Africa.)

AUSTRALIA

For seven decades Australia has been a world leader in radio astronomy. Government-funded radio astronomy in Australia sprang from the development of radar technologies during the Second World War. This early work was carried out at Dover Heights near Bondi Beach and at other field stations around Sydney. The pioneers who carried out this research were working in previously uncharted territory and made several major scientific discoveries.

They discovered the bright radio emission from the centre of our Galaxy, Sagittarius A*, which we now know is generated by material spinning around an enormous black hole. They studied our nearest star, the Sun, and made the fascinating observation that strong bursts of radio emission coincide with sunspot activity. They also mapped out the brightest radio sources in the sky, finding that many of them were very distant galaxies. This was a huge surprise at the time, since the brightest objects in the optical sky are usually the closest to us.

The discovery of bright radio galaxies was unexpected and led to many important streams of research. Their properties are unlike anything we see at optical wavelengths. They are incredibly luminous and have extremely powerful 'jets' of gas blasting from colossal black holes at their centres.

The next chapter in Australian radio astronomy began in 1961 when the sixty-four-metre Parkes Radio Telescope was opened. Its fantastic

Figure 9. Four of the six twenty-two-metre antennas of CSIRO's Australia Telescope Compact Array, near Narrabri, New South Wales. The three antennas in the foreground are arranged on the main East-West track and one in the background on the North spur. (Image courtesy of CSIRO.)

size and continual improvements to its receivers and computers have enabled astronomers to discover magnetic fields in space, identify new spiral arms in the Milky Way and to discover hundreds of fascinating rotating radio beacon stars called pulsars. The recent discovery of a gravitationally bound pair of pulsars was used to verify Einstein's theory of relativity (the curvature of space-time) to unprecedented accuracy.

Another jewel in the crown of Australian radio astronomy is the Australia Telescope Compact Array (ATCA), opened in 1988. The ATCA works at higher-radio frequencies than the Parkes Telescope and has sharper vision. The ATCA's achievements include photographing in real-time the growth of supernova remnants in the Milky Way and in nearby galaxies, mapping out the distributions of chemical elements in nebulous regions of star formation in the Milky Way, and investigating the structure and magnetic fields in giant clusters of galaxies.

A group of universities and research institutes in Australia and New Zealand regularly co-ordinate their observing schedules to form a giant radio telescope network called the Long Baseline Array that spans five thousand kilometres from Western Australia to the North Island of New Zealand and down to Tasmania in the south. This network allows astronomers to zoom in to very small regions of space and make radio images in exquisite detail.

THE SKA IN AUSTRALIA

Existing radio observatories on the east coast of Australia are increasingly affected by radio-frequency interference, the radio equivalent of light pollution, which is generated by electronic and communications devices that we all use today. In 2009, the Murchison Radio-astronomy Observatory was opened in the remote mid-west region of Western Australia, in response to the need to establish a remote site that would be suitable to host the SKA. The Wajarri Yamatji people are the traditional custodians of the land on which the observatory lies. The area has an extremely low population density (0.002 people per square kilometre) and was selected for its radio quietness and atmospheric stability.

There are currently two SKA pathfinder telescopes located at the observatory, the Australian SKA Pathfinder (ASKAP) and the Murchison Widefield Array (MWA).

THE AUSTRALIAN SKA PATHFINDER TELESCOPE

ASKAP comprises thirty-six dish radio telescopes, each with a diameter of twelve metres. The white metal dishes stand three storeys tall and are dotted across six kilometres of dry, red earth on an historical cattle station. At the focus of each ASKAP dish is a novel radio camera called a phased-array feed that enables astronomers to conduct fast radio surveys over wide areas of the sky. This design makes ASKAP thirty times more efficient at surveying the sky than an identical array fitted with traditional receivers.The ASKAP receiver's incredible power is that it combines information from different parts of the dish's focal plane to give a panoramic view of the sky. Phased arrays have been adopted widely for military and communications purposes but have not been used previously in radio astronomy.

The ASKAP telescope works as follows: signals from each ASKAP

Figure 10. Antennas of CSIRO's Australian SKA Pathfinder (ASKAP) telescope, at the Murchison Radio-astronomy Observatory in Western Australia. Phased array-feed technology, visible at the antenna focus, has been developed especially for radio astronomy to dramatically increase survey speed and sensitivity, and also offers enormous potential for other rapid-imaging applications. (Image courtesy of CSIRO.)

antenna are turned into digital information and transported via underground fibre-optic cables to a central control building on the observatory site. Data streams in from the telescopes at 72 trillion bits of information per second into a large, windowless room containing a very powerful computer. This computer (called a correlator) mathematically combines the data and compresses it by twenty thousand times to a more manageable forty billion bits per second.

This stream of information is transferred to a powerful supercomputing facility in Perth 750 kilometres away, where advanced data processing and imaging is carried out before the data are archived and stored for scientific use.

A campaign of commissioning and early science has been conducted with a six-antenna test array. This prototype has demonstrated the amazing scientific potential of ASKAP and produced high-quality images containing thousands of radio galaxies that span very large areas of the sky.

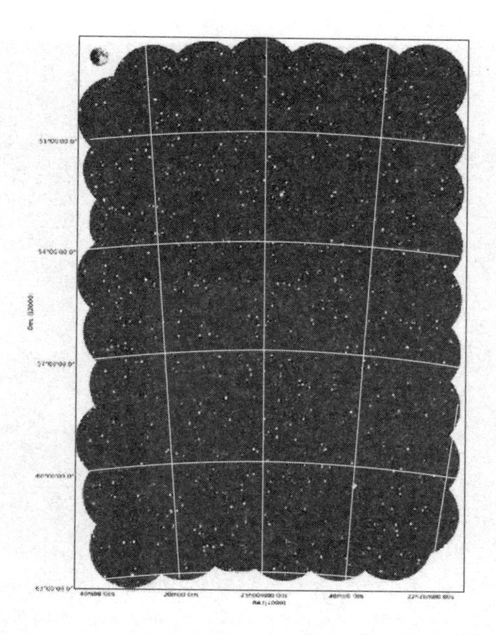

Figure 11. A radio image of 150 square degrees of the Tucana constellation taken with ASKAP's six-antenna engineering-test array. More than two thousand radio galaxies are visible. The full moon is shown for scale. (Image courtesy of CSIRO.)

At the time of writing, the first papers in refereed journals from the array have been published.

The full ASKAP array will be the fastest radio survey telescope in the world for studying centimetre wavelengths. Its ten science surveys, conducted by teams of over six hundred scientists from universities and research institutes across the world, promise great breakthroughs in many cutting-edge areas of astrophysics.

THE MURCHISON WIDEFIELD ARRAY

The MWA is the precursor to the low-frequency component of the SKA. Located at the Murchison Radio-astronomy Observatory, the MWA is made up of 128 'tiles' which are made up of a four-by-four array of ankle-high metal dipole antennas.

Each tile works as a single element of the array and has no moving parts. Each metal 'spider' is a dual polarization antenna, rather like a television aerial. The antenna converts radio waves into electrical signals.

Figure 12. One of the 128 receiver tiles of the Murchison Widefield Array. (Image courtesy of Natasha Hurley-Walker & the MWA Team.)

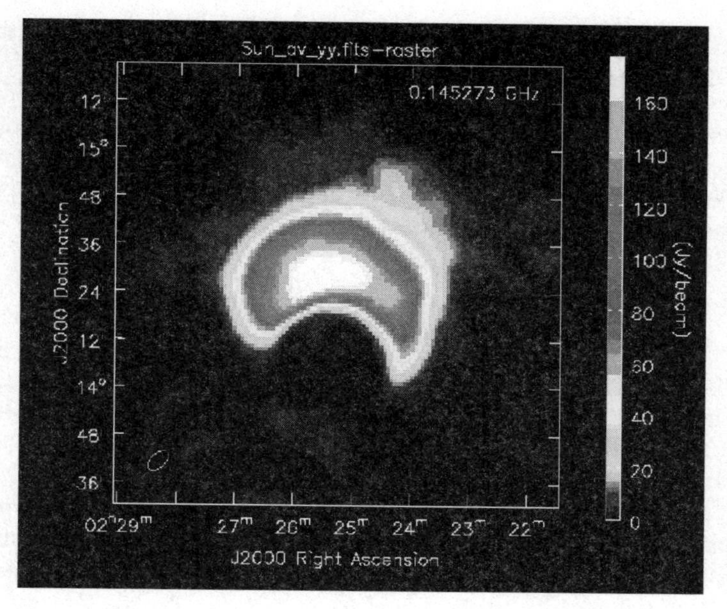

Figure 13. The partial solar eclipse, photographed with the MWA. (Image courtesy of Divvya Oberoi.)

The viewing angle of the telescope on the sky is 'virtually' steered by manipulating the way in which the signals are combined in an electronic device called a beamformer, which is attached to each tile.

The combination of tile signals takes place in a large on-site computing facility shared with ASKAP. Combined signals are then transported through a dedicated optical fibre link to the Pawsey Supercomputing Centre in Perth for further scientific analysis.

The MWA works at low frequencies, close to the FM radio band. There are very few telescopes operating at these frequencies and there have been rich pickings in this new discovery space.

The main science goals of the MWA are: 1) to detect fluctuations in the neutral hydrogen background from the era of reionization about five hundred million years after the Big Bang; 2) to carry out a full-sky survey of our Galaxy and of bright extragalactic objects; 3) to explore the time-variable sky for unusual and unknown transient radio emission; and 4) studying the sun and space weather.

At the time of writing, science operations are well underway. The

MWA commissioning team has already made discoveries that are having a significant impact in several of these scientific areas. With huge volumes of data now flowing from the telescope, we can expect many more breakthroughs in these areas over the coming years.

With the precursors already making important scientific advances, the excitement is building in the scientific community for the arrival of the SKA. So what areas of science will be the focus of the SKA?

SCIENCE WITH THE SKA

One of the key goals driving the SKA is to study how galaxies evolve through their several billion years of life. Such a historical study is only possible because our telescopes see distant galaxies as they appeared many millions of years ago. This is because it takes the radio waves many millions of years to reach Earth. Essentially, the information encoded in radio waves is a time capsule that is only opened when we receive the signal.

That is not to say that we can follow the evolution of an individual galaxy as it forms, grows and then fades through billions of years. This would require us to wait for the light to reach us in real time, which would also take billions of years. What we can do, however, is to make use of light travel time to look at the universe's history as a whole. By measuring the properties of many, many galaxies at various distances, we can get a census of galaxy demographics and see what the universe was like at different epochs in its history.

MEASURING DISTANCE IN SPACE

In the 1930s, the astronomer Edwin Hubble discovered that the light from almost every galaxy that we can see is stretched via the Doppler effect. Furthermore, the speed at which a galaxy is retreating depends on its distance from Earth. Hubble's observations proved that *all* galaxies are actually zooming apart from each other. From this, he made the startling discovery that the universe began expanding in a Big Bang around fourteen billion years ago.

The relationship between the velocity of a galaxy and its distance gives us a very powerful astronomical yardstick. It means that we can measure the distance to any galaxy in the universe as long as we have a sufficiently powerful telescope. With this little trick up our sleeve we can study what the universe looks like in any cosmic epoch we choose, simply by tuning our telescope to the right frequency. This enables astronomers (at least in principle) to see right back to the very first stars and galaxies that formed after the Big Bang.

SKA KEY SCIENCE PROJECTS

The SKA aims to solve some of the biggest questions in astronomy and fundamental physics. The unprecedented sensitivity provided by thousands of antennas promises to revolutionize our understanding of the formation and evolution of stars, galaxies and black holes after the Big Bang. The SKA will probe the role of magnetic fields in assembling and shepherding matter in stars, galaxies and planets. It will probe the evolution of large-scale structure on the universe and look at how galaxies evolve. It will investigate the fundamental nature of gravity and look into the age-old question: Are we alone in the universe?

We discuss these five key science projects, identified by the international astronomical community, below.

GALAXY EVOLUTION, COSMOLOGY AND DARK ENERGY

Current telescopes are only capable of detecting neutral gas in very local environs. The SKA will be ten to one hundred times more sensitive, enabling us for the first time to study the gas content of galaxies and even the gas in the 'empty' space between. By studying increasingly distant galaxies, we will learn how they form their stars, grow, merge and evolve through billions of years of cosmic history.

An equally important study planned for the SKA is to photograph and catalogue many tens or even hundreds of millions of galaxies across the sky. This giant cosmic census would enable astronomers to understand how variations in the environment of galaxies affect their mass, shape, properties and star formation. Information from this survey, applied through a number of slices through cosmic history, would also shed light on the influence of the mysterious substance known as dark matter on galaxy formation and evolution on cosmological scales.

THE ORIGIN AND EVOLUTION OF COSMIC MAGNETISM

Anyone who has used a compass to traverse the hills or navigate the open waters will know that the Earth is magnetic. Its north and south magnetic poles are generated and sustained by a dynamo generated by the rotation of iron-rich material deep within the earth's core. Magnetic materials align with the local field and tell us which way is north and south.

With the discovery of activity in the Sun and measurements of polarized emissions from its surface layers, astronomers have discovered that the Sun is also a giant magnet. What's more, the polarization of starlight from space has revealed that there is a large-scale alignment of dust grains in our Galaxy, rather like the alignment of iron filings around a bar magnet.

With increasingly sensitive observations of emissions from distant objects in space, we find that magnetic fields pervade the whole universe. Magnetism is one of the fundamental forces of nature and although it is relatively weak compared to the forces that glue fundamental particles together, it has a profound effect on the gas that fills the galaxies. Magnetic fields have a strong effect on channelling and redistributing materials in environments such as star-forming regions, the centres of active radio galaxies and supernova remnants. They may have played a vital role in the formation of the Sun and the Solar System.

One of the many unknowns about cosmic magnetic fields is how they were first generated, how they are amplified and how large-scale magnetic fields in vast 'empty' space between galaxies are sustained. With the SKA, astronomers will begin to study magnetism in very distant galaxies, in groups of galaxies, and on very large scales. This will enable us to answer some of the big mysteries relating to this highly influential force of nature.

THE CRADLE OF LIFE

As we have already mentioned, the SKA will explore the origins of our very existence and tackle the age-old question of whether or not we are alone in the universe.

For many years, scientists have been searching for radio signals as evidence for technological civilizations beyond the Earth. Scientists are somewhat divided on the chances of success of such a search, with scepticism arising largely from the fact that there is a miniscule cross-section between the age of life on Earth (3.6 billion years) and the age of electronic communications that could be picked up by nearby planet dwellers (100 years). Thus, if nearby alien civilizations had spent millions of years trying to detect radio transmissions from life on Earth, there is only a vanishingly small chance that they would have been successful!

One fact is inescapable, though – with the SKA we could readily detect airport radars on planets towards the closest few hundred stars to Earth, in only a few hours of observing time. If we were fortunate enough to observe during a planetary broadcast equivalent to the strongest radio transmissions that humans generate (during ionospheric research using RADAR), we could locate current technological civilizations anywhere within the closest few thousands of solar systems. Although it seems that the numbers are still stacked against us, with the SKA we may at least have a fighting chance.

Aside from eavesdropping on our cosmic neighbours, the SKA could make significant contributions in understanding how solar systems are born. If SKA dishes are separated by greater than 1500 kilometres, astronomers could potentially see individual protoplanets and the gaps that they clear in these disks as they orbit around their juvenile solar systems. Over the course of a few years, scientists could even make real-time movies of material orbiting these solar systems. As we study more and more planetary systems, we will gain a clearer understanding of how common is a solar system like our own with rocky planets in the habitable zone.

STRONG FIELD TESTS OF GRAVITY USING PULSARS AND BLACK HOLES

Physicists are ultimately seeking a theory of everything – a unified set of fundamental laws and principles that underpin the behaviour of everything in the universe. At present, physicists are unable to unify the laws of quantum mechanics, which describe the behaviour of particles and atoms, with the laws that describe gravity. Many hope that this would lead to a better understanding of the nature of matter and how it interacts.

Einstein's theory of general relativity is a description of gravity that involves space and time being distorted around each and every object. This theory is more than one hundred years old but it has stood the test of time. The supreme accuracy of Einstein's theory of general relativity underpins modern technologies such as GPS.

Despite this fact, physicists are keen to push Einstein's theory to the limit, to see whether (like Newton's simple description of gravity) or not it breaks down under extreme conditions. Measuring any deviation from Einstein's law could have profound changes to our understanding of physics.

In order to test the limits of relativity, we need to carry out an experiment in an environment with extremely strong gravity. Two astronomical environments where this may be possible are: 1) a pair of massive compact stars in a close orbit or 2) a massive compact star orbiting a black hole.

Pulsars are compact stars that rotate very rapidly. As they spin on their axis, they emit a beam of radio emission from each pole, like a lighthouse. These stars are ideal for measuring the laws of physics in extreme gravitational environments because their regular lighthouse beams mean that they produce very regular clock 'ticks'. As the pulsar moves around in its orbit, the distance between the pulsar and the Earth changes slightly and the ticks take a longer or shorter time to reach the Earth.

By measuring the orbital parameters of a pulsar in this way, we can measure how the laws of gravity are affecting its orbit. With the SKA we may find up to twenty thousand pulsars in our Galaxy. Astronomers predict that this will include up to a hundred compact binary objects in which the star is moving sufficiently close to the speed of light to test Einstein's theory to its limits.

PROBING THE COSMIC DAWN

As the universe expanded and cooled following the Big Bang, the most basic atoms (hydrogen) formed. At this time, the distribution of matter in the universe was incredibly uniform with nothing but a sea of hydrogen in every direction. If we imagine what this would look like to our modern-day radio telescopes, it would be a uniform signal emitted at 1420 MHz, stretched by the expansion of the universe to longer wavelengths. The wavelength of the emission tells us exactly how far we are looking back in time.

As the universe expanded and cooled further, the matter and dark matter began to form clumps and islands under the influence of gravity. This enabled the first stars and galaxies to form, which precipitated a light-bulb moment for the universe called the 'epoch of reionization'. During this epoch, the universe (which had previously been in a hot, dense and completely ionized state) had been reionized by ultraviolet radiation from the first stars.

The radio signal from this time would look like a background of 1420 MHz radiation from almost all directions, but with pieces missing from the background where the ionizing radiation had removed the neutral hydrogen atoms by turning them into electrons and protons. With a low-frequency (long wavelength) array as part of the SKA, astronomers will search for the global signal from this period referred to as the Cosmic Dawn.

A NEW ERA OF DISCOVERY

The SKA project has considerable momentum internationally. It has the backing of governments and research organizations across the globe. Many hundreds of astronomers, cosmologists and theoretical physicists are eager to use its unprecedented capabilities. The international research and development activities supporting the project have led to some remarkable advances in radio astronomy technology, including wide-field receivers, electronics, data transport, software and computing. The search for an ideal location for the telescope has led to the establishment of new astronomical observatories at two of the most radio-quiet locations on the planet. The SKA pathfinder telescopes are already producing new scientific discoveries and demonstrating the scientific benefits of recent technological innovations in radio astronomy.

With such a transformational radio telescope on the way, it is a truly exciting time to be a radio astronomer.

Sir John Frederick William Herschel (1792–1871): Astronomer, Cosmologist and 'Natural Philosopher' of the Victorian Age

ALLAN CHAPMAN

During the century that separated Sir William Herschel's first serious interest in what we now call 'deep-sky' astronomy and the death of his son Sir John in 1871, our understanding of the cosmos was transformed. And the two Herschels, William and John, along with William's sister Caroline, were to play a major part in that transformation. Indeed, for several decades, from about 1775 to 1835, the Herschels would be the only truly big players in the field: a role made possible by William's specially developed and largely secret techniques for figuring big speculum mirrors. These were mirrors which possessed a resolving or 'space penetrating' power that transcended that of any contemporary opticians, and which also enabled William to make many thousands of pounds from the commercial manufacture of high-quality reflecting telescopes.

It was only after the 1830s that new techniques of casting, steam-grinding and figuring speculum metal mirrors bigger than those of the Herschels became available in the burgeoning industrial age. And these techniques would, by the 1850s, be further applied to the newly developed glass-mirror technology, and, after 1860, would be complemented by the new sciences of astronomical photography and spectroscopy.

Sir John Herschel's sixty-odd-year career in front-rank science would encompass much of this transformation, extending from his family inheritance from William and Caroline to creative working

Figure 1. Sir John Herschel. Engraving after a portrait by Henry William Pickersgill, *c.*1845. (Image reproduced in Sir Robert Ball, *Great Astronomers* (Cambridge, 1895).)

friendships with people such as William Parsons, third Earl of Rosse, William Lassell, Mary Somerville, Sir William and Lady Margaret Huggins and many more, both at home and across Europe and the US. For both Herschels, father and son, were of congenial and generous dispositions, temperamentally inclined to make friends rather than enemies.

INHERITING THE FAMILY BUSINESS

It is clear from Sir John's private correspondence, diaries and personal notes that observational astronomy was not his first instinctive calling. For while he was growing up at Herschel House, Slough, surrounded by the world's biggest and most powerful deep-space telescopes, and meeting a galaxy of distinguished people who called in to see his father, young John's highly original intellect was drawn to other aspects of science. '*Light* was my first love! In an evil hour I quitted her for those brute & heavy bodies which tumbling thro' [the] ether, startle her from her deep recesses and drive her trembling and sensitive into our view',

Figure 2. Herschel House, Slough. (Image from Sir Robert Ball, *Great Astronomers.)*

obliging him, as he admitted to his wife Margaret in 1843, to go '*dallying* with the stars'. Indeed, optics would be a life-long passion, as we shall see, along with a passion for the pure intellectual beauty of higher mathematics and its application to the heavens. Of course, one can fully understand his youthful enchantment with light, living not only surrounded by giant telescopes, but also within what was his father's lucrative optical business. For mirrors and lenses of all kinds surrounded the young John as he grew up. But perhaps even more influential was the knowledge that his father had been the first scientist in history to make optical discoveries in those regions of the spectrum invisible to the human eye: the 'black light' or 'black heat' of infrared, which his father had made at home from laboratory experiments in February 1800, when John had been a precocious and curious eight-year-old. He also became captivated by gravitational physics as a boy.

His parents William and Mary, in fact, had their son educated at a local private school, and by private tutors, and except for a few unhappy months spent at nearby Eton College, he never went to a major school. Indeed, his health was delicate, and throughout his life his 'weak chest' was a topic of comment. Going up to St John's College, Cambridge, must have been a liberating, challenging and inspiring experience, for John now began to move among young gentlemen of his own age. In 1813, he graduated Smith's Prizeman and Senior Wrangler, winning the highest marks in the gruelling University Mathematical Tripos examinations. Like many people 'going up' to university, he would make friends who would stay with him for life, and during the years that he resided in Cambridge as a student, then as a mathematical Fellow of his college, he would lay the foundations of a prodigious reputation, not only as a physical scientist, but also as a classical and modern linguist, a philosopher and man of wide culture. His friends would include William Whewell, Sir Charles Babbage, George Peacock, and, in the 1820s, the young (later Sir) George Biddell Airy and Augustus de Morgan.

Yet in 1810, Cambridge mathematics was somewhat stuck in a rut. Vastly proud of its Newtonian legacy as the university was, it had proved reluctant to progress beyond Newton's 'sacred' methods of geometrical computation to espouse the new analytical 'higher' mathematics being pioneered on the Continent by Laplace and others. Herschel and his brilliant young friends began to petition the university to take on board the new mathematics of Paris and Continental Europe.

This new mathematics would transform planetary calculation and gravitational mechanics, and ironically, it would prove its worth when applied to another of William Herschel's discoveries: namely, the complex binary and compound star systems which he had discovered in the depths of the stellar universe. Stars, indeed, that rotated around each other, and came to act as a proof that Newton's Laws of Gravitation applied *beyond* the tried and trusted solar system, to operate in the very depths of space! Indeed, in the early 1830s, John Herschel would advance gravitational astronomy, as would his French contemporary Félix Savary, by developing a mathematical technique whereby the orbits of compound, rotating star systems could be analysed to determine the gravitational elements of the respective stars.

In the summer of 1816, however, John's mother and father were enjoying a holiday at Dawlish in south Devon. John came down to join them, and it was there that the 78-year-old William seems to have pointed out that he was getting no younger, and that there was still a great deal of 'big telescope' astronomy waiting to be done. And so, as a good son, John gave up Cambridge, and a half-hearted attempt at a legal career, and came back home to Slough, where his father began to teach him the techniques of big, reflecting telescope building, maintenance and use. Those 'brute & heavy bodies' had now tumbled into his life, and from 1816 onwards, John Herschel began his real career as an observational astronomer. For it was now his duty to take over the Herschel family business, or what his father had styled fathoming 'the length, breadth and depth . . . and profundity' of the Universe.

OBSERVING WITH BIG REFLECTORS

Although William had completed his iconic and royal-funded great forty-foot telescope with its four-foot-diameter mirror in 1789, the instrument never really lived up to expectations. Quite simply, it had

Figure 3. Jupiter and Saturn as seen with the twenty-foot telescope. (Image from Sir John F. W. Herschel, *Outlines of Astronomy* (London, 1849).)

pushed contemporary optical and mechanical technology just too far to be a regular research tool, and William warned his son about wasting his time on it. Instead, John, like his father before him, came to work with an upgraded version of William's 20-foot telescope with its 18¼-inch-diameter mirror.

The Herschel twenty-foot design had enabled William to do his groundbreaking cosmological work after 1783, being quite handy to use and having – by the standards of the day – excellent speculum optics. Indeed, in a letter to *The Times* in 1838, William Herschel's young surgeon-astronomer friend, Sir James South, would recall that William could direct the twenty-foot reflector to any object in the sky within five minutes!

It has long been my opinion that the twenty- and forty-foot Herschel telescopes owed key aspects of their design to ship-handling: for in their use of pulleys, blocks, capstans and rope-rigging to bring the great wooden tubes to bear on a celestial object, they represented a logical approach to moving heavy objects in the age of Nelson. One can easily see parallels – for example, between turning and adjusting a great telescope and the techniques routinely used to hoist the yard-arms and set the acre or so of canvas sail that powered a first-rate man-o'-war such as HMS *Victory*. (I have often wondered whether William had received advice on constructing his great wooden-framed telescopes from Admiralty friends. William, after all, was a good friend of Sir Joseph Banks, President of the Royal Society, who was himself interested in shipbuilding, had sailed around the world with Captain Cook, and knew Nelson.)

One can fully understand how the elderly William was desirous of bringing his son John into the family business in 1816, for using even the twenty-foot telescope took a great deal of physical agility, not to mention courage. Just imagine having to first ascend the fifteen or twenty feet of a steeply raked wooden scaffolding ladder, then swinging oneself around onto the observing gallery, and then operating the screw-cranks necessary to fine tune the position of the great tube, before even looking into the eyepiece. And all to be done in the dark, with no safety net in sight!

John also learned from his father the importance of working with optically matched sets of two or three 18¼-inch mirrors, often known as 'A', 'B' and 'C'. Because speculum metal tarnished relatively quickly, one would keep a mirror in the tube only until its reflectivity began to fade. It would then be removed and its twin fitted in its place, while the first mirror was repolished to restore its lustre, thereby ensuring that the telescope was always ready for optimum use.

Big timber-and-rope-framed Herschel telescopes, however, were altazimuth instruments, and not really suited to tracking objects across more than a few degrees of sky. They were best used in or around the meridian, where a celestial object's path was relatively flat, so that screw-operated lateral and vertical motions enabled Herschel to follow the right ascension and declination of an object across a narrow band of sky.

Both William and John Herschel also took advantage of the sky's own motion when sweeping great tracts of the heavens for nebulae, clusters and double stars, or for doing star-density studies within the Milky Way. One would select a zone or band of sky to be swept for a given evening, depending on the season of the year and time of night, then pre-set the tube's elevation to the appropriate angle. A bell could also be placed, the clang of which could indicate the maximum angle through which the telescope tube was to be elevated that night.

Let us suppose that a zone of sky some ten degrees in right ascension and two degrees in declination at forty degrees elevation was to be swept on a given night. The bell would be positioned so that when the telescope had been gently 'screwed' upwards, it rang after a two-degree rise had been achieved. The adjusting screw would now be reversed, until the tube had descended two degrees, at which point it would be made to rise again. Utilizing the natural motion of the heavens, and a gentle, methodical rising and falling of the tube, the observer would describe a sequence of overlapping 'A' shapes in the sky, which would enable him to make a methodical sweep of the selected zone.

By dividing the entire sky into a series of interconnecting geometrical zones, it was possible for William and John in their respective generations to sweep the entire northern heavens. And between 1834 and 1838, Sir John would use the same technique to sweep the southern skies at the Cape of Good Hope.

THE HERSCHEL COSMOS

The sweeps which William Herschel made of the northern heavens after 1780, made a sudden and dramatic addition to our ideas of deep space. Earlier that century, in 1716, Edmond Halley had discussed six 'lucid spots', or nebulae, which, to the simple, long-focus refractors of that time, were little more than glowing smudges that could not be brought into sharp focus, as could a star image. That number was to increase over the eighteenth century, until, in his third catalogue of 1784, Charles Messier published his list of 103 fuzzy-looking objects which might be confused with what were then seen as those much more interesting objects: comets. Some years earlier, in 1767, the Revd John Mitchell had speculated whether gravity might be causing stars to be attracted and drawn into the Pleiades and other clusters.

But for most of the eighteenth century, the deep sky was regarded as little more than a marginally interesting backdrop against which to measure and predict, using Newton's gravitational criteria, the motions of the moon, planets and comets. It was William Herschel, however, with his 6½-inch- then 18¼-inch-aperture reflecting telescopes, who was to change all that; he drew astronomical attention to the very nature of the nebulae, star clusters and binary star systems and their distribution in space, asking questions about what he styled the 'Construction of the Heavens' and the 'length, breadth, depth and profundity' of stellar space.

The universe within which both Herschels, father and son, were to work was what we might call 'steady state'. The Herschels were less concerned with the philosophical questions of 'infinity' than with fathoming physical structures.

Broadly speaking, theirs was a one, big galaxy universe, roughly discus-shaped, thicker in the middle and tapering off to the edge: a bit like two saucers placed edge to edge. And we, the Sun, Earth and Solar System, were at or very close to the geometrical centre, although, as we shall see below, Sir John was coming to modify this idea by the 1840s. Yet this was a perfectly reasonable conclusion to draw when the visible sky seemed to be evenly distributed around us, as we looked through the densest part of the Galaxy towards the edges and into the heart of the Milky Way.

In William Herschel's original model of the cosmos in around 1800, it was suggested that stars were probably of a generic size, just like people, cats and oak trees; some a bit bigger or smaller, but all roughly the same size. This meant that slightly bigger stars might attract smaller stars to eventually sweep a zone of space clean, as all local stars were gravitationally drawn into the growing mass of a star cluster – such as the Hercules and Pleiades clusters. (Yet by the 1840s, as we shall see, new evidences were causing Sir John to modify this 'generic size' star model.)

But either way, over-compressed clusters might eventually disintegrate, causing the stellar detritus to blast out into space. Could this starry detritus, or star-dust, be the stuff of which the nebulae were made? And could the fragments eventually condense into solid masses once more, begin to emit light, and become stars?

THE 'CONSTRUCTION OF THE HEAVENS': THE PUZZLE OF THE NEBULAE

We have seen that both Herschels saw the universe as a stable system in which change, gravitational attraction and a sort of 'recycling' of cosmic matter took place within a broadly changeless whole: rather like a long-term stable human or animal population, where things are born, grow, die and provide the basic chemicals from which the next generation comes. Indeed, they saw the telescopic sweeping of the cosmos for different specimens or types of objects as analogous to investigations into natural history. For the Herschels swept the skies for celestial specimens in much the same way as a conventional naturalist swept the countryside for plants, birds, insects or animals. By the 1830s, however, Sir John was further modifying his father's original 'natural history' approach to understanding deep space; for the terrestrial natural historian is capable of witnessing the whole life-cycle of plants, birds and animals, whereas no cosmologist will ever live long enough to see how a nebula or star cluster will develop.

The puzzle, though, lay in trying to fathom the relationship between individual self-luminous stars and the mysterious cloudy glow of the nebulae. Of what stuff could the nebulae be made? William Herschel had wrestled with the problem and, for over forty years, had changed his mind several times. Were the nebulae made up of glowing insubstantial luminous fluid or filmy *chevelures*, or were they made up of clouds of countless discrete particles, like the individual grains that make up a swirling sandstorm? Were they even like those reflecting airborne dust particles producing shafts of light when the sun catches them at the right angle when it shines into a room? In short, was there true nebulosity in space, or just streams of individual particles? Yet either way, a great puzzle remained; for how could any kind of matter in space glow and emit light, if there were no adjacent glowing stars to light it up?

Most baffling of all were those objects which the Herschels styled 'planetary nebulae': fixed, unchanging objects in deep stellar space which, nonetheless, appeared to present big planet-like discs to the telescopic observer – discs so vast that, Sir John computed, they were many times larger than the entire volume of the Solar System – which, by 1846, was known to extend as far as Neptune! In particular, the planetary nebulae, which have now been designated NGC 1514, had baffled William Herschel in the 1790s, for at the centre of the faintly glowing 'planetary' ring was a tiny star. Yet the light of that star was so feeble, comparatively speaking, that it could never have illuminated its distant ring! So where did the glowing ring get its light from?

This new cosmology, first discovered, explored and described by William and John Herschel, was also replete with philosophical problems about the nature of matter, light and gravitational action in deep space, about chance and divine design, and humanity's ability to grapple with the most profound questions regarding the cosmos. And there was no shortage of discussion in late Georgian and Victorian Great Britain, Europe and the US. In this age of global exploration, moreover, when exotic creatures were being discovered in Africa, Australasia and in the ocean depths, the Herschels and their friends (such as Lord Rosse, for example, who, by 1845, had brought his great 72-inch-aperture reflecting telescope into commission) were also discovering their cosmological equivalents in the deep space.

Nor should we forget that both Herschels were working in a cultural climate often referred to as the 'Romantic Age': an era in which the exotic, the bizarre and the mind-boggling were seizing the public imagination. Just think of J. W. M. Turner's paintings of wild, swirling storms and the contemporary craze for emotionally over-charged Gothic horror tales, such as the adventures of *Varney the Vampyre* (1845–7), especially when the stories contained a 'mad scientist' component, as did Mary Shelley's *Frankenstein* (1818). Indeed, on one level, the newly emerging Herschel universe, with is strange objects, vast distances and implicit cosmological infinities of time, space and magnitude, was the ideal cosmology for this Romantic Age.

And most controversial of all were the nebulae. Were they, as suggested above, vast congeries of glowing stars, the total mass of which appeared only as a faint, wispy mist in a giant telescope? Or were they composed of intangible shining fluid, as Emanuel Swedenborg, Pierre-Simon Laplace and others had speculated in the eighteenth century? Either way, their interpretation posed problems. If they were composed of stellar congeries, and several, such as the Orion and Andromeda nebulae (M42 and M31), extended over a degree or more of sky, then how physically *vast* were they, and how far away? Were some of them even outside the Galaxy – *our* Milky Way Galaxy, that is? For *all* cosmological objects were still assumed to be part of one great, singular, unified system.

Conversely, if they were made up of shining fluid, then what sort of physical laws did that fluid obey? For, in 1835, no one had any idea of how gravity, light or magnetism might act upon unknown forms of intangible stuff, *in vacuo*, in the remotest depths of the universe. For

Figure 4. Andromeda Nebula, M31, as seen through the twenty-foot telescope. (Image from Sir John F. W. Herschel, *A Treatise on Astronomy* (London, 1833).)

ionization, atomic radiation and similar concepts in physics still lay decades ahead in the future. And, too, how did the nebulae relate to the star clusters, the Milky Way, and to the planetary nebulae? Yet these mysterious objects were there in the sky to be seen by anyone who possessed a big enough telescope.

Quite simply, the Romantic imagination was struggling hard to get its head around a universe that seemed not only infinite, but also *infinitely* infinite, with star systems so inconceivably vast as to make the Solar System look like the merest speck of dust.

By the early 1830s, however, John Herschel had completed, refined and extended his father's survey of the Northern Hemisphere, to produce the first maps and tables of those strange objects which populated the deep sky. And he was also coming to formulate his own independent interpretations of the cosmos, based in part upon his Cambridge mathematical training, and also upon contemporary discoveries in physics and chemistry. After his aged mother Mary died in 1832, John, his wife Margaret and their growing family prepared to undertake a monumental upheaval that would extend over five years in all. They relocated to the Cape of Good Hope, so that John could undertake the first-ever in-depth survey of the skies of the Southern Hemisphere, and to complete William's northern sky surveys, no less, by extending them across two hemispheres. This was to make Sir John Herschel the first astronomer to examine the entire visible cosmos throughout 360 degrees in all directions. A feat that would not only win him universal accolades from across the international scientific community, but also see the already knighted (in 1832) Sir John elevated to the rank of baronet by the young Queen Victoria in July 1838.

SIR JOHN HERSCHEL AT THE CAPE OF GOOD HOPE

Sir John and Lady Margaret Herschel and their first three children, together with their servants, having sent their possessions along to be stowed in the hold of the *Mountstuart Elphinstone* as she lay moored in the Thames, travelled overland to Portsmouth. Then, when the great East Indiaman had sailed around from London to pick up passengers, they boarded as she called into Portsmouth before heading out to sea on 13 November 1833. These great three-masted East India Company wooden sailing ships were the liners of the age, providing as near as one could get to luxury afloat at the time. Sir John paid a hefty £500 fare to convey his household and equipment to the Cape of Good Hope, and it was a fare paid for entirely out of his own pocket, for, as a 'Grand Amateur' gentleman of science, he asked no one for money.

The voyage to South Africa took nine weeks and two days – quite fast – of largely plain sailing, and, as the vessel moved further south each night, Herschel spent hours on deck thoroughly familiarizing himself with the rising southern constellations. They made landfall at the Cape on 16 January 1834.

By that date, there was already a Royal Observatory at the Cape, established some years before to map the southern skies, regulate ships' chronometers, keep magnetic and meteorological records, and assist with cartographic enterprises in the Southern Hemisphere. Its Astronomer Royal was Thomas (later, Sir Thomas) Maclear, the former Bedford surgeon and Grand Amateur, whom Herschel had known in England, and who, on one occasion, obliged Lady Herschel by calling around to extract a troublesome tooth – without anaesthetic, for ether still lay a decade in the future. The extraction took place at Feldhausen, the elegant house and estate which Sir John and Lady Herschel leased just outside Cape Town. (Later, the financially astute Sir John was to make a deal with the owners of Feldhausen, and he purchased the property. In 1838, when the family were preparing to return to England – now with six children, he sold it, and the profit made on the deal paid for their passage home.)

John and Margaret Herschel would later recall their five years at the Cape as some of the happiest years of their lives. They had a lovely house, in a pleasantly temperate spot not far from the sea, and it was ideal for astronomical research. They enjoyed the company of their neighbours, who included, in addition to the Maclears, a local community of British and Dutch settlers, not to mention a stream of sea captains, visiting diplomats, and colonial administrators either *en route* for or returning home from India, for 'Herschel', as Lady Margaret always styled her husband, was already a world-famous scientist, and the number of people who wanted to meet him could sometimes be a nuisance.

Sir John's fame, and convenient two-month isolation from Europe while on the sea voyage to Cape Town, gave rise to the 'Great Lunar Hoax' of 1835. The New York journalist Richard Adams Locke composed a sensational tale about Herschel at the Cape using his great telescope to discover intelligent beings on the moon. The hoax caused a sensation in both the US (New York being only four or five weeks' sailing in a fast ship from Liverpool by 1835) and in Europe, and it is said that Sir John laughed when the news finally reached him in Cape Town.

Margaret Herschel's own letters and journals give us some fascinating details about life at the Cape. On one occasion, for instance, she used a poker to dispatch a poisonous puff adder which had slid into the drawing room, while even picnics on the beach were not without their hazards. 'Herschel' always took his rifle and cartridge box along, just in case marauding lions smelled their food and fancied eating something – or somebody!

The Herschels were at the Cape at the very time, 1834, when Parliament liberated all slaves within the British Empire. (The slave *trade* had been abolished since 1807, and the intervening years were spent in making provision for the economic implications of working plantations with free people, with wage-paid labour.) The Herschels were passionately anti-slavery, or abolitionist, and were furious to discover that some farmers were technically liberating their slaves, turning them out to face starvation, then re-employing the desperate people on pay and conditions that were scarcely better than their previous slavery. Outraged by this practice, the Herschels, who were well-connected and knew the 'right people' back in England and in Parliament, used all their social and moral clout to see fairer working conditions established.

Observing with the great twenty-foot telescope in South Africa was not without its additional hazards, either. Not only could there be snakes in the grass as Sir John walked to the great telescope and back home again, but feral dogs seemed to be everywhere: usually these were dogs living around the local farms. Just imagine climbing fifteen feet or so down the scaffolding ladder of the great telescope, in the pitch-dark, to hear a cacophony of barking and growling beneath your feet! Sometimes, we are told, Herschel would take a pair of smoothbore pistols loaded with small pellet shot when he went out for a night's observing. If the wild dogs sounded especially numerous and hungry as he climbed down, then – BANG, BANG! All the evidence suggests that Sir John was a kindly man who loved animals, but this did not mean he was willing to be eaten alive in his own garden!

In addition to astronomy, Sir John became fascinated by the geology of South Africa, and in particular, by that remarkable local feature, Table Mountain, at the very southern tip of the African Continent. At a time when European and American geologists were puzzling over the

Figure 5. The twenty-foot telescope at Herschel's 'Feldhausen' Observatory, near Cape Town, South Africa, *c.*1836. (Image from Sir Robert Ball, *Great Astronomers*, reproduced from Herschel, *Results of Astronomical Observations Made at the Cape of Good Hope* (1847).)

processes that had moulded the surface of the Earth – had there been spasmodic, violent 'catastrophes', or gentle long-term accumulative changes? – Sir John made a meticulous survey of Table Mountain, and, using his great skills as an artist and draughtsman, drew its complex rock formations and strata in great detail. This was just for the record, as it were, as little work had then been done on the geology of the Southern Hemisphere.

But what conclusions did he arrive at regarding the deep-sky cosmos during his years at the Cape?

THE CAPE RESULTS

One notable conclusion to which Sir John came concerned the different distribution of nebulae, double stars, star clusters and planetary nebulae from those which he and his father had so carefully recorded in the Northern Hemisphere. Sir John also embarked upon a meticulous new zonal survey of the great nebula in Orion, a repeat of the Orion survey which he had made from Slough in 1825–6. His purpose was to see whether or not changes had taken place in the nebula across the intervening decade. If visible changes had occurred, then this might suggest that the nebula was relatively 'local' and fairly close to us, cosmologically speaking. If not, it was probably remote and, no doubt, immensely large. In this quest, of course, Sir John was not alone, for astronomers had been drawing the Orion Nebula as far back as Christiaan Huygens in the 1650s. William Herschel had also made his own detailed studies, as, later, would Lord Rosse and William Lassell in the 1840s.

Initially, Sir John was inclined to think that he had detected changes between his two surveys, though he came to change his opinion, especially when he factored in the spectacularly clear skies of the Cape, which enabled the same twenty-foot reflector to reveal details of density, transparency and delicacy of structure that had been invisible under the damper skies of Slough.

But how did the Northern and Southern Hemisphere deep skies differ? For one thing, there were the Magellanic Clouds, resembling whole sections of the Milky Way that had somehow come adrift, for which there were no equivalents in the north. Both of these Magellanic Clouds, each a few degrees across, were veritably bursting with nebulae: Sir John counted 919 objects in the large cloud, and 226 in the small one!

His Cape observations, supplying as they did new data and insights into his own and his father's observations made in the north, led Sir John to make several fundamental changes to the original 'Herschel Universe' described earlier. These came about, moreover, not from new facts alone, but from styles of thinking and techniques of data analysis that were derived in part from his Cambridge mathematical training, and in part from the emerging concepts of strict intellectual rigour developing within science in general.

Strict *induction* had to be observed at every stage in the scientific process, Sir John emphasized, especially in the mathematical, physical sciences. *Speculation* was to be abhorred. For speculation about 'shining fluid', glowing *chevelures*, and unknown chemical and physical processes in the depths of space was simply not physics. It might be philosophical *metaphysics*, but it was not science, and it told us nothing about the real universe. Rather, the scientist should base his interpretations of celestial data upon things we can know, calculate and test and work things out from there. These guides to interpretation included such things as the action of Newton's Inverse Square Law of gravity and known facts about combustion and light.

Even before he sailed for the Cape in 1833, Sir John was coming to think that all matter in space must be *particulate*, or made up of tiny, physically quantifiable pieces of matter, rather than vague luminosities. For Newton's Laws can be applied to account mathematically for the movements of particles, no matter how tiny, whereas they could not be applied to hypothetical 'shining fluid'. Sir John's experience of the southern skies, with their abundance of nebulae, Magellanic Clouds, planetary nebulae and clusters, only confirmed him in his particulate concept of nebulous material.

He applied a similar inductive logic to the glowing, light-emitting properties of the nebulae, for no matter how much a laboratory chemist might heat a gas – as chemistry stood in the 1840s – it would never emit light. To make it do so, *carbonaceous* or solid particles had to be mixed with it, to facilitate a combustive chemical process and to create incandescence. John Herschel further argued that there might even be a material parallel between the nebulae and the zodiacal light. Modern light pollution probably makes this phenomenon unobservable in most of present-day Britain, including Slough and even rural Kent, to which he would relocate after returning from the Cape, but it was clearly conspicuous in the 1830s and 1840s.

The zodiacal light is best seen – under optimum conditions – in the evening following sunset, and appears as an elliptical band of light inclined to the western horizon at the same angle as the ecliptic. By 1830, astronomers were largely agreed that this faint light stream was made up of fine particles in space that caught the light of the sun and became luminous. It was considered likely that this *particulate* matter was related to the stuff of cometary tails, and even *meteorolites*, or meteorites. And as the glow of the zodiacal light was similar in appearance to what Sir John styled the 'flocky' streams and wisps of the nebulae, might they both be made up of a common physical, particulate space-stuff? Of course, one cannot deny that, to some extent, Sir John was breaking his own anti-speculation rule here, but at least – he might have claimed – he was speculating about *physical* materials, and not about vague 'shining fluid'. On the other hand, no matter how rigorous a scientist may aspire to be, it is impossible to avoid at least some element of speculation when it comes to interpreting the stuff of which remote celestial objects seem to be made. Even so, by the 1830s, Herschel felt confident that the nebulae *must*, on inductive grounds, be particle-based. His years of observation at the Cape only served to confirm him in this opinion, and then, after Lord Rosse brought his great 72-inch-aperture telescope into commission after 1845, Sir John felt this view was confirmed beyond all reasonable doubt when the 'Leviathan of Parsonstown', as Rosse's telescope came to be known, began to reveal structural details of several hitherto 'flocky' or misty objects. Most notably, these included M51 – now christened the 'Whirlpool' – an object which, in Herschel's 18¼-inch-aperture 20-footer, appeared as an asymmetrical glow, but which, in Rosse's 72-inch telescope, revealed fine structure and numerous small stars.

The rich harvest of new nebulae in the Southern Hemisphere also led Sir John to re-draw the geometry of the Galaxy. By the time of the publication of his *Results of Astronomical Observations Made at the Cape of Good Hope* (1847), and with the further apparent confirmation of *particulate* matter from Lord Rosse's discoveries, Herschel was pointing out that in the Northern Hemisphere there appeared to be a perhaps disproportionate aggregation of nebulae in the region of the northern galactic pole – Leo, Ursa Major, Virgo and adjacent constellations – whereas in the south 'a much great uniformity of distribution prevails', with the exception, that is, of the nebulae within the Magellanic Clouds. This geometrical asymmetry of distribution led Sir John to conclude that our Solar System, and hence our point of observation, was *not* in the centre of the Galaxy.

A VASTER, STRANGER AND EVEN MORE WONDERFUL UNIVERSE

By the time he published his *Cape Results* in 1847, John Herschel was thinking of a universe which was different in several key aspects to the one he had 'inherited' from his father in 1816. But that, after all, is what original scientific research is all about. For one thing, he had come to conclude that the 'nebulous system [was] distinct from the sidereal' – or, to put it another way, that the stars and the nebulae were probably made up of different types of matter, or, at least, matter in different chemical and physical states. For having abandoned the possible existence of shining or luminous fluid in space, he had come to accept that celestial objects were made up of a variety of different types and sizes of physical matter – such as glowing stars, particle streams, whirlpools, clouds, light-reflectors and light-emitters. All of them were physically connected, however, and all were governed by mathematical laws traceable back to Newtonian gravitation.

By this time, however, Sir John had also abandoned his father's model of all the stars being of an approximately generic size and luminosity, along with William's 'distance-luminosity' scale, in which a second-magnitude star must be twice as far away as one of the first magnitude. For it was at the Cape that the modern astronomical concept of 'absolute magnitude' was really born, in the wake of Sir John's experiments with his newly devised 'astrometer'. Using a ray of moonlight as his standard optical source, he contrived a prism device whereby that ray could be directed into the telescope eyepiece. Then, moving the prism to carefully measured distances from the eyepiece, he was able to establish a position at which the 'standard' moonbeam appeared to be of exactly the same brightness as a chosen star. By means of geometry and calculation, he was able, from these proportions, to establish the comparative brightness of 191 southern stars, and

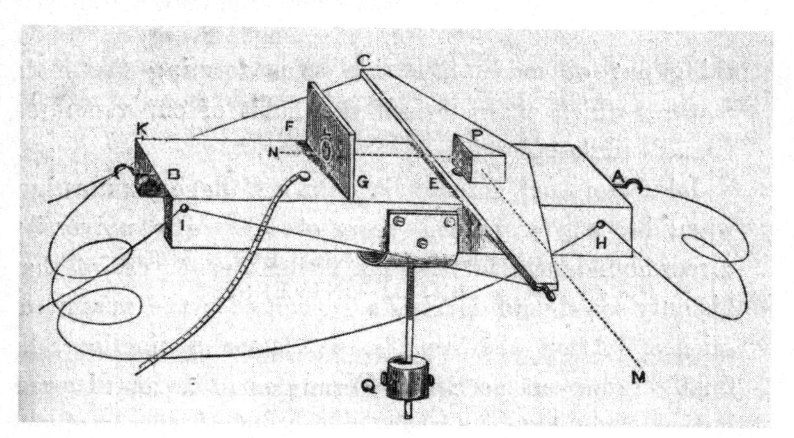

Figure 6. Herschel's 'astrometer'. A prism and a system of strings and adjusters enabled Herschel to compare the light of a star with that of a shaft of moonlight. (Image from Robert Ball, *Great Astronomers.*)

upon returning to England, to add some northern ones to their number as well. The results were quite amazing. Vega, for example, appeared forty times brighter than the sun on an absolute scale, while Arcturus was a breathtaking two hundred times brighter! Just how vast must the universe be with stars of such prodigious sizes, and how insignificant might our Solar System be?

By 1847, moreover, Herschel's view of the nebulae was changing. His own observations, and those of Lord Rosse, were suggesting that nebulae came in a variety of types and, maybe, sizes: about six in all. Some were elliptical, such as M31; others showed distinct structural details, like M51; while others were diffuse and peppered with dim stars, as was the Orion nebula, M42. Yet the *particulate* theory seemed to hold as Lord Rosse, William Cranch Bond at Harvard, and other big telescope astronomers started to glimpse individual tiny stars in previously bland glows, such as M31, with each improvement in optical power. In many ways, this particulate model still held after August 1864, when William Huggins (who was knighted in 1897), using the newly devised astronomical spectroscope in conjunction with a fine 8-inch, then 15-inch-aperture equatorial refractor from his south London back garden, made an epochal discovery in the Cat's Eye Nebula, in Draco, NGC 6543. For in the spectroscope, the Cat's Eye Nebula revealed not a stellar but a gaseous structure, for even by that date Huggins and others had discovered that nebulae such as M31 displayed complex spectra indicative of numerous emitting sources: namely, stars. But the Cat's Eye revealed the much simpler spectrum of a gas cloud. Yet a gas would not be the same as a shining fluid, for long before 1864, laboratory chemists had realized that a volume of gas was itself composed of millions of individual atoms – or physical particles.

The Cat's Eye Nebula – discovered by William Herschel in February 1786, and named from its visual similarity to a cat's eye – belongs to a class of nebulae which Herschel came to style *planetary* nebulae, because of the large circular, or spherical, diameters which they displayed.

When Sir John sailed for the Cape in November 1833, only six planetary nebulae were known in the northern skies, of which his father had discovered four or five, in the wake of Charles Messier's first recording of such an object, the Dumb Bell, M27, in 1764. By early April 1834, however, John recorded his discovery of the first of several southern planetaries, including one of a 'fine blue colour'. And by 1847, Sir John and other astronomers were placing the number of known planetary nebulae, in both hemispheres, at twenty-four or twenty-five.

The planetary nebulae posed a whole set of questions. What on earth could they be, with their spherical forms, large angular sizes and faint glows, sometimes with and sometimes without a central star? They posed questions about the nature of light, incandescence, matter, energy and the very structure of the universe, as such things were then understood. Sir John suggested that the planetary nebulae appeared 'in the nature of a hollow spherical shell', which, by modern standards, is not too far from the mark, since we now know (but which Herschel had no way of knowing) that they are expanding bodies of gas. Yet just as Sir John's astrometer measurements indicated that Vega and Arcturus were vastly larger than the sun, so the planetary nebulae appeared to be of breathtaking dimensions when he used geometrical criteria to estimate their sizes.

Herschel based his planetary nebula hypothetical size calculations upon a combination of the micrometric angles displayed by these nebulae, and the recently measured parallaxes of a small number of nearby stars. Take the planetary nebula β Ursae Majoris, for example, which displays an easily measurable angular diameter of $2'4''$ in the northern skies. Then let us *assume* it to be at the same distance as the star 61 Cygni, which in 1838 Bessel had determined at 10.3 light years. Ergo, the diameter of β Ursae Majoris must be seven times bigger than the diameter of the newly discovered planet Neptune! And this is only the size it would be if it were in our astronomical 'back garden', as it were, whereas, as Herschel knew from its appearance in powerful telescopes, in reality, as things stood in 1847, it must be immeasurably further away.

And whence did the nebulae derive their light, especially if they had no conspicuous central star to illuminate their 'shells', thereby occasioning some of them to glow rather like frosted-glass spheres? Indeed, the planetary nebulae truly were astronomical objects fit for the Romantic Age, with their vast sizes, immense distances, strange luminescent properties and aura of cosmological mystery!

SIR JOHN HERSCHEL: THE UNIVERSAL PHILOSOPHER OF EUROPE

By the time of his return from the Cape in 1838, when the newly created Knight Baronet was still only forty-six years old, Sir John Herschel was fast acquiring the mystique and the reputation of being one of the most learned men of the age. For was he not the only person in history to have surveyed the skies in both hemispheres with a telescope of prodigious power, to have seen wonders never glimpsed since the Creation, and to have applied his formidable intellect to unravelling the very mechanisms by which the universe had taken on its present form?

Not to mention his other talents and achievements: he was bilingual in English and German; he spoke fluent French and read other European languages; he was an internationally respected classical scholar who would later translate the ancient Greek poetry of Homer into German; he was a skilled draughtsman and painter; and he was also an accomplished musician. Moreover, he had a long and happy marriage to Margaret Brodie Stewart, the daughter of an eminent Edinburgh clergyman, fathered eleven gifted children, and, as indicated above, possessed a capacity to make lasting friendships, with no real enemies. And all this is before one thinks of Sir John's other scientific achievements.

On his return to England, Sir John discovered that he was an even more sought-after celebrity than he had been in 1833, and he did not entirely like it. He was congenial, but he was *not* a socialite. He needed space in which to think, read, conduct research, spend time with his family and just *be*; getting out of invitations to attend balls and prestigious state functions, or sitting on the committees of learned societies, required a strategy all of its own. Between Herschel's leaving Slough in 1833 and returning in 1838, Isambard Kingdom Brunel's Great Western Railway had reached the town, on its march to the far west – which made Herschel ever more vulnerable to unsolicited visits from admirers and metropolitan invitations. In 1840, therefore, John and Margaret bought a new home: the 'Collingwood' estate (formerly the residence of Admiral Cuthbert Collingwood, Lord Nelson's second in command at Trafalgar), in the Kentish village of Hawkhurst, and at that time several miles from the nearest railway station! Being well off, they didn't sell Herschel House, Slough, but leased it out, and the property remained in the possession of the Herschel family until the twentieth century, when Sir Patrick Moore visited it, shortly before 'developer' vandals demolished this birthplace of modern cosmology.

Before vacating Herschel House, Slough, however, Sir John dismantled the long-disused and semi-derelict timber framework of his father's great forty-foot telescope of 1789. He even composed a special anthem for the occasion of the formal de-commissioning, which the family sang, it is said, in the garden. Yet before dismantling the great structure, he gave it an immortality which probably he did not foresee at the time; for in 1839, Sir John took a *photograph* of the telescope, thereby making his picture of the great forty-footer the first photograph ever taken of a scientific instrument.

Photography was a sensational new technology in 1839, with the Frenchman Louis Daguerre and Herschel's friend, William Henry Fox Talbot, inventing their own optical-chemical processes. Yet in addition to photographing the great telescope, Sir John had an even greater photographic 'first' to his credit; for on 14 March 1839, he announced his method of using 'hyposulphite' (later immortalized as 'photographer's hypo') as a way of 'fixing' a photographic image and preventing it from going black when exposed to daylight.

Indeed, photography was only one of several affairs which Sir John was to have with his first love, light and optics, as mentioned above. He was also concerned with the mathematical theory involved in achromatic telescope object-glass design and manufacture, while he corresponded at length with Mrs Mary Somerville, the great mathematician, on optical as well as gravitational topics. In her *On the Connexion of the Physical Sciences* (1834), Mary Somerville reviewed in great detail the current state of knowledge in optics, including the physics and chemistry of the ultra-violet and infra-red bands of the spectrum. The letters which they exchanged over several decades, and which are now preserved in the Royal Society Library, cover a fascinating range of topics, including the connection between light and magnetism, and even the nature of phosphorescent light; to say nothing about their common interest in classical Greek and Latin literature. They also display an avid fascination with practical experimentation, which Herschel and Mary Somerville shared. For example, on one occasion in 1865, the 73-year-old Herschel told the still-intellectually productive 85-year-old Mary Somerville that one night he saw a dead lobster glowing conspicuously in the dark of the Hawkhurst kitchen, and wondered what was the spectral and chemical nature of its light! He said that he then passed the lobster's glow through a glass prism, only to see no discernible tints of colour, though he admitted that his visual acuity was not as good as it once had been. He added humorously, 'Is not the constellation Cancer a boiled Crab?'

At a time, moreover, when it was not expected that ladies would write books on intellectually demanding scientific, technical and mathematical subjects, Sir John Herschel had, since the 1820s, been one of Mary Somerville's warmest encouragers and promoters. Their friendship spanned five decades, and both John and Margaret Herschel were supporters of women's education.

Sir John Herschel spent his entire career in the 'Grand Amateur' tradition of self-funded scientific research, for apart from whatever he may have earned as a Fellow of St John's College, Cambridge as a young man, he had no properly salaried job until he was fifty-eight years old – and what a first job it was! In 1850, the Prime Minister, Lord John Russell, persuaded Sir John to accept the Mastership of the Royal Mint and to be responsible for the nation's currency. One of the attractions of the post was that it had once been held by none other than Sir Isaac Newton. An offer he could scarcely refuse.

Yet, unlike Sir Isaac, Sir John took no pleasure in administration or in the exercise of political power, and so, in a state of exhaustion, he resigned his post in 1856. For the life of the private gentleman scientist was the life for him.

The eyes that had glimpsed celestial wonders in both hemispheres finally closed, peacefully, at Hawkhurst, on 11 May 1871. He was seventy-nine.

BIBLIOGRAPHY

Manuscripts

The Royal Society Library houses many of Sir John Herschel's manuscripts. For his correspondence with Mrs Somerville, see R. Soc. MS HS16 pp. 337–82.

John F. W. Herschel: Key Publications

'Account of some observations made with a 20-feet Reflecting Telescope', *Memoirs of the Royal Astronomical Society* II (1826), pp. 459–97.

A Preliminary Discourse on Natural Philosophy (1830).
A Treatise on Astronomy (London, 1833).
'Observations of Nebulae and Clusters of Stars made at Slough, with a Twenty-Feet Reflecting Telescope between 1825 and 1833', *Philosophical Transactions of the Royal Society* 123 (1833), pp. 359–505.
Results of Astronomical Observations Made During the Years 1834, 1835, 1836, 1837, 1838, at the Cape of Good Hope (1847).
Outlines of Astronomy (London, 1849).
'Catalogue of Nebulae and Clusters of Stars', *Philosophical Transactions* 154 (1864–5), pp. 1–137.

John F. W. Herschel and Family: Published Correspondence

A Calendar of the Correspondence of Sir John Herschel, eds Michael J. Crowe, David R. Dyck and James R. Kevin (CUP, 1998).
Herschel at the Cape: Diaries and Correspondence of Sir John Herschel, eds T. J. Deeming, B. H. Evans, D. S. Evans and S. Goldfarb (University of Texas, Austin and London, 1969).
Constance A. Lubbock, *The Herschel Chronicle: The Life-Story of Sir William Herschel and His Sister Caroline* (Cambridge, 1933; reprinted by the William Herschel Society, Bath, 1997).
Brian Warner (ed.), *Lady Herschel: Letters From the Cape, 1834–1838*, (Cape Town, 1991).

Other Primary Sources

Caroline Herschel, *Caroline Herschel's Autobiographies*, ed. Michael Hoskin (Science History Publications, Cambridge, 2003).
Mrs John Herschel, *Memoir and Correspondence of Caroline Herschel* (1879; reprinted by the William Herschel Society, Bath, 2000).
Michael Hoskin, *William Herschel and the Construction of the Heavens* (1963).
Michael Hoskin, *The Herschel Partnership, as Viewed by Caroline* (Science History Publications, Cambridge, 2003).
Mary Somerville, *Personal Recollections of Mary Somerville* (1873); new edition, *Queen of Science, Personal Recollections of Mary Somerville*, ed. Dorothy McMillan (Canongate Classics 102, Edinburgh, 2001).

Secondary Sources

Günther Buttmann, *The Shadow of the Telescope: A Biography of John*

Herschel, transl. B. E. J. Pagel (Lutterworth, Guildford and London, 1974).

Allan Chapman, 'An Occupation for an Independent Gentleman: Astronomy in the Life of John Herschel', *Vistas in Astronomy* 36, No. 1 (1993), pp. 71–116. Reprinted as a separate monograph by Pergamon, 1994, Elsevier Science Ltd., pp. 1–25; also reprinted in A. Chapman, *Astronomical Instruments and their Users: Tycho Brahe to William Lassell* (Variorum, Ashgate, Aldershot, 1996), Chapter XIII.

Michael J. Crowe, 'Herschel, Sir John Frederick William, First Baronet (1792–1871), Mathematician and Astronomer', *Oxford Dictionary of National Biography* (OUP, pp. 200–14).

Hubble's Constant

DAVID M. HARLAND

INTRODUCTION

In this chapter we will explain the discovery of Hubble's Constant and the efforts to measure it, and then we will review its cosmological significance.

SURPRISING DISCOVERY

On being hired in 1919 by the Mount Wilson Observatory of the California Institute of Technology, Edwin P. Hubble set about using the newly commissioned 100-inch Hooker reflector, the largest

Figure 1. Edwin Hubble. (Image courtesy of the Carnegie Institution of Washington.)

telescope in existence, to study M31, a prominent nebula in the constellation of Andromeda whose oblique aspect displays a spiral form. When he pointed out that the glow of its inner regions was reminiscent of low-resolution pictures of the densest star fields on the Milky Way, supporting the theory that spiral nebulae were isolated 'island universes', other astronomers insisted that the glow was just gaseous nebulosity within the Milky Way. In 1920, in an effort to decide the issue, the Smithsonian Institution in Washington hosted a learned debate on *The Scale of the Universe*. The result was inconclusive.

Figure 2. The 100-inch Hooker telescope on Mount Wilson, commissioned in 1917. (Image courtesy of the Carnegie Institution of Washington.)

Hubble decided to settle the issue and, by 1923, employing an improved photographic emulsion, he managed to resolve individual stars in the periphery of M31. Some of the stars were variable, and by analysing their light curves he was able to identify one as being of a type known as a Cepheid. This was a momentous discovery.

The first such star to be recognized was Delta Cephei, by John Goodricke in England, an amateur astronomer who specialized in

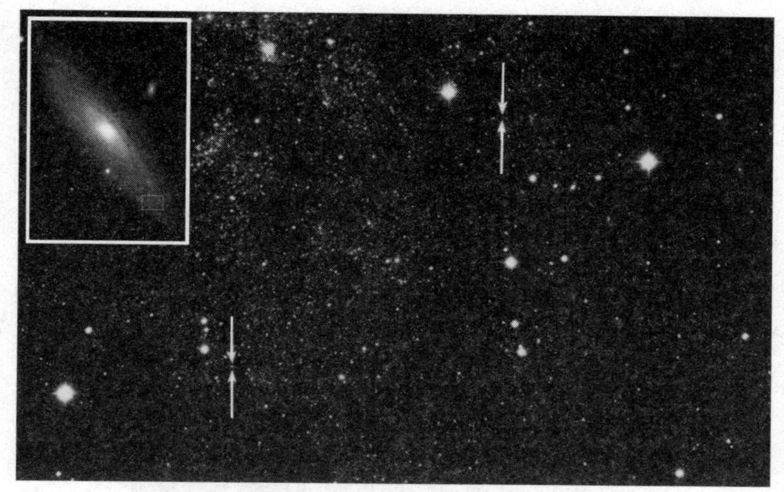

Figure 3. Two Cepheids discovered by Hubble in the periphery of the M31 spiral. (Image courtesy of the Carnegie Institution of Washington.)

variable stars. In 1784, shortly before his death at the age of only 21, he discovered this star to be varying in a well-defined manner over a period of 5.4 days. Over the years, others were identified, with periods ranging between about a day to several months, the mean being about a week. In 1894, Aristarkh A. Belopolski at the Pulkova Observatory took spectra of Delta Cephei and noticed a radial velocity cycle that was synchronized with its light curve, implying that the star was pulsating in size.

Between 1904 and 1908, Henrietta S. Leavitt at the Harvard College Observatory identified a number of Cepheids in the Small Magellanic Cloud and noticed that those with the longer periods appeared brighter. The significance of this observation was that, because all of the stars were at essentially the same distance from us, it was possible to infer a relationship between the period and the true luminosity of such a star. Leavitt published her results in 1912.

Although no 'local' Cepheid had a measurable parallax from which to calculate its distance using trigonometry, in 1913, Ejnar Hertzsprung at the University of Göttingen was able to apply a statistical technique to estimate the distances to some such stars and thereby obtain their true luminosities. This placed the Small Magellanic Cloud at a distance of 100,000 light years and served to calibrate Leavitt's relationship. In 1917, Harlow Shapley at Mount Wilson Observatory

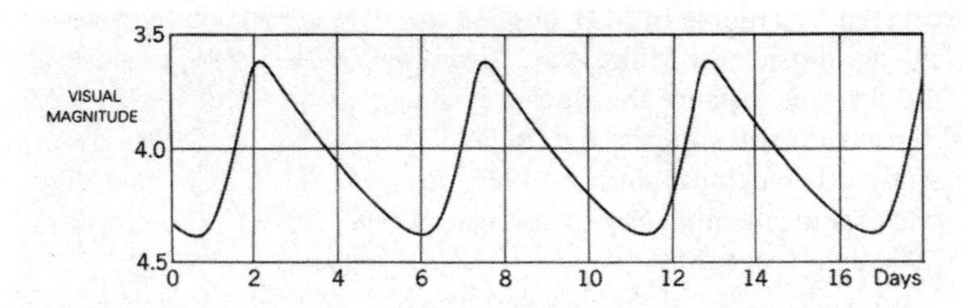

Figure 4. The light curve of the star Delta Cephei varies in a regular manner between well-defined maxima and minima over a period of 5.4 days.

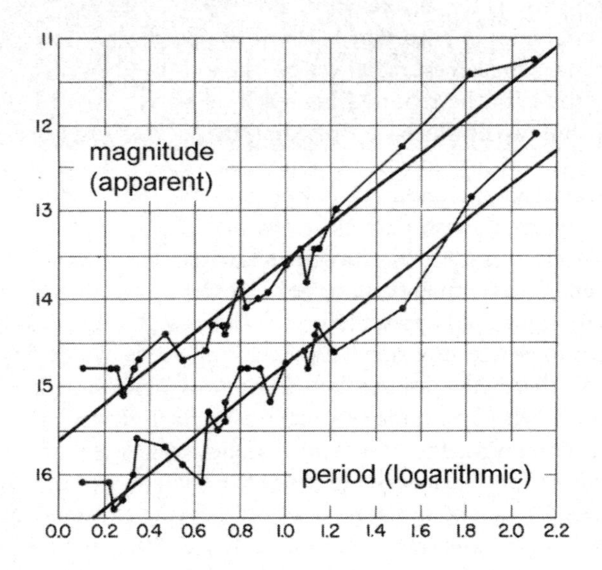

Figure 5. When Henrietta Leavitt plotted the periods of Cepheids in the Small Magellanic Cloud against their maximal and minimal apparent brightness, she realized that there was a simple relationship between the period and the luminosity of such stars. (Image courtesy of the Harvard College Observatory.)

estimated the Milky Way to be some 300,000 light years in diameter on the basis of a study of globular clusters, whose distances he calculated using Leavitt's period-luminosity relationship.* The Cepheids

* Later studies by other astronomers using better techniques would reduce this to 100,000 light years.

discovered by Hubble in M31 implied that it was 850,000 light years away – far outside the Milky Way. Given the large angular size of M31 in the sky, this distance meant that it must rival the Milky Way. The universe *really was* comprised of isolated galaxies. Hubble then set out to study other nebulae and, by 1926, he was able to announce that Cepheids indicated that M33, a face-on spiral in Triangulum, was also 850,000 light years away.

Meanwhile, Vesto M. Slipher had used the 24-inch refractor at the Lowell Observatory in Arizona to obtain the first spectrum of M31, an observation which required integrating on a photographic plate over several nights. The radial velocity calculated from the displacement in the spectral lines implied that M31 was approaching at 300 kilometres per second – the largest radial velocity of any celestial object measured to date. In 1914, he reported his results for fifteen spirals, including M31, all but two of which were receding at even higher speeds. A decade later, further measurements by Slipher and astronomers at other observatories had increased the sample to forty-five galaxies and the highest observed recessional velocity was 1,800 kilometres per second.

In parallel, Carl W. Wirtz in Germany had inspected some of these spirals and by assuming them to be of similar size, meaning that those furthest away would appear to be smallest, he pointed out that there appeared to be a relationship between distance and recessional velocity. To study this possibility further, it would be necessary to properly determine their distances; a task that was beyond his means.

Even employing the 100-inch, Hubble had been able to identify Cepheids in only half a dozen galaxies. In order to make progress, Hubble decided to make some assumptions. First, he presumed that the brightest stars in any galaxy were of similar luminosity. In the absence of evidence to the contrary, this was a fair assumption. Using the brightest stars as 'standard candles', he was able to infer the relative distances of nearby galaxies. To push further, he then presumed that all the galaxies in a vast cluster in Virgo occupied a spherical region of space and were of similar size, so that he could use their apparent sizes to calculate their relative distances.

Given the few individuals for which he was able to calculate distances based on 'standard candles', Hubble concluded that the centre of the Virgo cluster was 6.6 million light years away.*

* Herein, intergalactic distances are quoted in their contemporary context.

By 1928, Hubble had measured the distances to twenty-four of the spirals on Slipher's list and confirmed that there was indeed a redshift-distance relationship. On average, the radial velocity increased by about 160 kilometres per second per million light years of additional distance.*

Furthermore, the geometrical relationship meant that the spirals were racing away from one another at a constant rate. This implied that the universe had an *origin* and is in a *state of expansion*. This came as a surprise.

In 1917, when Albert Einstein explained gravitation in terms of his Theory of General Relativity, the universe was believed to be static, so he introduced a 'cosmological constant' to overcome the gravitational attraction which would otherwise result in universal collapse.† Nevertheless, in 1920, Alexander A. Friedmann in St Petersburg, Russia investigated a number of alternative solutions to Einstein's equations in which the universe would expand, so there was a theoretical basis for the observational discovery.

PROBING FURTHER

To extend the redshift-distance relationship to greater distances, it was necessary to obtain spectra of ever fainter galaxies, in doing so pushing the 100-inch to its limit. Milton L. Humason teamed up with Hubble. While Hubble estimated the distances to galaxies by a 'ladder' of different 'standard candles', Humason took spectra to measure their redshifts – in one case integrating a spectrum for ten nights. By 1931, they had extended the relationship to a recessional velocity of 20,000 kilometres per second and out to an estimated distance of about 100 million light years, which was a vast volume of space. The data suggested that although galaxies were congregated into dense clusters, the clusters were widely separated and randomly distributed in space.

As Hubble and Humason continued to push deeper into space, Walter Baade, who joined the Mount Wilson Observatory in 1931,

* A rate that corresponds to the separation between two objects 1,000 kilometres apart increasing by a fraction of a millimetre per year. M31's radial velocity arises from the fact that it is gravitationally bound within a group that includes our own galaxy.

† Upon hearing of Hubble's discovery that the universe was expanding, Einstein dismissed the cosmological constant as his 'greatest blunder'.

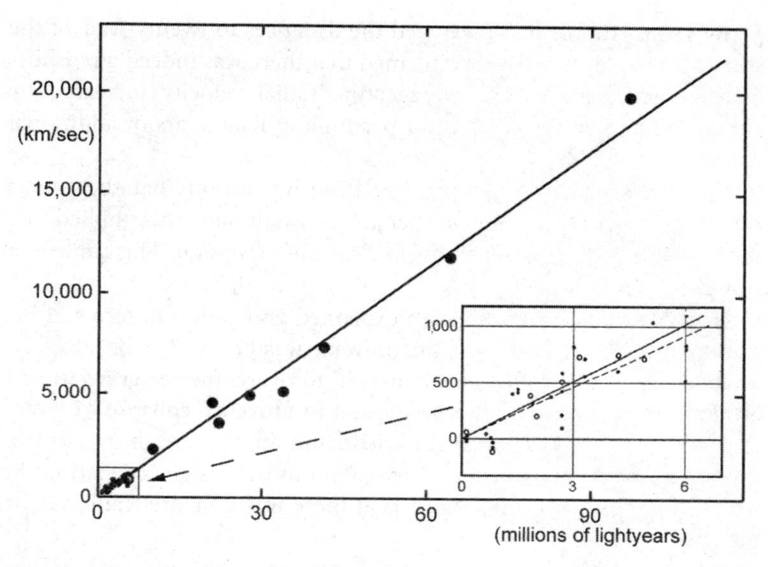

Figure 6. The redshift-distance relationship. The inset shows the data published by Hubble in 1929 as 'A Relation between Distance and Radial Velocity among Extragalactic Nebulae', with considerable scatter in the distance estimates. The main plot shows the extended relationship he published with Milton Humason in 1931 as 'The Velocity-Distance Relation among Extragalactic Nebulae'.

made a detailed study of M31 with a view to figuring out why its globular clusters seemed to be only half as luminous as those in our own Galaxy. The photographs taken by Hubble had resolved stars in the periphery of the spiral, but those in the nucleus had remained a diffuse glow. But when Los Angeles was blacked out during the Second World War, the much improved 'seeing' enabled the 100-inch to resolve stars in the nucleus. Baade realized that the spiral arms were predominantly 'hot blue' stars, and the nucleus and the globular clusters were composed of redder stars. Spectra indicated the stars within the spiral to be richer in 'metals' than those in the nucleus, a distinction that affected their luminosities.* There were, in fact, two types of Cepheid, with those in the spiral being about twice as luminous as those in the nucleus and the globular clusters. In 1952, applying this correction, Baade announced that M31 was actually 2.2 million light years away.

* To an astrophysical spectroscopist, a 'metal' in a star is any element heavier than helium.

Figure 7. The 200-inch Hale telescope on Mount Palomar, commissioned in 1949. (Image courtesy of the Carnegie Institution of Washington.)

This doubled the radius of the known universe and halved the slope of Hubble's redshift-distance relationship to about 80 kilometres per second per million light years.

With the introduction of the 200-inch Hale telescope on Mount Palomar in 1949, it became possible to image a galaxy during a ten-minute exposure that was barely able to be recorded by an all-night integration on the 100-inch. Having suffered a heart attack, Hubble assigned the telescope work needed to extend the redshift-distance relationship to postgraduate Allan R. Sandage, who, upon Hubble's death in 1953, continued the project.

Sandage exploited the light-gathering power of the 200-inch to search for Cepheids in nearby galaxies, to confirm the distances inferred by Hubble on the basis of the putative 'standard candles'. He found that many of the objects that his mentor had presumed to be very bright stars were, in fact, vast clouds of glowing hydrogen, and, being more luminous, were much more remote. On correcting this

Figure 8. Allan Sandage in front of the 200-inch in 1953. (Image courtesy of the Palomar Observatory.)

flaw, Sandage further reduced the slope of the redshift-distance relationship – the rate at which the universe is expanding, by then named Hubble's Constant in homage to its discoverer – to 55 kilometres per second per million light years. By the time Humason retired in 1957, they had measured the redshifts of 850 galaxies, calculating recessional velocities of up to 100,000 kilometres per second. This sample ranged out for several billion light years and encompassed eighteen rich clusters of galaxies. Upon noticing that there was a giant elliptical in each cluster that outshone its companions, Sandage plotted the apparent magnitudes of these 'first-ranked' ellipticals against their redshifts and obtained a straight line; this implied that they were equally luminous and could serve as 'standard candles' to probe deeper into space. By 1961, having noticed a discrepancy in the apparent brightness of globulars in galaxies in the Virgo cluster, Sandage announced that the value of Hubble's Constant might be much less than fifty.

In 1962, Sandage teamed up with Gustav A. Tammann of the University of Basel in Switzerland to analyse a decade's worth of photographs in search of Cepheids in galaxies in a small cluster in Ursa Major. Even using the 200-inch, it was possible to monitor Cepheids only during the brightest portions of their light curves. Nevertheless, in 1967, Sandage and Tammann were able to report that one member of this cluster was 10.6 million light years away, a distance that reinforced the suspicion that Hubble's Constant was much less than fifty. In 1967, after additional such studies, Sandage announced that it was in the range fifteen to seventeen.

At this point, Gerard de Vaucouleurs of the University of Texas claimed that Sandage was underestimating the value of Hubble's Constant by as much as a factor of two. This launched a vigorous debate over 'standard candles' and their calibration. The issue was exacerbated by the fact that although Sandage's latest value was just 10 per cent of Hubble's original estimate (in corrected terms), it was quoted with the same estimated error of plus/minus 15 per cent. Furthermore, there was no overlap between the ranges of the values of Sandage and of de Vaucouleurs. What was required was a significant improvement in the distances of the *first steps* on the distance scale.

By that time, the distances to 10,000 stars had been estimated by the measurement of their parallax and various indirect methods, but to achieve any significant increase in this sample was impracticable using terrestrial telescopes. This became the primary objective of the Hipparcos satellite, launched by the European Space Agency in 1989. As Earth travelled around the Sun, it provided a 300-million-kilometre baseline to measure the parallaxes in order to triangulate the distances to nearby stars, and, by being above the atmosphere, the satellite was able to achieve an accuracy a hundred times better than from the ground. The catalogue released in 1997 enabled Michael Feast at the University of Cape Town in South Africa to recalibrate the Cepheid period-luminosity relationship with an uncertainty in the distance of 5 per cent and, upon realizing that they were further away than supposed, he reduced the value of Hubble's Constant by 10 per cent. Although a step in the right direction, this in itself did nothing to reconcile the rival camps.

Determining the value of Hubble's Constant was one of the Key

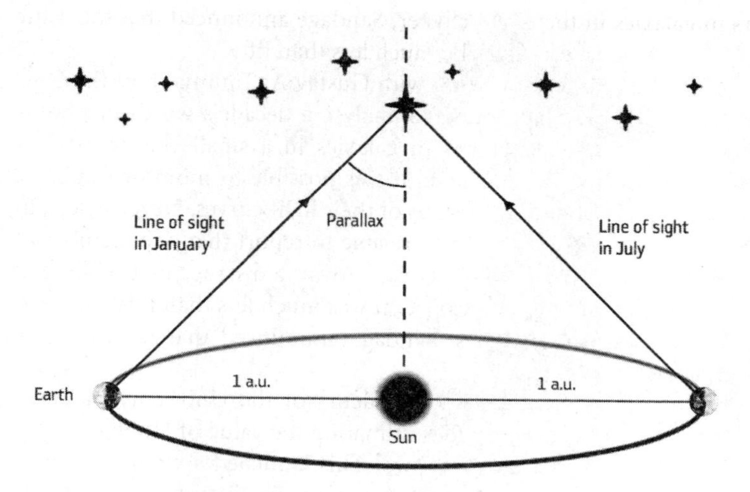

Figure 9. Using parallax to measure the distance of a star. (Image courtesy of ESA/Medialab.)

Projects of the Hubble Space Telescope. When launched by NASA in 1990, the optics of the telescope were discovered to be flawed by 'spherical aberration', but once this had been corrected by a Space Shuttle servicing mission in December 1993, long exposures were able to provide exceedingly sharp images of the faint outer regions of galaxies. The goal of the team led by Wendy L. Freedman of the Observatories of the Carnegie Institution in Washington, located in Pasadena, California, was to identify Cepheids in the Virgo cluster, the closest rich cluster. Over the ensuing years, the HST identified some 800 such stars in eighteen galaxies, spanning the period-luminosity relationship. This established the distance to the Virgo cluster and calibrated the 'standard candles' that had been used to reach much deeper. The value of Hubble's Constant was then calculated by a variety of methods. The 'best' estimate, announced by the team in 2001, was 21.5 kilometres per second per million light years to an accuracy of 10 per cent. By that time, the two values derived from terrestrial observations were just outside this range, one above and the other below, so the convergence was pleasing.

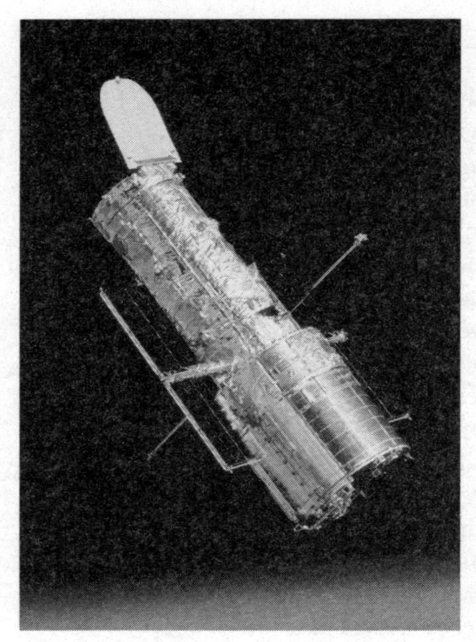

Figure 10. A view of the Hubble Space Telescope in orbit. (Image courtesy of NASA.)

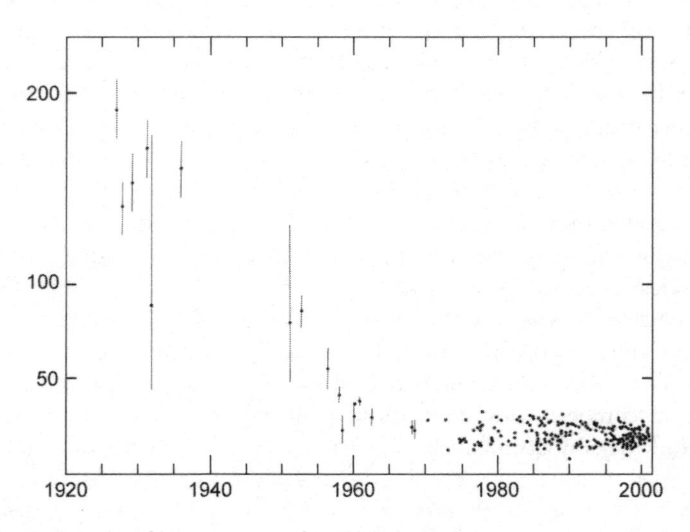

Figure 11. A plot of how estimates of Hubble's Constant have varied over the decades. (Image courtesy of John Huchra of the Harvard-Smithsonian Center for Astrophysics.)

MORE SURPRISES

Given the value of Hubble's Constant, it becomes possible to extrapolate back to a time when all the galaxies were together, and thereby calculate the *age* of the universe. However, this presumes that the expansion rate has been constant, which cannot have been the case because the mutual gravitational attraction of the mass of the galaxies must act to slow it. The rate of slowdown is measured by the Deceleration Parameter and its value determines the fate of the universe. If it were negative, then some force would have to be accelerating the expansion, but it seemed safe to reject this. If the Deceleration Parameter was zero, then gravity has had negligible effect and the age is obtained simply by extrapolating Hubble's Constant backwards. For a Deceleration Parameter greater than zero this calculation will give the *maximum age*, because the rate of expansion must have been greater in the past. In the case of a positive Deceleration Parameter less than 0.5, there is insufficient mass-energy for gravity to halt the expansion and the process will continue indefinitely and the universe is described as being 'open'. For a value exceeding 0.5, gravity will reverse the expansion and 'close' the universe. In the case of the critical value of 0.5, gravity will halt the expansion but doing so will take for ever, leaving no time for a contraction phase. In this case the universe is said to be 'flat'.

Any variation in the rate of expansion would reveal itself as a difference in the value of Hubble's Constant over time. Although Hubble himself had sought a departure from linearity at the far end of the redshift-distance relationship, there was considerable scatter in his data and, in any case, his survey did not extend deep enough into space. Even after extending the relationship to a recessional velocity of 250,000 kilometres per second, Sandage reported in 1974 that there was no clear evidence of divergence.

That same year, a rival team of J. Richard Gott and James E. Gunn at the California Institute of Technology and Beatrice M. Tinsley and David N. Schramm at the University of Texas decided that rather than try to compete with Sandage, they would measure the mean mass-energy density of the universe, which is known as Omega.* The value

*In his study of General Relativity, Friedmann had calculated the critical density for a 'flat' universe to be 10–28 g/cm^3, which is equivalent to only a few hydrogen atoms per cubic metre.

for a 'flat' universe would be 1.0. Estimating it involved thinking on the largest of scales.

The Coma cluster is a large conglomeration of 11,000 galaxies which spans about eight million light years with its members typically 300,000 light years apart, making it more densely packed together than the twenty members of our Local Group. In the early 1930s, soon after joining the California Institute of Technology, Fritz Zwicky noted that the individual galaxies in the Coma cluster were moving at such high velocities relative to the mean that for them to remain gravitationally bound there had to be more mass than was apparent. When Vera Rubin of the Carnegie Institution of Washington later studied the manner in which galaxies rotate, the 'rotation curves' revealed their gravitating mass to extend far beyond the visible structures. Further studies indicated that rather than each galaxy having its own 'halo', this 'non-luminous matter' or 'dark matter' pervaded the entire cluster, with the galaxies being merely dense patches that were luminous. Astronomers had presumed that the universe was comprised of what they could *see*, little suspecting that the luminous material was insignificant.*

Gott et al employed three methods to estimate Omega. The total luminosity of galaxies enabled the mass of the stars which emitted that light to be estimated. This gave a value for Omega of 0.01. By analysing the motions of galaxies in clusters, they obtained a value for Omega of 0.1. Finally, because the ratio of deuterium to helium in the universe was very sensitive to the conditions in the 'fireball' of the Big Bang, they calculated the deuterium abundance from the absorption lines in interstellar space, measured by the Copernicus Orbiting Astronomical Observatory launched by NASA in 1972.† In effect, deuterium measured the mass-energy density in the early universe, rather than as it is today, and this, too, yielded a value for Omega of 0.1. The team's results therefore indicated the universe to be 'open'.

Although 90 per cent of the gravitating mass-energy would have to be 'missing' for the universe to be 'flat', there was a compelling basis for believing this to be so. In 1979, Robert H. Dicke and Phillip J. E. Peebles at Princeton argued that the slightest deviation from unity in the

* A variety of candidates have been offered to explain 'dark matter', including red dwarfs, brown dwarfs, black holes, intergalactic hydrogen and a variety of exotic particles.

† Deuterium is an isotope of hydrogen which is chemically identical except its nucleus contains a neutron whereas that of hydrogen does not.

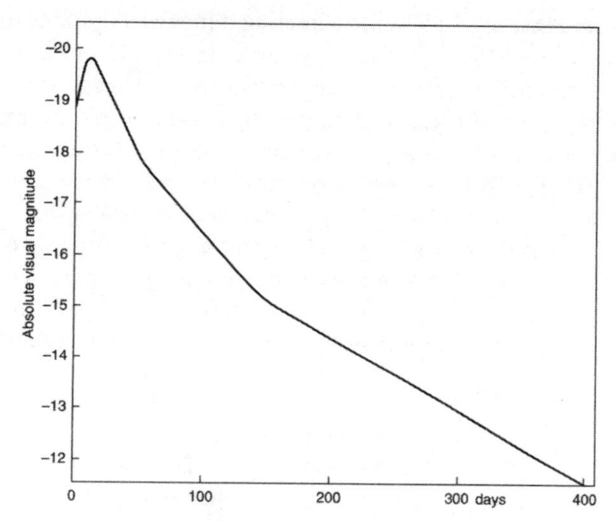

Figure 12. The light curve of a Type Ia supernova. When the mass of the accreting white dwarf reaches the Chandrasekhar limit, its core undergoes a thermonuclear runaway, ejecting its outer layer at 10,000 kilometres per second to form a nebulosity that radiates strongly in the visible spectrum for a time as its radioactive elements decay. Detailed models accurately explain the kinks in the light curve. The gas initially glows from the decay of nickel-56 to cobalt-56 with a half-life of 6 days and then, after a rapid decline, the slope follows the decay of cobalt-56 to iron with a 78-day half-life. At its peak, a supernova will outshine its parent galaxy.

value of Omega at the origin of the universe would cause it to rapidly become either infinitesimally small or infinitely large. For Omega today to be one tenth, it must have been *precisely one* at the time of the Big Bang. Further insight was provided by Alan Guth at Cornell, who discovered in 1979 that conditions during the Big Bang would have produced a period of *exponential expansion* that has been called *inflation*, and this would have 'flattened' the universe.

Thus, as the Hubble Space Telescope was making observations to determine the value of Hubble's Constant, cosmologists faced the dilemma of the majority of the gravitating mass-energy of the universe being somehow 'missing'. This issue was resolved in an unexpected manner.

In the early 1990s, two teams set out to use modern technology to extend the redshift-distance relationship for direct evidence of the

Deceleration Parameter. As 'standard candles' they used Type Ia supernovae because these are of uniform luminosity. Each is a close binary star system in which a white dwarf accretes hydrogen from its companion until its mass reaches the Chandrasekhar limit, at which time its core undergoes a thermonuclear runaway which liberates a well-defined burst of energy and produces a distinctive light curve.

The Supernova Cosmology Project was led by Saul Perlmutter in Berkeley, California, and the High-Z Supernova Search Team by one of his former postdocs, Brian Schmidt, then at Siding Springs in Australia. In 1998, they independently announced evidence that at very high redshifts a progressive departure from linearity in the redshift-distance

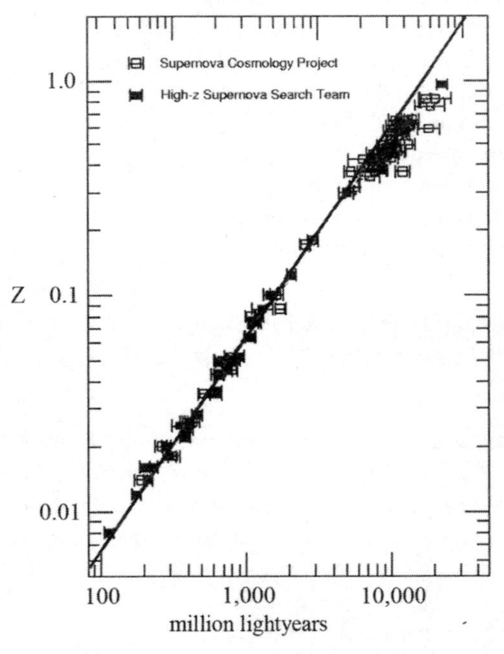

Figure 13. A simplified plot of the Type Ia supernovae data which extended the redshift-distance relationship sufficiently to reveal a deviation which indicated that the rate of expansion of the universe is accelerating. It has been redrawn to match the orientation of the earlier illustrations but with Z as a dimensionless redshift defined as $Z = (\lambda_{observed} - \lambda_{emitted}) / \lambda_{emitted}$, where λ is wavelength. (The data is derived from Riess et al. 1998, *Astronomical Journal,* Vol. 116, pp. 1009, and from Perlmutter et al. 1999, *Astrophysical Journal,* Vol. p. 517, p. 565.)

relationship meant that the rate of expansion was *accelerating* – a finding that was completely unexpected and led to the teams sharing the 2011 Nobel Prize for Physics.

This acceleration meant that the mutual gravitational attraction of the contents of the universe that acted to slow the expansion was being countered by a universal *repulsive force* reminiscent of the cosmological constant which Einstein added to General Relativity in order to hold the universe static against gravitational collapse. It had long been realized that whilst in General Relativity space is a continuum which is 'flat' on the smallest of scales and hence has zero energy, in Quantum Theory 'empty space' is *seething* with energy in the form of 'virtual particles' which flit in and out of existence. An energized 'false vacuum' imparts a pressure that *expands* space. Observations of the fine structure of the cosmic microwave background, the extremely redshifted glow of the Big Bang, by the Wilkinson Microwave Anisotropy Probe launched by NASA in 2001 have resolved the mystery. The value of Omega is indeed precisely 1.0 and the visible matter in galaxies accounts for just 4 per cent of the required density; the 'dark matter' whose gravity binds clusters together accounts for another 23 per cent; and the remaining 73 per cent is not present in the form of physical

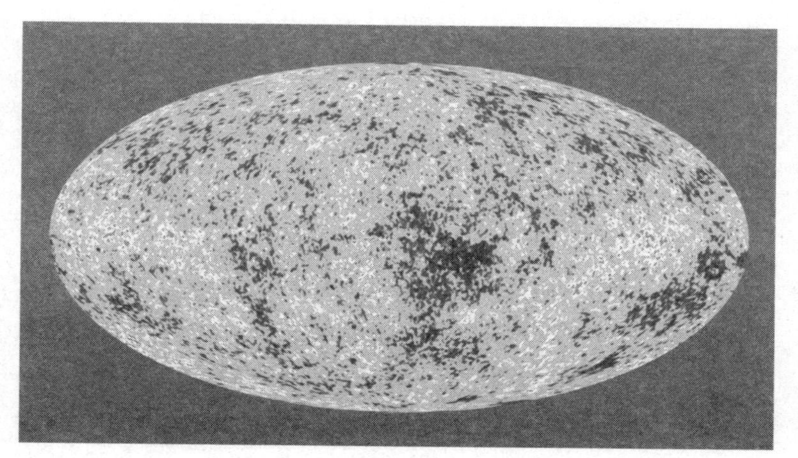

Figure 14. The fine structure of the cosmic microwave background measured by the Wilkinson Microwave Anisotropy Probe. (Image courtesy of the Goddard Space Flight Center, Princeton University, University of Chicago, University of California, Los Angeles, University of British Columbia and Brown University.)

matter, it is energy hidden in the vacuum and, by virtue of being a repulsive force, it is driving *runaway expansion*.

THE AGE OF THE UNIVERSE

With the Hipparcos recalibration of the Cepheids, the Hubble Space Telescope team's determination of Hubble's Constant as 21.5 kilometres per second per million light years implied an age for the universe of 14 billion years (in the absence of deceleration). By taking into account the value of Omega = 1.0 and allowing for deceleration (but not acceleration), the age would have been about 10 billion years. Charles Lineweaver of the University of New South Wales in Sydney, Australia, calculated how acceleration influenced the expansion rate and found that for Omega = 1.0, the rate of expansion initially decreased due to deceleration when the galaxies were tightly packed, but as the matter thinned out and the vacuum energy began to predominate, the expansion rate picked up some 5 billion years ago. Taking everything into account yields an age for the universe of 13.7 billion years.

New Horizons at Pluto: The First Results

JOHN MASON

On 14 July 2015, after a nine-and-a-half-year journey across the Solar System, and travelling some 5,000 million kilometres, the New Horizons spacecraft made a spectacular high-speed flyby of the dwarf planet Pluto and its moons. As this Yearbook went to press the very first images and scientific findings were being released. There will be a much fuller and more detailed summary in the 2017 Yearbook, but here is an overview of the first stunning results. Pluto's surface exhibits an amazing and quite unexpected diversity of geological features; there are young frozen plains, flowing ices, exotic surface chemistry, mountain ranges, and a thin atmosphere with layers of haze.

For a great many planetary scientists, the study of Pluto and other objects in the Edgeworth–Kuiper Belt – a relic of planetary formation beyond Neptune's orbit – marks a new era of Solar System exploration. In order to gain a complete understanding of our Solar System, they wanted to comprehend where the dwarf planet Pluto and its moons fitted in with other known objects. When considered alongside the four rocky inner planets (Mercury, Venus, Earth, and Mars) and the outer gas giants (Jupiter, Saturn, Uranus, and Neptune), Pluto and its largest moon, Charon, may be thought of as belonging to a third category of object – the 'ice dwarfs'. These have solid surfaces but, unlike the rocky terrestrial planets, a significant portion of their mass is icy material.

The baby-grand-piano-sized New Horizons spacecraft blasted off from Cape Canaveral Air Force Station atop its powerful Atlas V rocket on 19 January 2006, carrying with it the ashes of Clyde Tombaugh, the astronomer who discovered Pluto in 1930. The spacecraft was launched directly into an Earth-and-solar-escape trajectory, reaching a velocity of almost 58,540 kilometres per hour (36,375 miles per hour), and setting the record for the highest launch speed of a human-made object from Earth. By launching in January 2006, New Horizons could take advantage of a gravity assist from Jupiter, and on 28 February 2007 it made its closest approach to the giant planet, at a distance of 2.3 million kilometres.

The Jupiter flyby increased New Horizons' speed by 14,000 kilometres per hour and placed it on a trajectory that would see it reach Pluto in July 2015. The Jupiter encounter was also used to test New Horizons' scientific instruments, returning data about the planet's atmosphere, magnetosphere, and larger moons. Most of the post-Jupiter interplanetary cruise was spent in hibernation to preserve on-board systems, but there were brief annual checkouts. On 6 December 2014, New Horizons was brought back online in readiness for the Pluto encounter, and a thorough checkout of its instruments began. On 15 January 2015, almost nine years after launch, and six months before closest approach, the spacecraft began its approach phase to Pluto.

PLUTO'S CHAOTIC TUMBLING MOONS

Prior to the New Horizons flyby, the best images of Pluto had been obtained by NASA's Hubble Space Telescope (HST), and these images also revealed four previously unknown moons of Pluto: Nix, Hydra, Kerberos and Styx. In mythology, Charon was the ferryman who carried people across the River Styx to Hades (the underworld). Nix was the goddess of darkness and night, befitting a satellite orbiting distant Pluto, the god of the underworld. Nix was also the mother of Charon. Hydra was a terrifying monster with the body of a serpent and nine heads. Kerberos was the multi-headed 'hellhound' that guarded the gates of Hades and prevented ghosts of the dead from leaving, while Styx is named after the goddess who watched over the underworld river of the same name.

As the New Horizons spacecraft closed in for its historic flyby of Pluto, an analysis of data from HST by Mark Showalter of the SETI Institute in Mountain View, California and Doug Hamilton of the University of Maryland at College Park revealed that Nix and Hydra wobble chaotically. The orientation of Nix, in particular, changes quite unpredictably as it orbits Pluto.

Many astronomers describe Pluto and its innermost, largest moon, Charon, as a 'double planet' because they share a centre of mass located in the space between them. Their combined gravitational field shifts constantly and sends the smaller moons tumbling erratically. This effect is only magnified by the rugby ball-like, rather than spherical, shape of the moons Nix and Hydra. The consequences are that if you lived on either of these two tiny moons, you could not predict the time or direction the Sun would rise the next day! Clues to this chaotic motion first emerged when astronomers measured variations in the light reflected off Nix and Hydra. Analysing Hubble images of Pluto taken between 2005 and 2012, scientists compared the unpredictable changes in the moons' brightness to models of spinning bodies in complex gravitational fields. It is thought likely that Pluto's other two known moons, Kerberos and Styx, orbit in a similar chaotic fashion.

It has also been shown that three of Pluto's moons are at present locked together in resonance, meaning there is a precise ratio for their orbital periods. This means that if you were sitting on Nix, then you would see that Styx orbits Pluto twice for every three orbits made by Hydra.

Pluto's moons are thought to have been formed during a collision between the dwarf planet and a similar-sized body early in the history of our Solar System. The massive impact hurled out material that coalesced to produce the family of moons observed around Pluto today. The largest moon, Charon, is almost half the size of Pluto and was discovered by James Christy at the US Naval Observatory in Flagstaff in June 1978. Hubble discovered Nix and Hydra in 2005, Kerberos in 2011, and Styx in 2012. These little moons, measuring just a few tens of kilometres in diameter, were found during a Hubble search for objects that could be potential hazards to the New Horizons spacecraft as it passed the dwarf planet in July 2015.

The Hubble data also revealed that Kerberos is as black as asphalt, while the other frozen moons are as bright as sand. It had been predicted that dust blasted off the moons by meteorite impacts should coat the surfaces of all the moons, giving them a uniform look, which makes the very dark colouring of Kerberos a surprise.

CLOSING IN ON THE PLUTO SYSTEM

As New Horizons closed in on Pluto and its moons, images obtained by the spacecraft's telescopic Long Range Reconnaissance Imager

(LORRI) during late May and early June 2015 confirmed Pluto as a world with very bright and very dark terrain, and areas of intermediate brightness in between. These best ever views of Pluto revealed an increasingly complex surface with clear evidence of discrete equatorial bright and dark regions. Pluto's northern hemisphere displayed substantial dark terrain, though both Pluto's darkest and its brightest known terrain units were just south of or along its equator.

On 14 June, just one month out from Pluto, a 45-second thruster burst refined New Horizons' trajectory towards Pluto, targeting the optimal aim point for the spacecraft's flight through the Pluto system. This was only the second targeting manoeuvre of New Horizons' approach to Pluto; the burst adjusted the spacecraft's velocity by just 52 centimetres per second, aiming it toward the desired close-approach target point approximately 12,500 kilometres above Pluto's surface. The manoeuvre was based on the latest radio tracking data on the spacecraft and range-to-Pluto measurements made by optical-navigation imaging of the Pluto system taken by New Horizons in previous weeks. The New Horizons team continued to analyse spacecraft navigation and tracking data with an eye on 24 June, which would be the next opportunity to adjust course.

In their planning for the daring close approach that would take its spacecraft inside the orbits of all five of Pluto's known moons, the New Horizons team completed its analysis of the second and third sets of hazard-search observations of the Pluto system. These data were taken on 29–30 May and 5 June, using the LORRI camera. For these observations, LORRI was commanded to take hundreds of long-exposure (10-second) images, which were combined to enable a highly sensitive search for faint satellites, rings, or dust-sheets in the system – anything that might constitute a potential hazard to the spacecraft.

These hazard observations easily detected Pluto and all five known moons, but no rings, new moons, or hazards of any kind were found. The New Horizons hazard-detection team determined that satellites as faint as about four times dimmer than Pluto's faintest known moon, Styx, would have been seen if they existed beyond the orbit of Pluto's largest and closest moon, Charon. Limits on possible rings were unchanged since the first hazard observations in May: any undiscovered rings would have to be very faint or narrow – less than 1,600 kilometres wide or reflecting less than one 5-millionth of the incoming sunlight.

A subsequent hazard search began on 15 June, with the team reporting the results on about 25 June, after completing a detailed analysis of the new and still more sensitive data.

In the last week of June, the Pluto approach entered its third and final far-encounter science phase – called Approach Phase 3 (AP3), which ran until seven days before Pluto close approach. AP3 highlights included taking additional images of the Pluto system for final navigation purposes; mapping Pluto and Charon in increasing detail and watching for variability in colour, surface composition, and atmospheric patterns as the bodies rotate; and searching for new moons and rings with even greater sensitivity. New Horizons also continued sampling of the interplanetary environment – measuring both solar wind and high-energy particles, as well as dust-particle concentrations – approaching Pluto and its moons.

PLUTO AND CHARON'S ORBITAL DANCE

The first colour movies from NASA's New Horizons mission showed Pluto and Charon and the orbital motion of the two bodies. These near-true colour movies were assembled from images made in three colours – blue, red, and near-infrared – by the Multicolour Visible Imaging Camera on the instrument known as RALPH. The images were taken on nine different occasions from 29 May to 3 June. Even at this low resolution, it was possible to see that Pluto and Charon had different colours – Pluto was reddish-brown, while Charon was grey.

Pluto and Charon are tidally locked. Pluto spins on its axis every 6 days, 9 hours, and 17.6 minutes, the same time that it takes Charon and Pluto to orbit their barycentre – the shared centre of mass between the two bodies as they execute their orbital dance. As a result, an observer on Pluto's surface on the side facing Charon would always see Charon 'hanging' in the same part of the sky, but appearing seven times larger than Earth's Moon; an observer on the opposite side of Pluto would not see Charon at all. Because Pluto is substantially more massive than Charon, the barycentre is much closer to Pluto than to Charon. Looking closely at the images in the movies, it was possible to detect a regular shift in Pluto's brightness, due to the brighter and darker terrains on its differing faces.

At 1,208 km in diameter, Charon is half Pluto's size. However, it weighs only 12 per cent as much as Pluto. This suggests that Charon may be half ice and half rock; Pluto, by contrast, is about 70 per cent rock by mass. The pair form what planetary scientists call the only known 'binary planet' in the Solar System.

In a long series of images obtained by LORRI between 29 May and 19 June, Pluto and Charon appeared to more than double in size as New Horizons drew nearer. From this rapidly improving imagery, scientists found that the 'close approach hemisphere' of Pluto that New Horizons would fly over has the greatest variety of terrain types seen on the body so far. They also discovered that Charon has a 'dark pole' – a mysterious dark region that forms a kind of anti-polar cap.

THE SECOND 'RED PLANET'

With just two weeks to go before its historic 14 July flyby, the New Horizons spacecraft tweaked its path toward the Pluto system. The 23-second thruster burst was the third and final planned targeting manoeuvre of New Horizons' approach phase to Pluto; it was also the smallest of the nine course corrections since New Horizons launched in January 2006. It increased the spacecraft's velocity by just 27 centimetres per second, slightly adjusting its arrival time and position at the flyby close-approach target point approximately 12,500 km above Pluto's surface.

New colour images released on 1 July showed two very different faces of Pluto, one with a series of intriguing spots along the equator that were quite evenly spaced. Scientists had not seen anything quite like these dark spots; their presence captured the interest of the New Horizons science team, due to the remarkable consistency in their spacing and size. While the origin of the spots is a mystery for now, the answer may be revealed as more data becomes available.

New Horizons team members combined black-and-white images of Pluto and Charon from LORRI with lower-resolution colour data from the RALPH instrument to produce approximately true colour views.

Pluto is reddish brown in colour. Although this is reminiscent of Mars, the cause is almost certainly very different. On Mars the colouring agent is iron oxide, commonly known as rust. On Pluto, the reddish colour is likely caused by hydrocarbon molecules that are formed when cosmic rays and solar ultraviolet light interact with methane in Pluto's atmosphere and on its surface.

'Pluto's largest dark spot is clearly more red than the majority of the surface, while the brightest area appears closer to neutral gray,' said Alex Parker, a member of the New Horizons Composition team, of the Southwest Research Institute, Boulder, Colorado.

'Pluto's reddish colour has been known for decades, but New Horizons is now allowing us to correlate the colour of different places on the surface with their geology and soon, with their compositions,' added New Horizons Principal Investigator Alan Stern, also from South-West Research Institute. 'This will make it possible to build sophisticated computer models to understand how Pluto has evolved to its current appearance.'

Experts have long thought that reddish substances are generated as a particular wavelength of ultraviolet light from the Sun, called Lyman-alpha, strikes molecules of the gas methane (CH_4) in Pluto's atmosphere, powering chemical reactions that create complex compounds called tholins, which drop to the surface to form a sort of reddish 'gunk'.

Recent measurements with New Horizons' ALICE instrument reveal that the diffuse Lyman-alpha radiation falling on Pluto from all directions in interplanetary space is strong enough to produce almost as much tholin as the direct rays of the Sun.

'This means Pluto's reddening process occurs even on the night side where there's no sunlight, and in the depths of winter when the Sun remains below the horizon for decades at a time,' said New Horizons co-investigator Michael Summers, of George Mason University, Fairfax, Virginia.

Tholins have been found on other bodies in the outer Solar System, including Titan and Triton, the largest moons of Saturn and Neptune, respectively, and made in laboratory experiments that simulate the atmospheres of those bodies.

THE 4TH JULY ANOMALY!

Quite unexpectedly, just ten days before its scheduled arrival at Pluto, the New Horizons mission operations centre at the Johns Hopkins University Applied Physics Laboratory in Laurel, Maryland, lost contact with the spacecraft at 5:54 p.m. UTC on 4 July, but regained communications at 7:15 p.m.

During that time the autonomous autopilot on board the spacecraft recognized a problem and switched from the main to the backup computer. The autopilot placed the spacecraft in 'safe mode', and commanded the backup computer to reinitiate communication with Earth. New Horizons then began to transmit telemetry to help engineers diagnose the problem. The team then worked non-stop to return New Horizons to its original flight plan. Due to the 9-hour, round-trip communication delay that results from operating a spacecraft almost 4,900 million kilometres from Earth, full recovery was expected to take from one to several days.

The investigation into the anomaly that caused New Horizons to enter safe mode on 4 July concluded that no hardware or software fault occurred on the spacecraft. It was simply that the main computer was overloaded due to a timing conflict in the spacecraft command sequence. The computer was tasked with receiving a large command load at the same time it was engaged in compressing previous science data. The main computer responded as it was programmed to do, by entering safe mode, switching to the backup computer, stopping science, and calling Earth for help. Fortunately, back on Earth, engineers quickly understood the problem and the mission leadership chose to suspend science activities to focus on recovery efforts. Science operations resumed on 7 July and only thirty planned science observations were lost during the three-day recovery period – none of them important to the top-level science goals of the mission.

The really good news was the problem was not one that could happen during the Pluto encounter as no similar operations were planned!

DARK AND BRIGHT REGIONS ON PLUTO

Between 1 and 3 July, LORRI obtained three new images of Pluto. Together, the three images showed the full extent of a continuous swath of dark terrain that wrapped around much of Pluto's equatorial region. The western end of the swath broke up into a series of striking dark regularly-spaced spots (Figure 1), which were first detected in New Horizons' images taken in late June. Intriguing details were beginning to emerge in the bright material north of the dark region, in

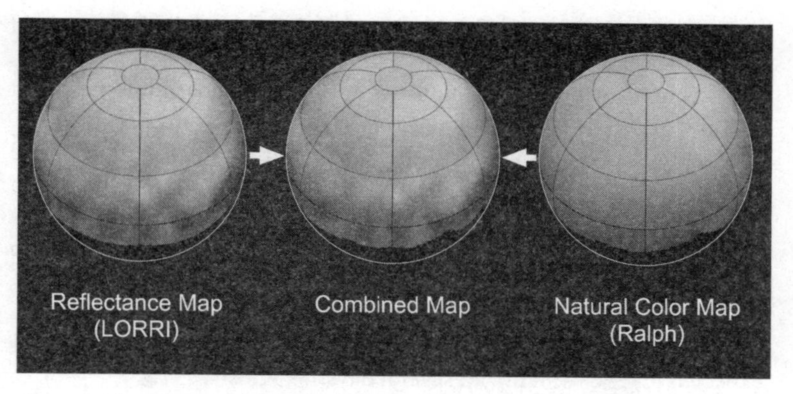

Figure 1. Pluto displays a continuous swath of dark terrain that wraps around much of the equatorial region. The western end of the swath breaks up into a series of striking dark regularly spaced spots, which were first detected in New Horizons' images taken in late June 2015. Such images were used to compile the first maps of Pluto's surface features from New Horizons' data. In the centre is a combined map, produced by merging the high-resolution LORRI data (left) and the lower-resolution RALPH data which provided colour information in the original (right). (All images in this article are reproduced courtesy of NASA/Johns Hopkins University Applied Physics Laboratory/Southwest Research Institute.)

particular a series of bright and dark patches that were conspicuous just below the centre of the disk in the right image. In one of the images, on the right side of the disk, a large bright area on the hemisphere of Pluto that would be seen close-up by New Horizons on 14 July could be discerned.

From images acquired by LORRI between 27 June and 3 July, combined with lower-resolution colour data from the spacecraft's RALPH instrument, scientists created a map of Pluto. The centre of the map corresponded to the 'encounter side' of Pluto that would be seen close up during New Horizons' 14 July flyby. This map provided mission scientists with an important tool to decipher the complex and intriguing pattern of bright and dark markings on Pluto's surface. Features from all sides of Pluto could now be seen at a glance and from a consistent perspective, making it much easier to compare their shapes and sizes.

The elongated dark area informally known as 'the whale', along the equator on the left side of the map, was one of the darkest regions visible to New Horizons. It measures about 3,000 km in length. Directly to the right of the whale's 'snout' was the brightest region visible on the dwarf planet. This was thought to be a region where relatively fresh deposits of ice or frost – perhaps including frozen methane, nitrogen, and/or carbon monoxide – form a bright coating. Continuing to the right, along the equator, it was possible to discern the four intriguing dark spots (Figure 2). Meanwhile, the whale's 'tail', at the left end of the

Figure 2. This New Horizons' LORRI image of Pluto was taken on 3 July 2015 from a range of 12.5 million kilometres, with a central longitude of 19°. It shows the intriguing line of evenly spaced, dark circular patches which lie along the equator of the Charon-facing hemisphere of Pluto.

dark feature, cradled a bright donut-shaped feature about 350 km across. At first glance it resembled circular features seen elsewhere in the Solar System, from impact craters to volcanoes.

On 8 July, mission scientists received a new view of Pluto – the most detailed yet returned by LORRI. The image had been taken on 7 July, when the spacecraft was just under 8 million kilometres from Pluto. The view was centred roughly on the area that would be seen close up during New Horizons' 14 July closest approach (Figure 3). This side of Pluto is dominated by three broad regions of varying brightness. Most prominent were 'the whale', and a large heart-shaped bright area

Figure 3. In the early morning of 8 July, mission scientists received this new view of Pluto – the most detailed then returned by the LORRI camera. The image was taken on 7 July when the NASA spacecraft was just under 8 million kilometres from Pluto, and was the first to be received since the 4 July anomaly that sent the spacecraft into safe mode. The view is centred roughly on the area that would be seen close-up during New Horizons' 14 July closest approach.

measuring almost 2,000 km across to the right. Above those features lay a polar region that was intermediate in brightness.

By 9 July, New Horizons was only 5.4 million kilometres away from Pluto, and at this range, Pluto was beginning to reveal the first signs of discrete geologic features. One image showed the side of Pluto that always faces Charon, and included the so-called 'tail' of the dark whale-shaped feature along its equator. The large, bright feature shaped like a heart had rotated out of view when this image was captured.

New Horizons' last look at Pluto's Charon-facing hemisphere, taken early in the morning of 11 July, revealed intriguing geological details that were of great interest to mission scientists. The image showed newly resolved linear features above the equatorial region that intersect, suggestive of polygonal shapes. The spots appear on the side of Pluto that always faces Charon – the face that would be invisible to New Horizons when the spacecraft made its close flyby on the morning of 14 July. The spots are connected to a dark belt that circles Pluto's equatorial region (Figures 4a and 4b). These large dark areas were estimated to be 480 km across. In comparison with earlier images, it was

Figures 4a and 4b. Two views of Pluto acquired by New Horizons on 11 July (Figure 4a, left) and on 12 July (Figure 4b, right), which show the slow west-to-east rotation of the dwarf planet. For the first time on Pluto, these views reveal linear features that may be cliffs, as well as a circular feature that could be an impact crater. Just rotating into view in Figure 4b (right) is the bright heart-shaped feature that would be seen in far greater detail during New Horizons' closest approach on 14 July.

possible to now see that the dark areas were more complex than they initially appeared, while the boundaries between the dark and bright terrains were irregular and sharply defined.

When New Horizons made its closest approach to Pluto on 14 July, it would focus on the opposite or 'encounter hemisphere' of the dwarf planet. On the morning of 14 July, New Horizons would pass about 12,500 km from the face with the large, bright heart-shaped feature.

CHARON'S CHASMS AND CRATERS

Charon is 1,208 km across, about half the diameter of Pluto – making it the Solar System's largest moon relative to its parent body. Its smaller size and lower surface contrast initially made it harder for New Horizons to capture its surface features from afar, but the new, closer images of Charon's surface showed intriguing fine details. Newly revealed were brighter areas on Charon that members of the mission's Geology, Geophysics, and Imaging (GGI) team suspected might be impact craters. If so, the scientists would put them to good use.

'If we see impact craters on Charon, it will help us see what's hidden beneath the surface,' said Jeff Moore, leader of the GGI team at NASA's Ames Research Center in Moffett Field, California. 'Large craters can excavate material from several kilometres down and reveal the composition of the interior.'

As New Horizons drew nearer, its images revealed Charon to be a world of chasms and craters. The most pronounced chasm, which lies in the southern hemisphere, is longer and far deeper than Earth's Grand Canyon.

'This is the first clear evidence of faulting and surface disruption on Charon,' said William McKinnon, deputy lead scientist with the GGI team, who is based at the Washington University in St Louis, Missouri.

The most prominent crater, which lies near the south pole of Charon in an image taken on 11 July and radioed to Earth the following day, is about 100 km across (Figure 5). The brightness of the rays of material blasted out of the crater suggests it formed relatively recently in geologic terms, during a collision with a small Kuiper Belt object some time in the last billion years.

The darkness of the crater's floor is especially intriguing. One explanation is that the crater has exposed a different type of icy material than the more reflective ices that lie on the surface. Another possibility is that the ice in the crater floor is the same material as its surroundings but has a larger ice-grain size, which reflects less sunlight. In this

Figure 5. Possible chasms (far right and lower right), probable impact craters (centre right and lower centre), and a dark north polar region are revealed in this image of Pluto's largest moon, Charon, taken by New Horizons on 11 July.

scenario, the impactor that gouged the crater melted the ice in the crater floor, which then refroze into larger grains.

A high-contrast array of bright and dark features covers Pluto's surface, while on Charon, only a dark polar region interrupts a generally more uniform light grey terrain. The reddish materials that colour Pluto seem to be absent on Charon. Pluto has a significant atmosphere; Charon does not. On Pluto, exotic ices like frozen nitrogen, methane, and carbon monoxide have been found, while Charon's surface is made of frozen water and ammonia compounds. The interior of Pluto is mostly rock, while Charon contains equal measures of rock and water ice.

PLUTO – 'KING OF THE EDGEWORTH–KUIPER BELT'

Images acquired with LORRI were also used to determine that Pluto is 2,370 km in diameter, somewhat larger than many prior estimates. This result confirms what was already suspected: Pluto is larger than all other known Solar System objects beyond the orbit of Neptune, deservedly earning it the title 'King of the Edgeworth–Kuiper Belt'.

Pluto's newly estimated size meant that its density is slightly lower than previously thought, and the fraction of ice in its interior is slightly higher. Also, the lowest layer of Pluto's atmosphere, called the troposphere, is shallower than previously believed.

Measuring Pluto's size has been a decades-long challenge due to complicating factors from its atmosphere. Charon lacks a substantial atmosphere, and its diameter was easier to determine using ground-based telescopes. New Horizons observations of Charon confirmed previous estimates, at 1,208 km.

LORRI also zoomed in on two of Pluto's smaller moons, Nix and Hydra. Nix had been estimated to be about 35 km across, and Hydra to be roughly 45 km across. These sizes led mission scientists to conclude that their surfaces were quite bright, possibly due to the presence of ice. What about Pluto's two smallest moons, Kerberos and Styx? Fainter than Nix and Hydra, they are harder to measure. Mission scientists should be able to determine their sizes with observations New Horizons will make during the flyby and will transmit to Earth at a later date.

CLOSEST APPROACH

At about 8 p.m. UTC on 13 July – some sixteen hours before closest approach – New Horizons captured a stunning image of one of Pluto's most dazzling and dominant features. The 'heart', estimated to be 1,600 km across at its widest point, rests just above the equator (Figure 6). The angle of view in all of the images displayed mostly the northern hemisphere, because the north pole of Pluto was tilted towards the Sun. The image revealed, for the first time, that some surfaces on Pluto are peppered with impact craters and are therefore relatively ancient, perhaps several billion years old. Other regions, such as the interior of the heart, show no obvious craters and thus are probably younger,

Figure 6. Pluto nearly fills the frame in this image from New Horizons' LORRI camera, taken on 13 July, when the spacecraft was 768,000 km from the surface. This is the last and most detailed image sent to Earth before the spacecraft's closest approach to Pluto the following day. The original high-resolution black and white image was combined with lower-resolution colour information from the RALPH instrument acquired earlier on 13 July. This view is dominated by the large, bright heart-shaped feature informally named 'Tombaugh Regio', which measures approximately 1,600 km across.

indicating that Pluto has experienced a long and complex geological history. Some craters appear partially destroyed, perhaps by erosion. There are also hints that parts of Pluto's crust have been fractured, as indicated by the series of linear features to the left of the heart.

Below the heart are dark terrains along Pluto's equator, including, on the left, the large dark feature known as the 'whale'. Craters pockmark part of the whale's head; areas that appear smooth and featureless may, however, be a result of image compression.

On 14 July 2015, at 11:49 UTC, New Horizons hurtled past Pluto at 13.8 kilometres per second (30,870 miles per hour), and passing just 12,500 kilometres (7,800 miles) above the surface (Figure 7). Thirteen hours later, at 00:52:37 UTC, NASA received the first communication from the probe following the flyby, at the time expected. Engineering data indicated that it had been completely successful and that the probe had operated in all respects as expected.

Figure 7. Artist's impression of the New Horizons spacecraft during its planned close encounter with Pluto and its moon Charon. The craft's miniature cameras, radio science experiment, ultraviolet and infrared spectrometers, and space plasma experiments would characterize the global geology and geomorphology of Pluto and Charon, map their surface compositions and temperatures, and examine Pluto's atmosphere in detail. The spacecraft's most prominent design feature is a nearly 2.1-metre dish antenna, through which it would communicate with Earth from as far as 7.5 billion kilometres away.

MOUNTAINS, ICY PLAINS AND ICE FLOWS

The first close-up images of Pluto released following the flyby were stunning, especially the full disk view (Figure 8) showing the smooth western half of the large, bright heart-shaped region – informally named 'Tombaugh Regio'. A close-up image of an equatorial region near the base of the bright heart-shaped feature showed a mountain range with peaks jutting as high as 3,500 metres (11,000 feet) above the surface of the icy body (Figure 9). These mountains on Pluto – subsequently called Norgay Montes – possibly formed no more than 100 million years ago. This suggests the close-up region, which covers about one per cent of Pluto's surface, may still be geologically active today. Unlike the icy moons of giant planets, Pluto cannot be heated by gravitational interactions with a much larger planetary body. Some other process must be generating the mountainous landscape.

In the centre left of Pluto's heart-shaped feature lies a vast, craterless

Figure 8. Four images from New Horizons' LORRI camera were combined with colour data from the RALPH instrument to create this global view of Pluto, released on 25 July. The images, taken when the spacecraft was 450,000 kilometres from Pluto, show features as small as 2.2 kilometres, twice the resolution of the single-image view taken on 13 July and shown in Figure 6.

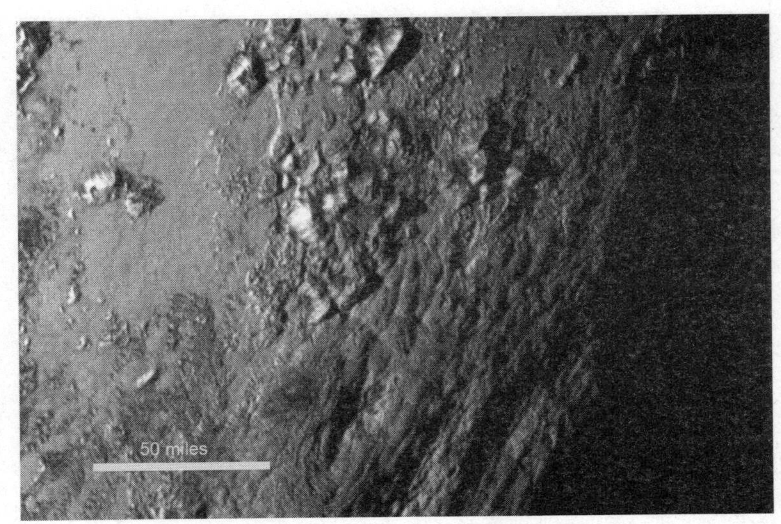

Figure 9. Close-up images of a region near Pluto's equator revealed a giant surprise: a range of mountains rising 3.5 km above the surface. The mountains likely formed no more than 100 million years ago – mere youngsters compared with the 4.56-billion-year age of the Solar System – and may still be in the process of building. That suggests the close-up region may still be geologically active today. Planetary scientists base the youthful age estimate on the lack of craters in this scene. Like the rest of Pluto, this region would presumably have been pummelled by space debris for billions of years and would have once been heavily cratered – unless recent activity had given the region a facelift, erasing those pockmarks. The mountains are probably composed of Pluto's water-ice 'bedrock'. Although methane and nitrogen ice cover much of the surface of Pluto, these materials are not strong enough to build the mountains. Instead, a stiffer material, most likely water ice, created the peaks. This close-up image was taken about 1.5 hours before New Horizons' closest approach to Pluto, when the craft was 77,000 km from the surface.

plain – informally dubbed 'Sputnik Planum' – that appears to be no more than 100 million years old, and is possibly still being shaped by geologic processes. It has a broken surface of irregularly shaped segments, roughly 20 km across, bordered by what appear to be shallow troughs (Figure 10). Some of these troughs have darker material within them, while others are traced by clumps of hills that appear to rise above the surrounding terrain. Elsewhere, the surface appears to be etched by fields of small pits that may have formed by a process called

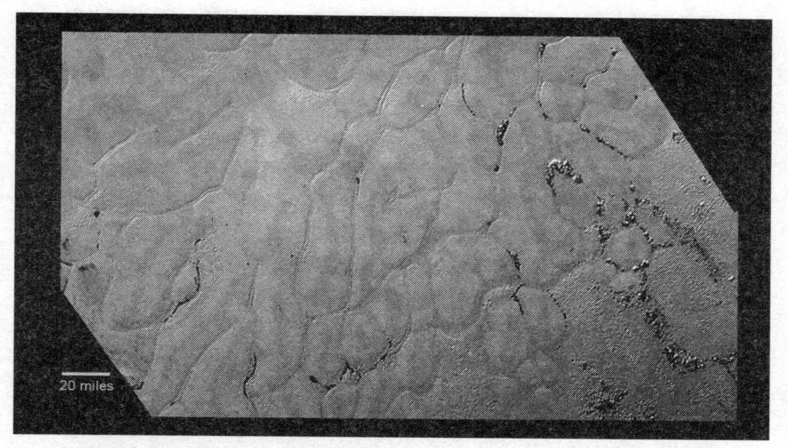

Figure 10. In the centre left of Tombaugh Regio lies a vast, craterless plain, named Sputnik Planum, after Earth's first artificial satellite, that appears to be no more than 100 million years old, and is possibly still being shaped by geologic processes. The surface appears to be divided into irregularly shaped segments that are ringed by narrow troughs. Features that appear to be groups of mounds and fields of small pits are also visible. This image was acquired by LORRI on 14 July from a distance of 77,000 km. Features as small as 1 km across are visible.

sublimation, in which ice turns directly from solid to gas, just as dry ice does on Earth.

Scientists have two working theories as to how these segments were formed. The irregular shapes may be the result of the contraction of surface materials, similar to what happens when mud dries. Alternatively, they may be a product of convection, similar to wax rising in a lava lamp. On Pluto, convection would occur within a surface layer of frozen carbon monoxide, methane, and nitrogen, driven by the scant warmth of Pluto's interior.

Pluto's icy plains also display dark streaks that are a few kilometres long. These streaks appear to be aligned in the same direction and may have been produced by winds blowing across the frozen surface.

Just a week after the flyby, it was announced that New Horizons had discovered a second, apparently less lofty, mountain range on the lower-left edge of Tombaugh Regio. These newly discovered frozen peaks were estimated to be 1,000–1,500 metres (3,300–5,000 feet) high, about the same height as the Appalachian Mountains in the United

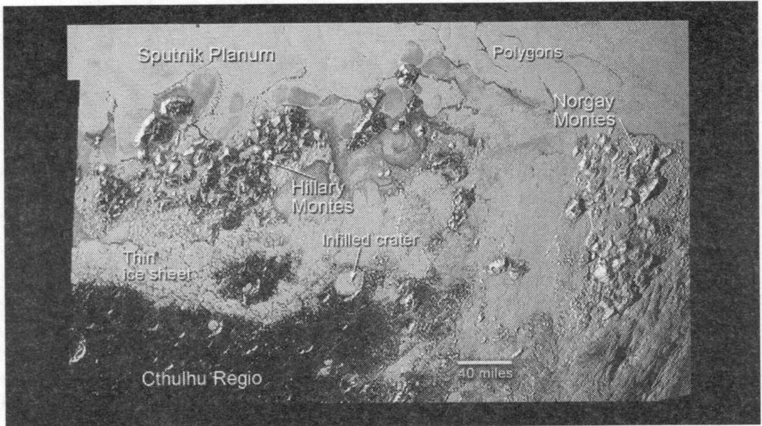

Figures 11a and 11b. A newly discovered mountain range, Hillary Montes, lies near the southwestern margin of Tombaugh Regio, situated between bright, icy plains and dark, heavily cratered terrain as shown in Figure 11a. The close-up image in Figure 11b (the region shown boxed in Figure 11a) was acquired by New Horizons' LORRI camera on 14 July from a distance of 77,000 km and sent back to Earth on 20 July. Features as small as 1 km across are visible. These frozen peaks are estimated to be 1–1.5 km high.

States. The taller Norgay Montes (discovered on 15 July) more closely approximate the height of the Alps in Switzerland. The new range, Hillary Montes, is just west of the Sputnik Planum region within Pluto's heart. These peaks lie some 110 km northwest of Norgay Montes (Figures 11a and 11b) and further illustrate the remarkably well-defined topography along the western edge of Tombaugh Regio.

'There is a pronounced difference in texture between the younger, frozen plains to the east and the dark, heavily cratered terrain to the west,' said GGI leader Jeff Moore, of NASA's Ames Research Center. 'There's a complex interaction going on between the bright and the dark materials that we're still trying to understand.'

Data still stored on New Horizons' recorders at the time of going to press includes much higher-resolution views of Sputnik Planum, along with spectral readings that will reveal what types of ices comprise the ice field. The information will help narrow down the age of Sputnik Planum, which tapers toward a boundary with rugged mountains of water ice in the north.

The best images of Sputnik Planum on the ground show the region with a resolution of about 400 metres, but New Horizons' black-and-white camera took pictures with resolutions as high as 70 metres, good enough to see something as small as the ponds in New York City's Central Park. Additionally, compositional data from New Horizons' RALPH instrument indicate the centre of Sputnik Planum is rich in nitrogen, carbon monoxide, and methane ices. At Pluto's surface temperature of –234 degrees Celsius, these ices can flow like a glacier.

AND MORE ON PLUTO'S MOONS

A new view of Charon released post flyby (Figure 12) revealed a youthful and varied terrain. Scientists were surprised by the apparent lack of craters. A swath of cliffs and troughs stretching about 1,000 km suggests widespread fracturing of Charon's crust, likely the result of internal geological processes. Images revealed a canyon estimated to be 7–9 km deep. In Charon's north polar region, informally named 'Mordor Macula', the dark surface markings have a diffuse boundary, suggesting a thin deposit or stain on the surface.

While Charon grabbed most of the spotlight, two of Pluto's smaller and lesser-known satellites started to come into focus via new images

Figure 12. Remarkable new features of Charon are revealed in this image from the LORRI camera, taken late on 13 July from a distance of 466,000 km. A swath of cliffs and troughs stretches about 1,000 km from left to right, suggesting widespread fracturing of Charon's crust, likely a result of internal processes. At upper right, along the moon's curving edge, is a canyon estimated to be 7–9 km deep. Mission scientists were surprised by the apparent lack of craters on Charon, indicating a relatively young surface that has been reshaped by geologic activity. In Charon's north polar region, a dark marking prominent in New Horizons' approach images is now seen to have a diffuse boundary, suggesting it is a thin deposit of dark material. Underlying it is a distinct, sharply bounded, angular feature.

from New Horizons (Figure 13). Nix and Hydra – the second and third moons to be discovered – are approximately the same size, but their similarity ends there. New Horizons' first colour image of Nix revealed an intriguing region on the jelly-bean-shaped satellite, which is estimated to be 42 km long and 36 km wide. Although the overall surface colour of Nix is neutral grey in the image, the new-found region had a distinct reddish tint. Hints of a bullseye pattern led scientists to speculate that the reddish region is a crater.

The sharpest images of Pluto's satellite Hydra show that it has an irregular shape, and observations indicate that its surface is probably coated with water ice. There appear to be at least two large craters, one

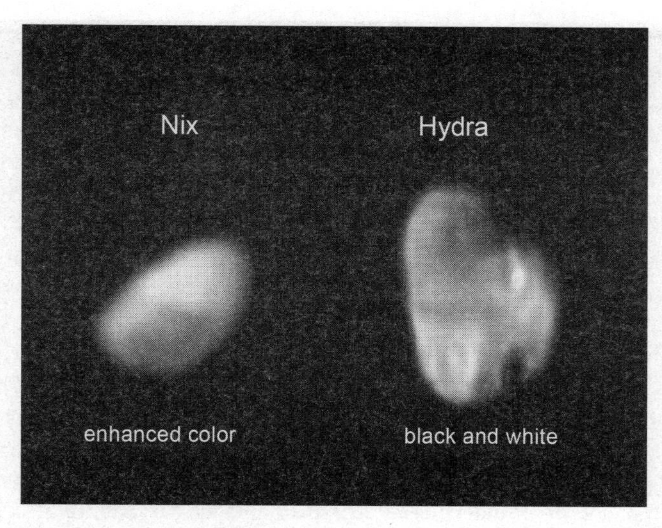

Figure 13. Pluto's moon Nix (left), as imaged by the New Horizons' RALPH instrument, has a reddish spot (the dark grey region here) that has attracted the interest of mission scientists. The data was obtained on the morning of 14 July, and received on the ground on 18 July. At the time the observations were taken New Horizons was about 165,000 km from Nix. The image shows features as small as approximately 3 km across on Nix. Pluto's small, irregularly shaped moon Hydra (right) is revealed in this black and white image taken from New Horizons' LORRI instrument on 14 July, from a distance of about 231,000 km. Features as small as 1.2 kilometres are visible on Hydra.

of which is mostly in shadow. The upper portion looks darker than the rest, suggesting a possible difference in surface composition. From this image, mission scientists have estimated that Hydra is 55 km long and 40 km wide.

Images of Pluto's most recently discovered moons, Styx and Kerberos, were expected to be transmitted to Earth no later than mid-October 2015.

ATMOSPHERE, HAZE LAYERS AND ESCAPING NITROGEN

The New Horizons Atmospheres team observed Pluto's nitrogen-rich atmosphere as far as 1,600 km above the surface, demonstrating that it is quite extended. This is the first observation at altitudes higher than 270 km above the surface. The new information was gathered by New Horizons' ALICE imaging spectrograph during a carefully designed alignment of the Sun, Pluto, and the spacecraft starting about an hour after its closest approach to the planet on 14 July. During this event, known as a solar occultation, New Horizons passed through Pluto's shadow while the Sun backlit Pluto's atmosphere.

The New Horizons Particles and Plasma team has discovered a region of cold, dense ionized gas tens of thousands of miles beyond Pluto – the planet's atmosphere being stripped away by the solar wind and lost to space. Beginning an hour and half after closest approach, the Solar Wind Around Pluto (SWAP) instrument observed a cavity in the solar wind – the outflow of electrically charged particles from the Sun – between 77,000 km and 109,000 km downstream of Pluto. SWAP data revealed this cavity to be populated with nitrogen ions forming a 'plasma tail' of undetermined structure and length extending behind the planet. Similar plasma tails are observed at planets like Venus and Mars. Prior to closest approach, nitrogen ions were detected far upstream of Pluto by the Pluto Energetic Particle Spectrometer Science Investigation (PEPSSI) instrument, providing a foretaste of Pluto's escaping atmosphere.

Just seven hours after closest approach, New Horizons had aimed its LORRI camera back at Pluto, capturing sunlight streaming through the atmosphere and revealing hazes extending as high as 130 km above the surface (Figure 14). A preliminary analysis of the image shows two distinct layers of haze – one about 80 km above the surface and the other at about 50 km. Studying Pluto's atmosphere should provide clues as to what is happening below.

'The hazes detected in this image are a key element in creating the complex hydrocarbon compounds that give Pluto's surface its reddish hue,' said Michael Summers, New Horizons co-investigator at George Mason University in Fairfax, Virginia.

Models suggest the hazes form when ultraviolet sunlight breaks up methane-gas particles – a simple hydrocarbon in Pluto's atmosphere. The breakdown of methane triggers the buildup of more complex hydrocarbon gases, such as ethylene and acetylene, which were also discovered in Pluto's atmosphere by New Horizons. As these hydrocarbons fall to the lower, colder parts of the atmosphere, they condense into ice particles that create the hazes. Ultraviolet sunlight chemically

Figure 14. Backlit by the light of the Sun, Pluto's thin and very tenuous atmosphere rings its silhouette like a luminous halo in this breathtaking image taken by the New Horizons spacecraft on 15 July, some thirty-nine hours after closest approach. This global portrait of the atmosphere was captured when the spacecraft was about 2 million kilometres from Pluto and shows structures as small as 20 km across. The image was delivered to Earth on 23 July.

converts hazes into tholins, the dark hydrocarbons that colour Pluto's surface.

The latest data from New Horizons reveals diverse features on Pluto's surface and an atmosphere dominated by nitrogen gas. Pluto's atmosphere is similar to Earth's in that it is predominantly composed of nitrogen (N). But Pluto's atmosphere is approximately 98 per cent N, while Earth's is only approximately 78 per cent N. Pluto's atmosphere is also considerably thinner than Earth's with approximately 10,000 times lower pressure at the surface.

The nitrogen in Pluto's atmosphere (in the form of N_2 gas) is actually flowing away and escaping the planet at an estimated rate of hundreds of tonnes per hour. We also see what looks like flowing ice on Pluto's surface in high-resolution images made by New Horizons (Figures 15a and 15b). The water ice (H_2O) that we are familiar with on Earth would be completely rigid and stiff at Pluto's surface temperatures, but ice made out of N_2 would be able to flow like a glacier. So where does all of this nitrogen come from?

'More nitrogen has to come from somewhere to resupply both the nitrogen ice that is moving around Pluto's surface in seasonal cycles,

Figures 15a and 15b. Probable nitrogen ice flows, located near the northern periphery of Sputnik Planum (the region shown boxed in Figure 15a) are revealed here in close up (Figure 15b). Although at Pluto's surface temperature of −234 degrees Celsius the water ice that we are familiar with on Earth would be completely rigid and stiff, ice made out of nitrogen would be able to flow like a glacier.

and the nitrogen that is escaping off the top of the atmosphere as the result of heating by ultraviolet light from the Sun,' said Kelsi Singer, a postdoctoral researcher at Southwest Research Institute.

Singer and her mentor Alan Stern, New Horizons principal investigator and SwRI associate vice president, looked at a number of different ways that nitrogen might be resupplied. They wondered if comets could deliver enough nitrogen to Pluto's surface to resupply what is escaping its atmosphere. They also looked at whether craters made by the comets hitting the surface could excavate enough nitrogen – but that would require a very deep layer of nitrogen ice at the surface, which is not proven. The team also studied whether craters could expose more surface area, by punching through surface deposits that would likely be built up over time.

'We found that all of these effects, which are the major ones from cratering, do not seem to supply enough nitrogen to supply the escaping atmosphere over time,' continued Singer. 'While it's possible that the escape rate was not as high in the past as it is now, we think geologic activity is helping out by bringing nitrogen up from Pluto's interior.'

'Our pre-flyby prediction, made when we submitted the paper, is that it's most likely that Pluto is actively resupplying nitrogen from its interior to its surface, possibly meaning the presence of ongoing geysers or cryovolcanism,' said Stern.

AND SO MUCH MORE TO COME . . .

As planned, following an initial release of Pluto and Charon images in July 2015, New Horizons took something of a post-flyby break, using the time to send back lower data-rate information collected by the energetic-particle, solar-wind and space-dust instruments. It was scheduled to resume sending back new flyby images and other data in early September. Then the team was expected to dive back into data analysis with the pace of news and image releases picking up on the team's activity. New Horizons – which is healthy and operating normally – will continue to send data stored in its onboard recorders back to Earth through to the autumn of 2016.

The first images to be received from the New Horizons spacecraft since late July were due back at Earth on 5 September 2015. As this Yearbook went to press, only about 5 per cent of the 50 gigabits of data

New Horizons collected during its nine-day encounter with Pluto had been received. Only seven close-up images of Pluto from New Horizons' telescopic camera had been downlinked to Earth, and those files were compressed to expedite their transmission back home. All of the best images were still on the spacecraft, all the high-resolution mapping, almost all the high-resolution composition mapping, the vast majority of radio occultation data, along with global high-resolution uncompressed images of Charon and observations of the smaller moons.

Since New Horizons is now more than 4,800 million kilometres from Earth, it can only send back data at the agonizingly slow rate of about 2 kilobits per second, a fraction of the speed of dial-up Internet. That is why it will take more than a year for the spacecraft to broadcast all its detailed measurements (in both compressed and uncompressed form), including hundreds of images, to the eager scientists (and the interested general public) waiting on the ground. One can only imagine what amazing discoveries will come to light when that data finally arrives.

Note: The information and images in this article have been gleaned from press releases and various media resources provided by NASA at www.nasa.gov/mission_pages/newhorizons/overview and by the team at the Johns Hopkins University Applied Physics Laboratory at pluto.jhuapl.edu. To all of these people the author extends his grateful thanks.

Part Three

Miscellaneous

Some Interesting Variable Stars

JOHN ISLES

All variable stars are of potential interest, and hundreds of them can be observed with the slightest optical aid – even with a pair of binoculars. The stars in the list that follows include many that are popular with amateur observers, as well as some less well-known objects that are, nevertheless, suitable for study visually. The periods and ranges of many variables are not constant from one cycle to another, and some are completely irregular.

Finder charts are given after the list for those stars marked with an asterisk. These charts are adapted with permission from those issued by the Variable Star Section of the British Astronomical Association. Apart from the eclipsing variables and others in which the light changes are purely a geometrical effect, variable stars can be divided broadly into two classes: the pulsating stars, and the eruptive or cataclysmic variables.

Mira (Omicron Ceti) is the best-known member of the long-period subclass of pulsating red-giant stars. The chart is suitable for use in estimating the magnitude of Mira when it reaches naked-eye brightness – typically from about a month before the predicted date of maximum until two or three months after maximum. Predictions for Mira and other stars of its class follow the section of finder charts.

The semi-regular variables are less predictable, and generally have smaller ranges. V Canum Venaticorum is one of the more reliable ones, with steady oscillations in a six-month cycle. Z Ursae Majoris, easily found with binoculars near Delta, has a large range, and often shows double maxima owing to the presence of multiple periodicities in its light changes. The chart for Z is also suitable for observing another semi-regular star, RY Ursae Majoris. These semi-regular stars are mostly red giants or supergiants.

The RV Tauri stars are of earlier spectral class than the semi-regulars, and in a full cycle of variation they often show deep minima and double maxima that are separated by a secondary minimum. U Monocerotis is one of the brightest RV Tauri stars.

Among eruptive variable stars is the carbon-rich supergiant R Coronae Borealis. Its unpredictable eruptions cause it not to brighten, but to fade. This happens when one of the sooty clouds that the star throws out from time to time happens to come in our direction and blots out most of the star's light from our view. Much of the time R Coronae is bright enough to be seen in binoculars, and the chart can be used to estimate its magnitude. During the deepest minima, however, the star needs a telescope of 25-cm or larger aperture to be detected.

CH Cygni is a symbiotic star – that is, a close binary comprising a red giant and a hot dwarf star that interact physically, giving rise to outbursts. The system also shows semi-regular oscillations, and sudden fades and rises that may be connected with eclipses.

Observers can follow the changes of these variable stars by using the comparison stars whose magnitudes are given below each chart. Observations of variable stars by amateurs are of scientific value, provided they are collected and made available for analysis. This is done by several organizations, including the British Astronomical Association, the American Association of Variable Star Observers (49 Bay State Road, Cambridge, Massachusetts 02138, USA), and the Royal Astronomical Society of New Zealand (PO Box 3181, Wellington, New Zealand).

Star	RA		Declination		Range	Type	Period	Spectrum
	h	m	°	′			(days)	
R Andromedae	00	24.0	+38	35	5.8–14.9	Mira	409	S
W Andromedae	02	17.6	+44	18	6.7–14.6	Mira	396	S
U Antliae	10	35.2	–39	34	5–6	Irregular	—	C
Theta Apodis	14	05.3	–76	48	5–7	Semi-regular	119	M
R Aquarii	23	43.8	–15	17	5.8–12.4	Symbiotic	387	M+Pec
T Aquarii	20	49.9	–05	09	7.2–14.2	Mira	202	M
R Aquilae	19	06.4	+08	14	5.5–12.0	Mira	284	M
V Aquilae	19	04.4	–05	41	6.6–8.4	Semi-regular	353	C
Eta Aquilae	19	52.5	+01	00	3.5–4.4	Cepheid	7.2	F–G
U Arae	17	53.6	–51	41	7.7–14.1	Mira	225	M
R Arietis	02	16.1	+25	03	7.4–13.7	Mira	187	M
U Arietis	03	11.0	+14	48	7.2–15.2	Mira	371	M

Some Interesting Variable Stars

Star	RA h	RA m	Declination °	Declination ′	Range	Type	Period (days)	Spectrum
R Aurigae	05	17.3	+53	35	6.7–13.9	Mira	458	M
Epsilon Aurigae	05	02.0	+43	49	2.9–3.8	Algol	9892	F+B
R Boötis	14	37.2	+26	44	6.2–13.1	Mira	223	M
X Camelopardalis	04	45.7	+75	06	7.4–14.2	Mira	144	K–M
R Cancri	08	16.6	+11	44	6.1–11.8	Mira	362	M
X Cancri	08	55.4	+17	14	5.6–7.5	Semi-regular	195?	C
R Canis Majoris	07	19.5	–16	24	5.7–6.3	Algol	1.1	F
VY Canis Majoris	07	23.0	–25	46	6.5–9.6	Unique	—	M
S Canis Minoris	07	32.7	+08	19	6.6–13.2	Mira	333	M
R Canum Ven.	13	49.0	+39	33	6.5–12.9	Mira	329	M
*V Canum Ven.	13	19.5	+45	32	6.5–8.6	Semi-regular	192	M
R Carinae	09	32.2	–62	47	3.9–10.5	Mira	309	M
S Carinae	10	09.4	–61	33	4.5–9.9	Mira	149	K–M
l Carinae	09	45.2	–62	30	3.3–4.2	Cepheid	35.5	F–K
Eta Carinae	10	45.1	–59	41	-0.8–7.9	Irregular	—	Pec
R Cassiopeiae	23	58.4	+51	24	4.7–13.5	Mira	430	M
S Cassiopeiae	01	19.7	+72	37	7.9–16.1	Mira	612	S
W Cassiopeiae	00	54.9	+58	34	7.8–12.5	Mira	406	C
Gamma Cas.	00	56.7	+60	43	1.6–3.0	Gamma Cas.	—	B
Rho Cassiopeiae	23	54.4	+57	30	4.1–6.2	Semi-regular	—	F–K
R Centauri	14	16.6	–59	55	5.3–11.8	Mira	546	M
S Centauri	12	24.6	–49	26	7–8	Semi-regular	65	C
T Centauri	13	41.8	–33	36	5.5–9.0	Semi-regular	90	K–M
S Cephei	21	35.2	+78	37	7.4–12.9	Mira	487	C
T Cephei	21	09.5	+68	29	5.2–11.3	Mira	388	M
Delta Cephei	22	29.2	+58	25	3.5–4.4	Cepheid	5.4	F–G
Mu Cephei	21	43.5	+58	47	3.4–5.1	Semi-regular	730	M
U Ceti	02	33.7	–13	09	6.8–13.4	Mira	235	M
W Ceti	00	02.1	–14	41	7.1–14.8	Mira	351	S
*Omicron Ceti	02	19.3	–02	59	2.0–10.1	Mira	332	M
R Chamaeleontis	08	21.8	–76	21	7.5–14.2	Mira	335	M
T Columbae	05	19.3	–33	42	6.6–12.7	Mira	226	M
R Comae Ber.	12	04.3	+18	47	7.1–14.6	Mira	363	M
*R Coronae Bor.	15	48.6	+28	09	5.7–14.8	R Coronae Bor.	—	C
S Coronae Bor.	15	21.4	+31	22	5.8–14.1	Mira	360	M
T Coronae Bor.	15	59.6	+25	55	2.0–10.8	Recurrent nova	—	M+Pec
V Coronae Bor.	15	49.5	+39	34	6.9–12.6	Mira	358	C
W Coronae Bor.	16	15.4	+37	48	7.8–14.3	Mira	238	M
R Corvi	12	19.6	–19	15	6.7–14.4	Mira	317	M
R Crucis	12	23.6	–61	38	6.4–7.2	Cepheid	5.8	F–G

Star	RA		Declination		Range	Type	Period	Spectrum
	h	m	°	′			(days)	
R Cygni	19	36.8	+50	12	6.1–14.4	Mira	426	S
U Cygni	20	19.6	+47	54	5.9–12.1	Mira	463	C
W Cygni	21	36.0	+45	22	5.0–7.6	Semi-regular	131	M
RT Cygni	19	43.6	+48	47	6.0–13.1	Mira	190	M
SS Cygni	21	42.7	+43	35	7.7–12.4	Dwarf nova	50±	K+Pec
*CH Cygni	19	24.5	+50	14	5.6–9.0	Symbiotic	—	M+B
Chi Cygni	19	50.6	+32	55	3.3–14.2	Mira	408	S
R Delphini	20	14.9	+09	05	7.6–13.8	Mira	285	M
U Delphini	20	45.5	+18	05	5.6–7.5	Semi-regular	110?	M
EU Delphini	20	37.9	+18	16	5.8–6.9	Semi-regular	60	M
Beta Doradûs	05	33.6	–62	29	3.5–4.1	Cepheid	9.8	F–G
R Draconis	16	32.7	+66	45	6.7–13.2	Mira	246	M
T Eridani	03	55.2	–24	02	7.2–13.2	Mira	252	M
R Fornacis	02	29.3	–26	06	7.5–13.0	Mira	389	C
R Geminorum	07	07.4	+22	42	6.0–14.0	Mira	370	S
U Geminorum	07	55.1	+22	00	8.2–14.9	Dwarf nova	105±	Pec+M
Zeta Geminorum	07	04.1	+20	34	3.6–4.2	Cepheid	10.2	F–G
Eta Geminorum	06	14.9	+22	30	3.2–3.9	Semi-regular	233	M
S Gruis	22	26.1	–48	26	6.0–15.0	Mira	402	M
S Herculis	16	51.9	+14	56	6.4–13.8	Mira	307	M
U Herculis	16	25.8	+18	54	6.4–13.4	Mira	406	M
Alpha Herculis	17	14.6	+14	23	2.7–4.0	Semi-regular	—	M
68, u Herculis	17	17.3	+33	06	4.7–5.4	Algol	2.1	B+B
R Horologii	02	53.9	–49	53	4.7–14.3	Mira	408	M
U Horologii	03	52.8	–45	50	6–14	Mira	348	M
R Hydrae	13	29.7	–23	17	3.5–10.9	Mira	389	M
U Hydrae	10	37.6	–13	23	4.3–6.5	Semi-regular	450?	C
VW Hydri	04	09.1	–71	18	8.4–14.4	Dwarf nova	27±	Pec
R Leonis	09	47.6	+11	26	4.4–11.3	Mira	310	M
R Leonis Minoris	09	45.6	+34	31	6.3–13.2	Mira	372	M
R Leporis	04	59.6	–14	48	5.5–11.7	Mira	427	C
Y Librae	15	11.7	–06	01	7.6–14.7	Mira	276	M
RS Librae	15	24.3	–22	55	7.0–13.0	Mira	218	M
Delta Librae	15	01.0	–08	31	4.9–5.9	Algol	2.3	A
R Lyncis	07	01.3	+55	20	7.2–14.3	Mira	379	S
R Lyrae	18	55.3	+43	57	3.9–5.0	Semi-regular	46?	M
RR Lyrae	19	25.5	+42	47	7.1–8.1	RR Lyrae	0.6	A–F
Beta Lyrae	18	50.1	+33	22	3.3–4.4	Eclipsing	12.9	B
U Microscopii	20	29.2	–40	25	7.0–14.4	Mira	334	M
*U Monocerotis	07	30.8	–09	47	5.9–7.8	RV Tauri	91	F–K

Some Interesting Variable Stars

Star	RA		Declination		Range	Type	Period	Spectrum
	h	m	°	′			(days)	
V Monocerotis	06	22.7	–02	12	6.0–13.9	Mira	340	M
R Normae	15	36.0	–49	30	6.5–13.9	Mira	508	M
T Normae	15	44.1	–54	59	6.2–13.6	Mira	241	M
R Octantis	05	26.1	–86	23	6.3–13.2	Mira	405	M
S Octantis	18	08.7	–86	48	7.2–14.0	Mira	259	M
V Ophiuchi	16	26.7	–12	26	7.3–11.6	Mira	297	C
X Ophiuchi	18	38.3	+08	50	5.9–9.2	Mira	329	M
RS Ophiuchi	17	50.2	–06	43	4.3–12.5	Recurrent nova	—	OB+M
U Orionis	05	55.8	+20	10	4.8–13.0	Mira	368	M
W Orionis	05	05.4	+01	11	5.9–7.7	Semi-regular	212	C
Alpha Orionis	05	55.2	+07	24	0.0–1.3	Semi-regular	2335	M
S Pavonis	19	55.2	–59	12	6.6–10.4	Semi-regular	381	M
Kappa Pavonis	18	56.9	–67	14	3.9–4.8	W Virginis	9.1	G
R Pegasi	23	06.8	+10	33	6.9–13.8	Mira	378	M
X Persei	03	55.4	+31	03	6.0–7.0	Gamma Cas.	—	O9.5
Beta Persei	03	08.2	+40	57	2.1–3.4	Algol	2.9	B
Zeta Phoenicis	01	08.4	–55	15	3.9–4.4	Algol	1.7	B+B
R Pictoris	04	46.2	–49	15	6.4–10.1	Semi-regular	171	M
RS Puppis	08	13.1	–34	35	6.5–7.7	Cepheid	41.4	F–G
L^2 Puppis	07	13.5	–44	39	2.6–6.2	Semi-regular	141	M
T Pyxidis	09	04.7	–32	23	6.5–15.3	Recurrent nova	7000±	Pec
U Sagittae	19	18.8	+19	37	6.5–9.3	Algol	3.4	B+G
WZ Sagittae	20	07.6	+17	42	7.0–15.5	Dwarf nova	1900±	A
R Sagittarii	19	16.7	–19	18	6.7–12.8	Mira	270	M
RR Sagittarii	19	55.9	–29	11	5.4–14.0	Mira	336	M
RT Sagittarii	20	17.7	–39	07	6.0–14.1	Mira	306	M
RU Sagittarii	19	58.7	–41	51	6.0–13.8	Mira	240	M
RY Sagittarii	19	16.5	–33	31	5.8–14.0	R Coronae Bor.	—	G
RR Scorpii	16	56.6	–30	35	5.0–12.4	Mira	281	M
RS Scorpii	16	55.6	–45	06	6.2–13.0	Mira	320	M
RT Scorpii	17	03.5	–36	55	7.0–15.2	Mira	449	S
Delta Scorpii	16	00.3	–22	37	1.6–2.3	Irregular	—	B
S Sculptoris	00	15.4	–32	03	5.5–13.6	Mira	363	M
R Scuti	18	47.5	–05	42	4.2–8.6	RV Tauri	146	G–K
R Serpentis	15	50.7	+15	08	5.2–14.4	Mira	356	M
S Serpentis	15	21.7	+14	19	7.0–14.1	Mira	372	M
T Tauri	04	22.0	+19	32	9.3–13.5	T Tauri	—	F–K
SU Tauri	05	49.1	+19	04	9.1–16.9	R Coronae Bor.	—	G
Lambda Tauri	04	00.7	+12	29	3.4–3.9	Algol	4.0	B+A
R Trianguli	02	37.0	+34	16	5.4–12.6	Mira	267	M

Star	RA		Declination		Range	Type	Period	Spectrum
	h	m	°	′			(days)	
R Ursae Majoris	10	44.6	+68	47	6.5–13.7	Mira	302	M
T Ursae Majoris	12	36.4	+59	29	6.6–13.5	Mira	257	M
*Z Ursae Majoris	11	56.5	+57	52	6.2–9.4	Semi-regular	196	M
*RY Ursae Majoris	12	20.5	+61	19	6.7–8.3	Semi-regular	310?	M
U Ursae Minoris	14	17.3	+66	48	7.1–13.0	Mira	331	M
R Virginis	12	38.5	+06	59	6.1–12.1	Mira	146	M
S Virginis	13	33.0	–07	12	6.3–13.2	Mira	375	M
SS Virginis	12	25.3	+00	48	6.0–9.6	Semi-regular	364	C
R Vulpeculae	21	04.4	+23	49	7.0–14.3	Mira	137	M
Z Vulpeculae	19	21.7	+25	34	7.3–8.9	Algol	2.5	B+A

V CANUM VENATICORUM 13h 19.5m +45° 32′ (2000)

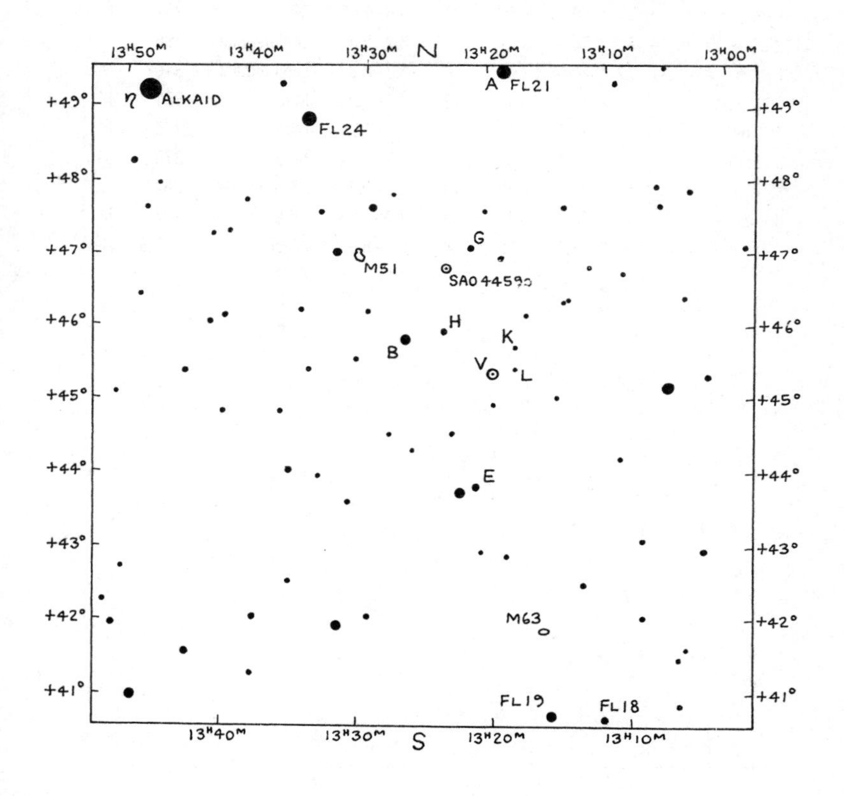

A 5.1	H 7.8
B 5.9	K 8.4
E 6.5	L 8.6
G 7.1	

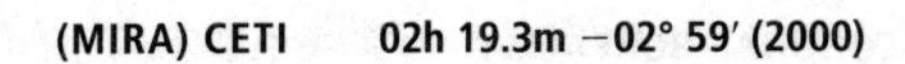

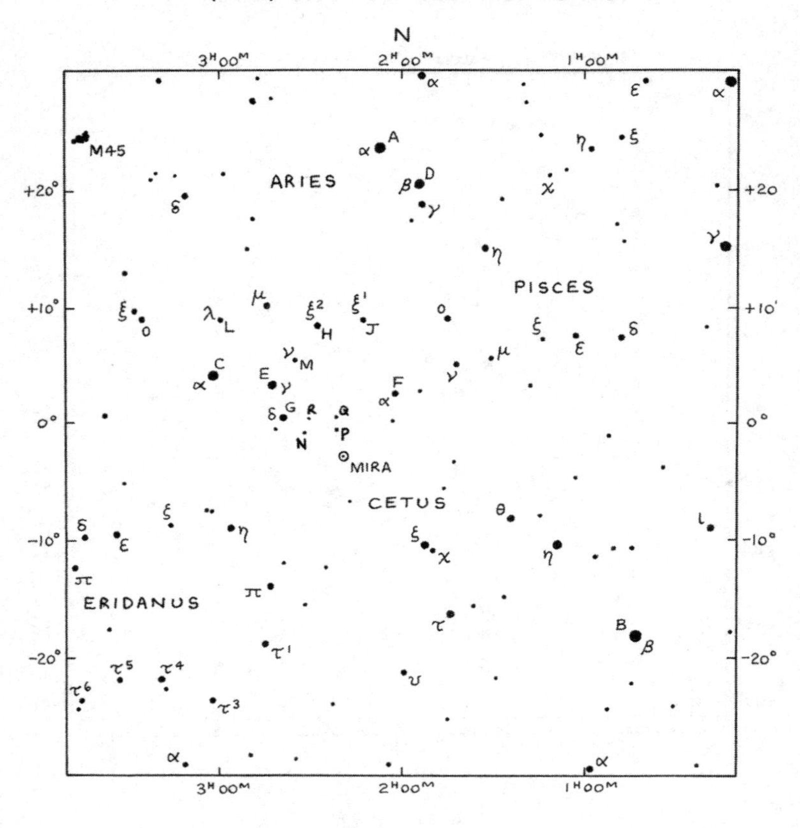

A 2.2	J 4.4
B 2.4	L 4.9
C 2.7	M 5.1
D 3.0	N 5.4
E 3.6	P 5.5
F 3.8	Q 5.7
G 4.1	R 6.1
H 4.3	

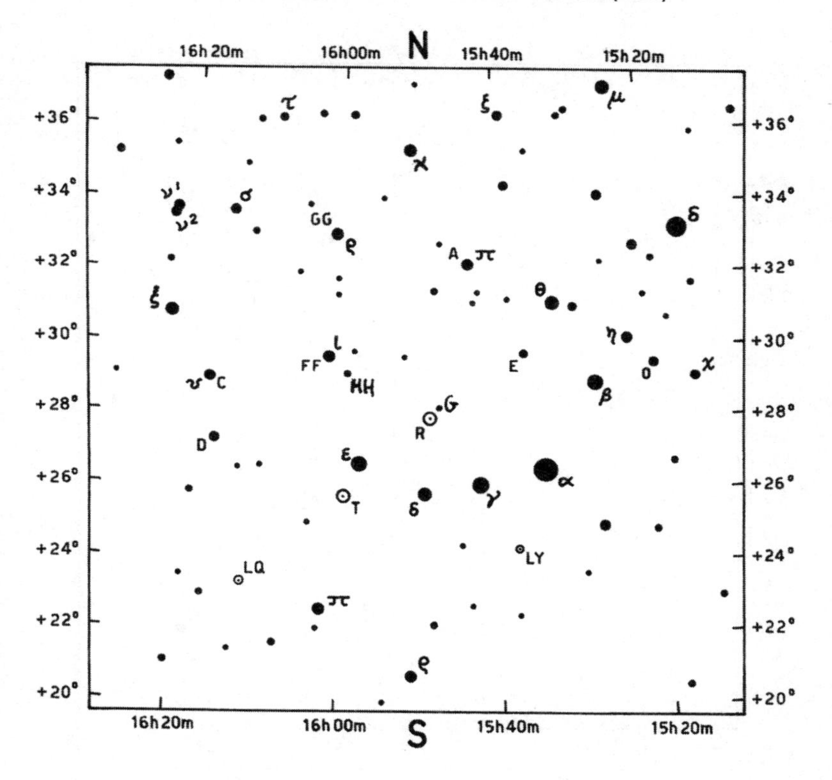

FF	5.0	C	5.8
GG	5.4	D	6.2
A	5.6	E	6.5
		HH	7.1
		G	7.4

CH CYGNI 19h 24.5m +50° 14′ (2000)

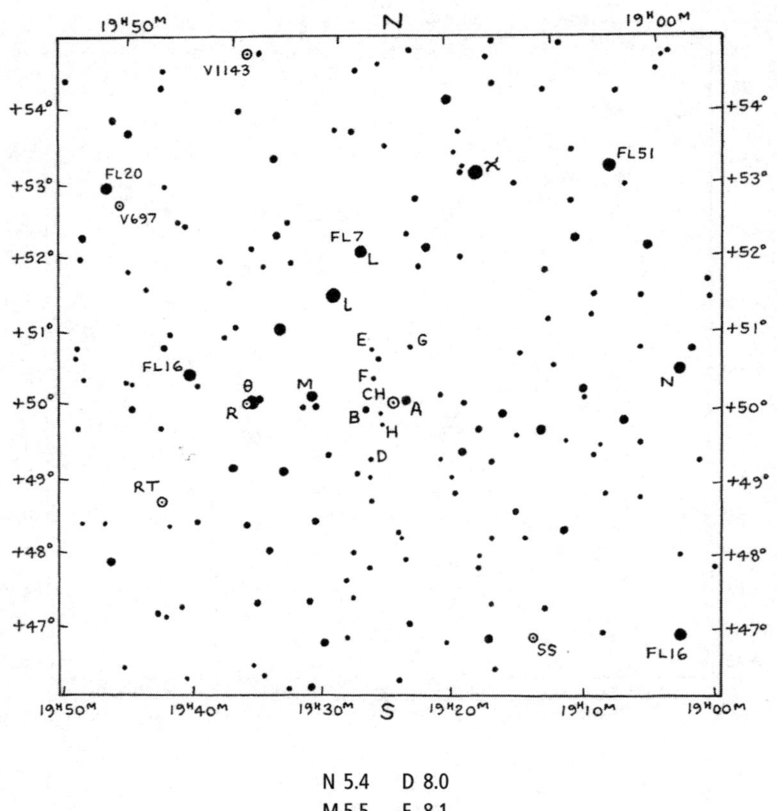

N 5.4	D 8.0
M 5.5	E 8.1
L 5.8	F 8.5
A 6.5	G 8.5
B 7.4	H 9.2

U MONOCEROTIS 07h 30.8m −09° 47′ (2000)

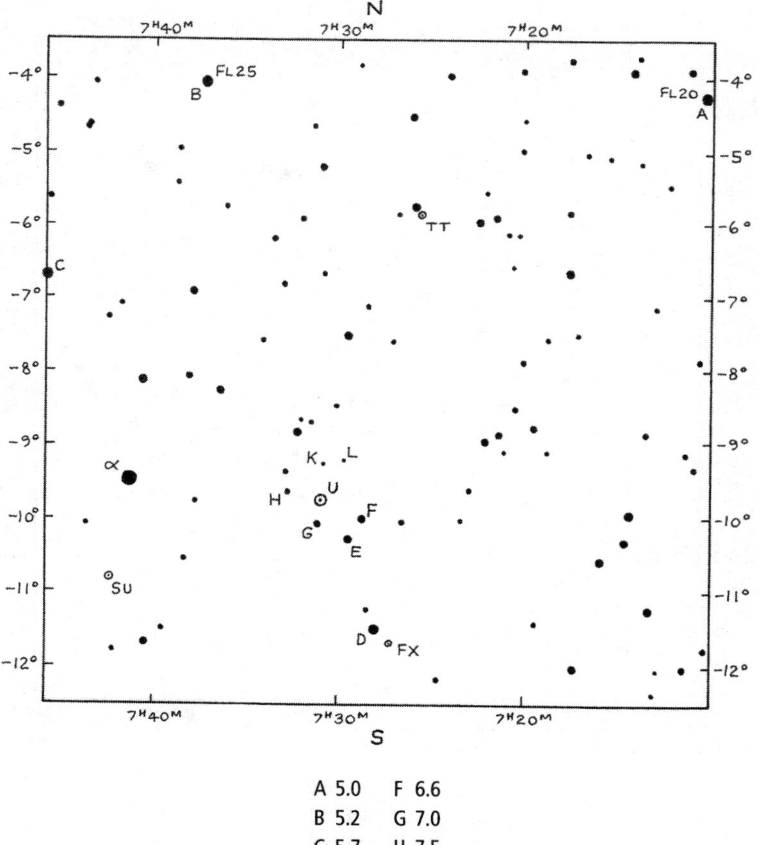

A	5.0	F	6.6
B	5.2	G	7.0
C	5.7	H	7.5
D	5.9	K	7.8
E	6.0	L	8.0

RY URSAE MAJORIS 12h 20.5m +61° 19′ (2000)
Z URSAE MAJORIS 11h 56.5m +57° 52′ (2000)

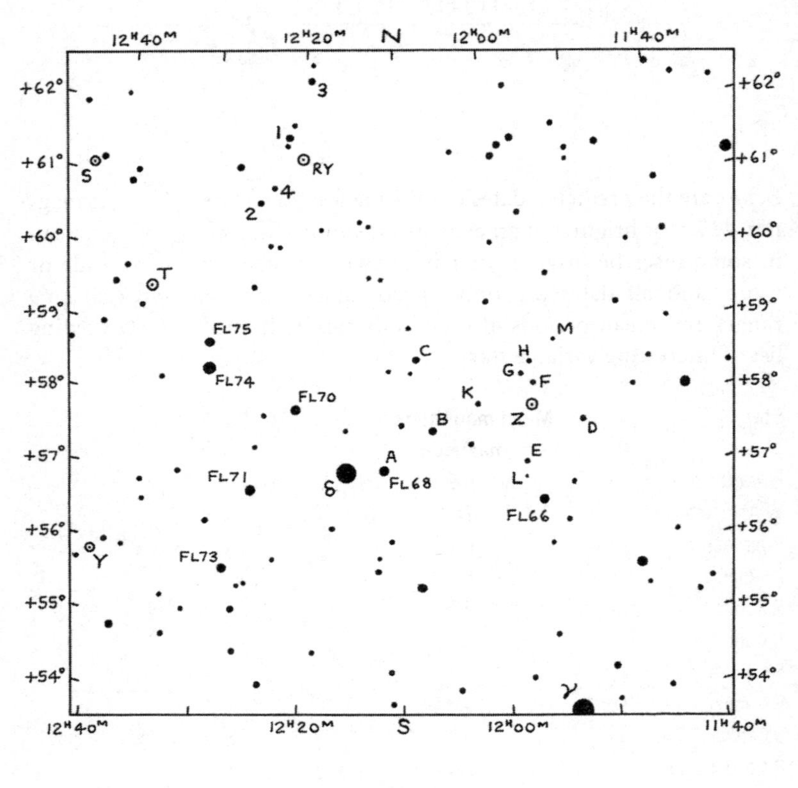

A 6.5	F 8.6	M 9.1
B 7.2	G 8.7	1 6.9
C 7.6	H 8.8	2 7.4
D 8.0	K 8.9	3 7.7
E 8.3	L 9.0	4 7.8

Mira Stars: Maxima, 2016

JOHN ISLES

Below are the predicted dates of maxima for Mira stars that reach magnitude 7.5 or brighter at an average maximum. Individual maxima can in some cases be brighter or fainter than average by a magnitude or more, and all dates are only approximate. The positions, extreme ranges and mean periods of these stars can be found in the preceding list of interesting variable stars.

Star	Mean magnitude at maximum	Dates of maxima
R Andromedae	6.9	29 Apr
W Andromedae	7.4	19 Jan
R Aquarii	6.5	29 May
R Aquilae	6.1	8 June
R Boötis	7.2	23 Apr, 2 Dec
R Cancri	6.8	24 July
S Canis Minoris	7.5	26 May
R Carinae	4.6	7 Oct
S Carinae	5.7	25 Jan, 22 June, 18 Nov
R Cassiopeiae	7.0	16 Feb
R Centauri	5.8	25 Sept
T Cephei	6.0	18 May
U Ceti	7.5	15 Jan, 5 Sept
Omicron Ceti	3.4	1 Apr
T Columbae	7.5	5 Aug
S Coronae Borealis	7.3	28 Aug
V Coronae Borealis	7.5	1 Oct
R Corvi	7.5	18 Jan, 30 Nov
R Cygni	7.5	20 Jan

Star	Mean magnitude at maximum	Dates of maxima
U Cygni	7.2	29 July
RT Cygni	7.3	7 May
Chi Cygni	5.2	8 Oct
R Geminorum	7.1	15 Feb
U Herculis	7.5	6 June
R Horologii	6.0	2 Oct
R Hydrae	4.5	26 Dec
R Leonis	5.8	1 July
R Leonis Minoris	7.1	5 Apr
R Leporis	6.8	3 Mar
RS Librae	7.5	25 Jan, 29 Aug
V Monocerotis	7.0	24 Aug
T Normae	7.4	12 Apr, 9 Dec
V Ophiuchi	7.5	29 June
X Ophiuchi	6.8	20 May
U Orionis	6.3	31 Mar
R Sagittarii	7.3	12 July
RR Sagittarii	6.8	13 Apr
RT Sagittarii	7.0	21 Mar
RU Sagittarii	7.2	7 June
RR Scorpii	5.9	14 Feb, 21 Nov
RS Scorpii	7.0	28 Feb
S Sculptoris	6.7	24 Dec
R Serpentis	6.9	8 July
R Trianguli	6.2	12 Mar
R Ursae Majoris	7.5	27 Aug
R Virginis	6.9	14 May, 7 Oct
S Virginis	7.0	26 Mar

Some Interesting Double Stars

BOB ARGYLE

The positions, angles and separations given below correspond to epoch 2016.0.

No.	RA	Declination	Star	Magnitudes	Separation	PA	Catalogue	Comments
	h m	° ′			arcsec	°		
1	00 31.5	−62 58	β Tuc	4.4,4.8	27.1	169	LCL 119	Both stars again difficult doubles
2	00 49.1	+57 49	η Cas	3.4,7.5	13.4	324	Σ60	Easy. Creamy, bluish. P = 480 years.
3	00 55.0	+23 38	36 And	6.0,6.4	1.1	330	Σ73	P = 168 years. Both yellow. Slowly opening.
4	01 13.7	+07 35	ζ Psc	5.6,6.5	23.1	63	Σ100	Yellow, reddish-white.
5	01 39.8	−56 12	p Eri	5.8,5.8	11.7	187	Δ5	Period = 484 years.
6	01 53.5	+19 18	γ Ari	4.8,4.8	7.5	1	Σ180	Very easy. Both white.
7	02 02.0	+02 46	α Psc	4.2,5.1	1.8	260	Σ202	Binary, period = 933 years.
8	02 03.9	+42 20	γ And	2.3,5.0	9.6	63	Σ205	Yellow, blue. Relatively fixed. BC now beyond
			γ2 And	5.1,6.3	0.0	162	OΣ38	range of amateur instruments (0″.02!)
9	02 29.1	+67 24	ι Cas AB	4.9,6.9	2.6	228	Σ262	AB is long-period binary. P = 620 years.
			ι Cas AC	4.9,8.4	7.1	117		
10	02 33.8	−28 14	ω For	5.0,7.7	10.8	245	HJ 3506	Common proper motion.
11	02 43.3	+03 14	γ Cet	3.5,7.3	2.3	298	Σ299	Not too easy.
12	02 58.3	−40 18	θ Eri	3.4,4.5	8.3	90	PZ 2	Both white.

No.	RA	Declin-ation	Star	Magni-tudes	Separa-tion	PA	Cata-logue	Comments
	h m	° ′			arcsec	°		
13	02 59.2	+21 20	ε Ari	5.2,5.5	1.3	210	Σ333	Closing very slowly. PZ=1216 years? Both white.
14	03 12.1	–28 59	α For	4.0,7.0	5.3	300	HJ 3555	P = 269 years. B variable?
15	03 48.6	–37 37	f Eri	4.8,5.3	8.2	215	Δ16	Pale yellow. Fixed.
16	03 54.3	–02 57	32 Eri	4.8,6.1	6.9	348	Σ470	Fixed. Deep yellow and white.
17	04 32.0	+53 55	1 Cam	5.7,6.8	10.3	308	Σ550	Fixed.
18	04 50.9	–53 28	ι Pic	5.6,6.4	12.4	58	Δ18	Good object for small apertures. Fixed.
19	05 13.2	–12 56	κ Lep	4.5,7.4	2.0	357	Σ661	Visible in 7.5 cm. Slowly closing.
20	05 14.5	–08 12	β Ori	0.1,6.8	9.5	204	Σ668	Companion once thought to be close double.
21	05 21.8	–24 46	41 Lep	5.4,6.6	3.4	93	HJ 3752	Deep yellow pair in a rich field.
22	05 24.5	–02 24	η Ori	3.8,4.8	1.8	77	DA 5	Binary nature doubtful. Slowly getting wider.
23	05 35.1	+09 56	λ Ori	3.6,5.5	4.3	44	Σ738	Fixed. Both stars white.
24	05 35.3	–05 23	θ Ori AB	6.7,7.9	8.6	32	Σ748	Trapezium in M42.
			θ Ori CD	5.1,6.7	13.4	61		
25	05 40.7	–01 57	ζ Ori	1.9,4.0	2.2	167	Σ774	Can be split in 7.5 cm. Long-period binary.
26	06 14.9	+22 30	η Gem	var,6.5	1.6	252	β1008	Well seen with 20 cm. Primary orange.
27	06 46.2	+59 27	12 Lyn AB	5.4,6.0,	1.9	66	Σ948	AB is binary, P = 908 years.
			12 Lyn AC	5.4,7.3	8.8	308		
28	07 08.7	–70 30	γ Vol	3.9,5.8	14.1	298	Δ42	Very slow binary.
29	07 16.6	–23 19	HJ3945 CMa	4.8,6.8	26.8	51	–	Contrasting colours. Yellow and blue.

Some Interesting Double Stars

No.	RA	Declin-ation	Star	Magni-tudes	Separa-tion	PA	Cata-logue	Comments
	h m	° ′			arcsec	°		
30	07 20.1	+21 59	δ Gem	3.5,8.2	5.5	229	Σ1066	Not too easy. Yellow, pale blue.
31	07 34.6	+31 53	α Gem	1.9,2.9	5.1	55	Σ1110	Binary. 460 years. Widening. Easy with 7.5 cm.
32	07 38.8	–26 48	κ Pup	4.5,4.7	9.8	318	H III 27	Both white.
33	08 09.5	–47 20	γ Vel	1.8,4.1	40.3	229	Δ 65	Spectacular pair in lovely field.
34	08 12.2	+17 39	ζ Cnc AB	5.6,6.0	1.2	17	Σ1196	Period (AB)= 59.6 years. Near maximum separation.
			ζ Cnc AB–C	5.0,6.2	5.9	66	Σ1196	Period (AB-C) = 1115 years.
35	08 46.8	+06 25	ε Hyd	3.3,6.8	2.8	308	Σ1273	PA slowly increasing. A is a very close pair.
36	09 18.8	+36 48	38 Lyn	3.9,6.6	2.6	226	Σ1334	Almost fixed.
37	09 47.1	–65 04	υ Car	3.1,6.1	5.0	129	RMK 11	Fixed. Fine in small telescopes.
38	10 20.0	+19 50	γ Leo	2.2,3.5	4.6	126	Σ1424	Binary, period = 510 years. Both orange.
39	10 32.0	–45 04	s Vel	6.2,6.5	13.5	218	PZ 3	Fixed. Both white.
40	10 46.8	–49 25	μ Vel	2.7,6.4	2.4	57	R 155	P = 149 years. Near widest separation.
41	10 55.6	+24 45	54 Leo	4.5,6.3	6.6	111	Σ1487	Slowly widening. Pale yellow and white.
42	11 18.2	+31 32	ξ UMa	4.3,4.8	1.8	177	Σ1523	Binary, 59.9 years. Needs 7.5 cm.
43	11 23.9	+10 32	ι Leo	4.0,6.7	2.1	95	Σ1536	Binary, period = 186 years.
44	11 32.3	–29 16	N Hya	5.8,5.9	9.4	210	H III 96	Both yellow. Long-period binary.
45	12 14.0	–45 43	D Cen	5.6,6.8	2.8	243	RMK 14	Orange and white. Closing.
46	12 26.6	–63 06	α Cru	1.4,1.9	4.0	114	Δ 252	Glorious pair. Third star in a low power field.

No.	RA	Declin-ation	Star	Magni-tudes	Separa-tion	PA	Cata-logue	Comments
	h m	° ′			arcsec	°		
47	12 41.5	–48 58	γ Cen	2.9,2.9	0.1	109	HJ 4539	Period = 83.1 years. Just past periastron. Both yellow.
48	12 41.7	–01 27	γ Vir	3.5,3.5	2.4	4	Σ1670	Now widening quickly. Beautiful pair for 10 cm.
49	12 46.3	–68 06	β Mus	3.7,4.0	1.0	53	R 207	Both white. Closing slowly. Orbit needs revision.
50	12 54.6	–57 11	μ Cru	4.3,5.3	34.9	17	Δ126	Fixed. Both white.
51	12 56.0	+38 19	α CVn	2.9,5.5	19.3	229	Σ1692	Easy. Yellow, bluish.
52	13 22.6	–60 59	J Cen	4.6,6.5	60.0	343	Δ133	Fixed. A is a close pair.
53	13 24.0	+54 56	ζ UMa	2.3,4.0	14.4	152	Σ1744	Very easy. Naked-eye pair with Alcor.
54	13 51.8	–33 00	3 Cen	4.5,6.0	7.7	102	H III 101	Both white. Closing slowly.
55	14 39.6	–60 50	α Cen	0.0,1.2	4.0	302	RHD 1	Finest pair in the sky. P = 80 years. Closing.
56	14 41.1	+13 44	ζ Boo	4.5,4.6	0.4	289	Σ1865	Both white. Closing. Needs at least 30 cm.
57	14 45.0	+27 04	ε Boo	2.5,4.9	2.9	344	Σ1877	Yellow, blue. Fine pair.
58	14 46.0	–25 27	54 Hya	5.1,7.1	8.3	122	H III 97	Closing slowly. Yellow and reddish.
59	14 49.3	–14 09	μ Lib	5.8,6.7	1.8	6	β106	Becoming wider. Fine in 7.5 cm.
60	14 51.4	+19 06	ξ Boo	4.7,7.0	5.6	302	Σ1888	Fine contrast. Easy. P = 151.6 years.
61	15 03.8	+47 39	44 Boo	5.3,6.2	0.8	70	Σ1909	Period = 210 years. Becoming considerably more difficult..
62	15 05.1	–47 03	π Lup	4.6,4.7	1.6	64	HJ 4728	Widening. Both pale yellow.

Some Interesting Double Stars

No.	RA	Declination	Star	Magnitudes	Separation	PA	Catalogue	Comments
	h m	° ′			arcsec	°		
63	15 18.5	−47 53	μ Lup AB	5.1,5.2	0.8	117	HJ 4753	AB closing. Long-period binary, 772 years.
			μ Lup AC	4.4,7.2	22.7	127	Δ180	AC almost fixed.
64	15 23.4	−59 19	γ Cir	5.1,5.5	0.8	356	HJ 4757	Closing. Needs 20 cm. Long-period binary.
65	15 34.8	+10 33	δ Ser	4.2,5.2	4.0	172	Σ1954	Long-period binary.
66	15 35.1	−41 10	γ Lup	3.5,3.6	0.8	276	HJ 4786	Binary. Period = 190 years. Needs 20 cm.
67	15 56.9	−33 58	ξ Lup	5.3,5.8	10.2	49	PZ 4	Fixed. Both pale yellow?
68	16 14.7	+33 52	σ CrB	5.6,6.6	7.2	238	Σ2032	Long period binary. Both white.
69	16 29.4	−26 26	α Sco	1.2,5.4	2.6	277	GNT 1	Red, green. Difficult from mid-northern latitudes.
70	16 30.9	+01 59	λ Oph	4.2,5.2	1.4	42	Σ2055	P = 129 years. Fairly difficult in small apertures.
71	16 41.3	+31 36	ζ Her	2.9,5.5	1.2	131	Σ2084	Period 34.5 years. Now widening. Needs 20 cm.
72	17 05.3	+54 28	μ Dra	5.7,5.7	2.5	2	Σ2130	Period 812 years. Both stars white.
73	17 14.6	+14 24	α Her	var,5.4	4.6	103	Σ2140	Red, green. Long-period binary.
74	17 15.3	−26 35	36 Oph	5.1,5.1	5.0	141	SHJ 243	Period = 471 years.
75	17 23.7	+37 08	ρ Her	4.6,5.6	4.1	319	Σ2161	Slowly widening.
76	17 26.9	−45 51	HJ 4949 AB	5.6,6.5	2.1	251	HJ 4949	Beautiful coarse triple. All white.
			Δ216 AC	,7.1	105.0	310		
77	18 01.5	+21 36	95 Her	5.0,5.1	6.5	257	Σ2264	Colours thought variable in C19.
78	18 05.5	+02 30	70 Oph	4.2,6.0	6.4	125	Σ2272	Opening. Easy in 7.5 cm. P = 88.4 years.

No.	RA	Declin-ation	Star	Magni-tudes	Separa-tion	PA	Cata-logue	Comments
	h m	° ′			arcsec	°		
79	18 06.8	–43 25	h5014 CrA	5.7,5.7	1.7	0	–	Period = 450 years. Needs 10 cm.
80	18 25.4	–20 33	21 Sgr	5.0,7.4	1.7	279	JC 6	Slowly closing binary, orange and green.
81	18 35.9	+16 58	OΣ358 Her	6.8,7.0	1.5	146	—	Period = 380 years.
82	18 44.3	+39 40	ε1 Lyr	5.0,6.1	2.3	346	Σ2382	Quadruple system with epsilon2. Both pairs
83	18 44.3	+39 40	ε2 Lyr	5.2,5.5	2.4	76	Σ2383	visible in 7.5 cm.
84	18 56.2	+04 12	θ Ser	4.5,5.4	22.4	104	Σ2417	Fixed. Very easy. Both stars white.
85	19 06.4	–37 04	γ CrA	4.8,5.1	1.4	345	HJ 5084	Beautiful pair. Period = 122 years.
86	19 30.7	+27 58	β Cyg AB	3.1,5.1	34.3	54	STFA 43	Glorious. Yellow, blue-greenish.
			β Cyg Aa	3.1,5.2	0.4	84	MCA 55	Aa. Needs 40 cm. Period = 214 years.
87	19 45.0	+45 08	δ Cyg	2.9,6.3	2.7	217	Σ2579	Slowly widening. Period = 780 years.
88	19 48.2	+70 16	ε Dra	3.8,7.4	3.0	19	Σ2603	Slow binary. Yellow and blue.
89	19 54.6	–08 14	57 Aql	5.7,6.4	36.0	170	Σ2594	Easy pair. Contrasting colours.
90	20 46.7	+16 07	γ Del	4.5,5.5	9.0	265	Σ2727	Easy. Yellowish. Long-period binary.
91	20 59.1	+04 18	ε Equ AB	6.0,6.3	0.2	281	Σ2737	Fine triple. AB a test for 40 cm. P = 101.5 years.
			ε Equ AC	6.0,7.1	10.3	66		
92	21 06.9	+38 45	61 Cyg	5.2,6.0	31.6	152	Σ2758	Nearby binary. Both orange. Period = 678 years.
93	21 19.9	–53 27	θ Ind	4.5,7.0	7.0	271	HJ 5258	Pale yellow and reddish. Long-period binary.
94	21 44.1	+28 45	μ Cyg	4.8,6.1	1.5	322	Σ2822	Period = 789 years.

Some Interesting Double Stars

No.	RA	Declination	Star	Magnitudes	Separation	PA	Catalogue	Comments
	h m	° ′			arcsec	°		
95	22 03.8	+64 37	ξ Cep	4.4,6.5	8.4	274	Σ2863	White and blue. Long-period binary.
96	22 14.3	−21 04	41 Aqr	5.6,6.7	5.1	113	H N 56	Yellowish and purple?
97	22 26.6	−16 45	53 Aqr	6.4,6.6	1.3	65	SHJ 345	Long-period binary; periastron in 2023.
98	22 28.8	−00 01	ζ Aqr	4.3,4.5	2.3	164	Σ2909	Period = 487 years. Slowly widening.
99	23 19.1	−13 28	94 Aqr	5.3,7.0	12.3	351	Σ2988	Yellow and orange. Probable binary.
100	23 59.5	+33 43	Σ3050 And	6.6,6.6	2.4	340	–	Period = 717 years. Visible in 7.5 cm.

Some Interesting Nebulae, Clusters and Galaxies

Object	RA		Declina-tion		Remarks
	h	m	°	′	
M31 Andromedae	00	40.7	+41	05	Andromeda Galaxy, visible to naked eye.
H VIII 78 Cassiopeiae	00	41.3	+61	36	Fine cluster, between Gamma and Kappa Cassiopeiae.
M33 Trianguli	01	31.8	+30	28	Spiral. Difficult with small apertures.
H VI 33–4 Persei, C14	02	18.3	+56	59	Double cluster; Sword-handle.
Δ142 Doradûs	05	39.1	−69	09	Looped nebula round 30 Doradûs. Naked eye. In Large Magellanic Cloud.
M1 Tauri	05	32.3	+22	00	Crab Nebula, near Zeta Tauri.
M42 Orionis	05	33.4	−05	24	Orion Nebula. Contains the famous Trapezium, Theta Orionis.
M35 Geminorum	06	06.5	+24	21	Open cluster near Eta Geminorum.
H VII 2 Monocerotis, C50	06	30.7	+04	53	Open cluster, just visible to naked eye.
M41 Canis Majoris	06	45.5	−20	42	Open cluster, just visible to naked eye.
M47 Puppis	07	34.3	−14	22	Mag. 5.2. Loose cluster.
H IV 64 Puppis	07	39.6	−18	05	Bright planetary in rich neighbourhood.
M46 Puppis	07	39.5	−14	42	Open cluster.
M44 Cancri	08	38	+20	07	Praesepe. Open cluster near Delta Cancri. Visible to naked eye.
M97 Ursae Majoris	11	12.6	+55	13	Owl Nebula, diameter 3′. Planetary.
Kappa Crucis, C94	12	50.7	−60	05	'Jewel Box'; open cluster, with stars of contrasting colours.
M3 Can. Ven.	13	40.6	+28	34	Bright globular.
Omega Centauri, C80	13	23.7	−47	03	Finest of all globulars. Easy with naked eye.
M80 Scorpii	16	14.9	−22	53	Globular, between Antares and Beta Scorpii.
M4 Scorpii	16	21.5	−26	26	Open cluster close to Antares.

Some Interesting Nebulae, Clusters and Galaxies

Object	RA h m		Declination ° ′		Remarks
M13 Herculis	16	40	+36	31	Globular. Just visible to naked eye.
M92 Herculis	16	16.1	+43	11	Globular. Between Iota and Eta Herculis.
M6 Scorpii	17	36.8	−32	11	Open cluster; naked eye.
M7 Scorpii	17	50.6	−34	48	Very bright open cluster; naked eye.
M23 Sagittarii	17	54.8	−19	01	Open cluster nearly 50′ in diameter.
H IV 37 Draconis, C6	17	58.6	+66	38	Bright planetary.
M8 Sagittarii	18	01.4	−24	23	Lagoon Nebula. Gaseous. Just visible with naked eye.
NGC 6572 Ophiuchi	18	10.9	+06	50	Bright planetary, between Beta Ophiuchi and Zeta Aquilae.
M17 Sagittarii	18	18.8	−16	12	Omega Nebula. Gaseous. Large and bright.
M11 Scuti	18	49.0	−06	19	Wild Duck. Bright open cluster.
M57 Lyrae	18	52.6	+32	59	Ring Nebula. Brightest of planetaries.
M27 Vulpeculae	19	58.1	+22	37	Dumb-bell Nebula, near Gamma Sagittae.
H IV 1 Aquarii, C55	21	02.1	−11	31	Bright planetary, near Nu Aquarii.
M15 Pegasi	21	28.3	+12	01	Bright globular, near Epsilon Pegasi.
M39 Cygni	21	31.0	+48	17	Open cluster between Deneb and Alpha Lacertae. Well seen with low powers.

(M = Messier number; NGC = New General Catalogue number; C = Caldwell number.)

Our Contributors

Dr Natalie Starkey is a space scientist at The Open University where she analyses comet and asteroid samples collected in the Earth's stratosphere, and returned by space missions such as NASA Stardust and JAXA Hayabusa. She has written about science for the *Guardian* newspaper as a British Science Association Media Fellow, and for The Conversation. In 2014 she received a SEPnet award for Public Engagement in the Media and Communications category.

Dr Stephen Webb is the author of several books, including *New Eyes on the Universe: Twelve Cosmic Mysteries and the Tools We Need To Solve Them* (which provides an overview of the many observatories that are planned or are already in construction) and *Measuring the Universe: The Cosmological Distance Ladder* (which gives a detailed account of how astronomers measure distances in the universe). His most recent book, which contains a foreword by Martin Rees, the Astronomer Royal, is an expanded and updated edition of his award-winning *Where is Everyone?* He blogs on various astronomy- and physics-related topics at http://stephenswebb.info/.

Richard Myer Baum is a former Director of the Mercury and Venus Section of the British Astronomical Association, an amateur astronomer and an independent scholar. He is author of *The Planets: Some Myths and Realities* (1973), (with W. Sheehan) *In Search of Planet Vulcan: The Ghost in Newton's Clockwork Universe* (1997) and *The Haunted Observatory* (2007). He has contributed to the *Journal of the British Astronomical Association*, the *Journal for the History of Astronomy*, *Sky & Telescope* and many other publications including *The Dictionary of Nineteenth-Century British Scientists* (2004), and *The Biographical Encyclopedia of Astronomers* (2007).

Martin Mobberley is one of the UK's most active imagers of comets, planets, asteroids, variable stars, novae and supernovae and served as President of the British Astronomical Association from 1997 to 1999.

In 2000, he was awarded the Association's Walter Goodacre Award. He is the sole author of ten popular astronomy books published by Springer as well as three children's 'Space Exploration' books published by Top That Publishing. In addition he has authored hundreds of articles in *Astronomy Now* and numerous other astronomical publications.

Dr Lisa Harvey-Smith is a radio astronomer researching the birth and death of stars and cosmic magnetic fields. Lisa is the Project Scientist for the Australian Square Kilometre Array Pathfinder (ASKAP), part of an international project to build the world's largest radio telescope. She is an acclaimed science communicator, bringing astronomy to large audiences through television, radio and the written press. Lisa runs grassroots mentoring and educational programmes in schools across Australia. In 2012 Lisa was named in the 'Top 100: Most Influential People' by the *Sydney Morning Herald*. Regular updates on her work can be found at www.lisaharveysmith.com.

Dr Allan Chapman, of Wadham College, Oxford, is probably Britain's leading authority on the history of astronomy. He has published many research papers and several books, as well as numerous popular accounts. He is a frequent contributor to the *Yearbook*.

Dr David M. Harland gained his BSc in astronomy in 1977 and a doctorate in computational science. Subsequently, he has lectured in computer science, worked in industry and managed academic research. In 1995, he 'retired' and has since published many books on space themes.

Dr John Mason is the Editor of the *Yearbook of Astronomy*. He is a past President of the British Astronomical Association and Director of the BAA's Meteor Section. He is currently Principal Lecturer at the South Downs Planetarium and Science Centre in Chichester. He appeared many times with Sir Patrick Moore on BBC TV's *The Sky at Night*. For over thirty years he has been leading overseas expeditions to observe and record annular and total solar eclipses, the polar aurora and major meteor showers. He was made an MBE in the 2009 New Year's Honours List for his services to science education.

Note from the Editor

Previous editions of the Yearbook have included a list of Astronomical Societies in the British Isles. It has become apparent that much of this information is rather out-of-date and is constantly changing, and so it has been decided to omit this section from the Yearbook this time.